Investigations and Applications of Severe Plastic Deformation

NATO Science Series

A Series presenting the results of activities sponsored by the NATO Science Committee. The Series is published by IOS Press and Kluwer Academic Publishers, in conjunction with the NATO Scientific Affairs Division.

A. Life Sciences	IOS Press
B. Physics	Kluwer Academic Publishers
C. Mathematical and Physical Sciences	Kluwer Academic Publishers
D. Behavioural and Social Sciences	Kluwer Academic Publishers
E. Applied Sciences	Kluwer Academic Publishers
F. Computer and Systems Sciences	IOS Press

1. Disarmament Technologies	Kluwer Academic Publishers
2. Environmental Security	Kluwer Academic Publishers
3. High Technology	Kluwer Academic Publishers
4. Science and Technology Policy	IOS Press
5. Computer Networking	IOS Press

NATO-PCO-DATA BASE

The NATO Science Series continues the series of books published formerly in the NATO ASI Series. An electronic index to the NATO ASI Series provides full bibliographical references (with keywords and/or abstracts) to more than 50000 contributions from international scientists published in all sections of the NATO ASI Series.
Access to the NATO-PCO-DATA BASE is possible via CD-ROM "NATO-PCO-DATA BASE" with user-friendly retrieval software in English, French and German (WTV GmbH and DATAWARE Technologies Inc. 1989).

The CD-ROM of the NATO ASI Series can be ordered from: PCO, Overijse, Belgium

Series 3. High Technology – Vol. 80

Investigations and Applications of Severe Plastic Deformation

edited by

Terry C. Lowe
Materials Science and Technology Division,
Los Alamos National Laboratory,
Los Alamos, NM, U.S.A.

and

Ruslan Z. Valiev
Institute of Physics of Advanced Materials,
Ufa State Aviation Technical University,
Ufa, Russia

Kluwer Academic Publishers

Dordrecht / Boston / London

Published in cooperation with NATO Scientific Affairs Division

Proceedings of the NATO Advanced Research Workshop on
Investigations and Applications of Severe Plastic Deformation
Moscow, Russia
2-7 August 1999

A C.I.P. Catalogue record for this book is available from the Library of Congress.

ISBN 0-7923-6280-2 (HB)
ISBN 0-7923-6281-0 (PB)

Published by Kluwer Academic Publishers,
P.O. Box 17, 3300 AA Dordrecht, The Netherlands.

Sold and distributed in North, Central and South America
by Kluwer Academic Publishers,
101 Philip Drive, Norwell, MA 02061, U.S.A.

In all other countries, sold and distributed
by Kluwer Academic Publishers,
P.O. Box 322, 3300 AH Dordrecht, The Netherlands.

Printed on acid-free paper

TABLE OF CONTENTS

II. MICROSTRUCTURAL CHARACTERIZATION AND MODELING OF SEVERE PLASTIC DEFORMATION MATERIALS

III. MICROSTRUCTURE EVOLUTION DURING SEVERE PLASTIC DEFORMATION PROCESSING

IV. PHYSICAL AND MECHANICAL PROPERTIES OF SEVERE PLASTIC DEFORMATION MATERIALS

PREFACE

In the spring of 1997 Professors Ruslan Valiev and Julia Weertman visited Los Alamos National Laboratory to discuss their research collaborations with Drs. Terry Lowe and Yuntian Zhu. These individuals were co-investigators on a project entitled "Properties and Processing of Nanocrystalline Materials" sponsored by the U.S. Department of Energy. On this same day, Ross Lemons, Director of the Materials Science and Technology Division at Los Alamos had planned a barbecue for his leadership team at his home in White Rock, New Mexico. He invited Ruslan and Julia to participate in this evening event.

White Rock is located at the edge of an ancient volcanic mesa on the eastern slope of the Jemez Mountains. The outdoor patio area at the Lemons' home offers a stunning view looking across the Rio Grande valley to the majestic 13,000 foot (4000 m) peaks of the Sangre de Cristo range of the Rocky Mountains. In the evening, the setting sun paints the tops of the peaks a brilliant red, an especially enjoyable sight at the end of a warm spring day. The setting is truly inspiring.

Throughout the evening, there were many interesting discussions, including some on the topic of nanostructured materials. At the very end of the evening when the guests were all bidding each other farewell, Ruslan Valiev pointed out that though there had been various recent symposia on the topic of nanocrystalline materials, none yet had brought together the research community that focused on studying such materials produced by severe plastic deformation. He suggested to Terry Lowe that they jointly submit a proposal to NATO to co-organize a workshop on severe plastic deformation – a meritorious idea, but a non-trivial undertaking. Somehow, onerous endeavors seem less so following a pleasant evening in an inspirational New Mexico setting. Terry Lowe consented, agreeing to a partnership that has led to the creation of this volume.

During the following two years, many individuals were helpful in the planning and preparations for the workshop. First, the workshop would not have been possible without the generous support of many sponsoring organizations and their key representatives. Our greatest thanks go to the founding and primary sponsor, NATO. Nancy Schulte, Director of the NATO Program on High Technology within the NATO Science Program, was key to the initiating the workshop. Next, we were fortunate to receive the joint sponsorship of the European branches of U.S. Armed Services research organizations, including the European Research Office of the U.S. Army, the European Office of Aerospace Research and Development, and the Office of Naval Research, International Field Office. Thanks are due to Joyce Illinger, Jerry Sellars, Ben Wilcox, and Hugh Casey from these offices. Similarly, financial support was provided by the Materials Science and Technology Division and the Center for International Security Affairs at Los Alamos National Laboratory and the Chemistry and Materials Science Directorate of Lawrence Livermore National Laboratory. Special thanks are due to the leaders of these organizations, Drs. Ross Lemons, John Shaner, and Larry Newkirk. Special financial support was also provided by the Russian Foundation for Basic Research. Additional participation of more than 20 Russian scientists was supported by the Russian Academy of Sciences and various Russian universities. We are also grateful for the endorsements provided by the Shaping and Forming Committee of The

Minerals, Metals & Materials Society (TMS) and the Computational Materials Science and Engineering Committee of ASM International and TMS.

The majority of the hard work of preparing the workshop can be attributed to teams of workers at Ufa State Aviation Technical University, Moscow Steel and Alloys Institute, and Los Alamos National Laboratory, most ably coordinated by Ireck Zamilov and Christine Roybal. Guiding the entire process and the technical program were the workshop organizing committee members, Dr. Terry Lowe, and Profs. Ruslan Valiev, Haël Mughrabi, and Sergei Dobatkin. The result of much detailed planning was a remarkably successful technical workshop, where 77 leading experts from 17 countries worked in six thematic presentation and poster sessions, and had active round table discussions. This success was made all the more enjoyable by the outstanding hospitality of the Russian hosts from the International Convention Centre in Golitsino, Moscow. The workshop experience itself created memories and inspiration for the participants that were comparable in some ways to inspirational setting at sunset in New Mexico in which the workshop was originally conceived.

The success of the workshop inspired an extraordinary effort to document its content. The Organizing Committee selected 48 papers for the proceedings. A formal peer review was conducted for all these papers. The editors and their associates in Russia and the United States worked diligently with the authors and the publisher for months following the workshop to create proceedings of the highest quality. Special thanks are due to Prof. Igor Alexandrov, Dr. Yuntian Zhu, and Ms. Annelies Kersbergen for their dedicated contributions to the preparation of this volume.

Finally, it is the intent of the editors that this volume set a high standard for subsequent volumes on severe plastic deformation that will surely follow in future years. It is also our hope that this first-ever book dedicated to this topic serve as a valuable reference for researchers working in the emerging field of severe plastic deformation.

Terry C. Lowe and Ruslan Z. Valiev
Workshop Co-Directors

INTRODUCTION

This book contains the principal research papers presented at the first international scientific workshop devoted to investigations and applications of severe plastic deformation (SPD). SPD is an emerging field of materials science that encompasses aspects of mechanical metallurgy, the physics and mechanics of materials, plasticity, and processing of nanostructured materials. As in any new field of research, the development and use of commonly accepted terminology is very important and thus, this topic was actively discussed at the workshop. The editors and other reviewers further noted this issue during the editing of the contributions included in this volume, all of which were prepared prior to the workshop. In this introduction we propose standardized terminology and briefly outline the main topics of each chapter.

As is typical for workshops, much time was dedicated to collective discussion of ideas and issues surrounding the workshop topics. Central to the workshop discussions were questions relating to application of severe plastic deformation as an innovative procedure for obtaining unique microstructures in materials and, consequently, for attaining new unique materials properties. There was a high level of interest and interactivity at the workshop, driven by the exciting prospect of creating ultrafine-grained nanostructures in metals and alloys, leading to many new applications of SPD-processed materials.

In recent years, bulk nanostructured materials (NSM) processed by methods of SPD have attracted the interest of experts in material science [1]. This interest continues to grow, not only because of the unique physical and mechanical properties inherent to various nanostructured materials, *e.g.* processed by gas condensation [2, 3] or ball milling with subsequent consolidation [4, 5], but also because of the increasingly apparent advantages of SPD materials as compared to other NSM processing methods. In particular, SPD methods overcome difficulties that exist in other methods, for example residual porosity in compacted samples, impurities from ball milling, inability to process large-scale billets, and impracticality for reduction to industrial practice. Processing of bulk nanostructured materials using SPD methods is also emerging as an attractive alternative to the existing methods of nanopowder compaction.

Terminology

Formation of nanostructures requires attaining very high plastic strains (true strains are higher than 6–8, as a rule) at relatively low temperatures through the application of special deformation schemes, such as severe torsion under pressure or equal channel angular pressing (ECAP) [6]. Severe torsional straining must be accompanied by pressures of several GPa to cause concurrent compressive strains and create the conditions necessary to generate refined microstructures. During the workshop discussions it was recommended to adopt the term High Pressure Torsion (HPT) as the most accurate identification of this SPD method. There were also discussions on the use of the terms "ECA pressing" or "ECA extrusion" for the same processing method, introduced by Segal and co-workers [7]. "Extrusion" in metal working usually means pressing of material through two channels where the second channel is smaller than the first. This is not the case in ECA since both channels have the same cross section. Therefore we believe the term "pressing" and the associated acronym ECAP to be more general.

For the ECAP methods, researchers have distinguished four distinct processing routes which are distinguished by the how the billet is rotated with respect to the die for each subsequent pass. The first paper in Chapter III of this book by Langdon *et al.* illustrates these routes. Figure 2 on page 150 contains schematics of the routes along with the corresponding designations A, B_A, B_C, and C. We find the nomenclature used by Langdon *et al.* to be the most compelling, and most commonly used in the literature. Thus, we propose that others adopt these designations.

Another very important question regards the formation of nanostructures during SPD. It is well known that heavy deformations, for example, by cold rolling or drawing, can also result in significant refinement of microstructure at low temperatures [8–11]. However, the substructures formed are usually of a cellular type having boundaries with low angle misorientations. In contrast, the structures formed by SPD can be ultrafine-grained nanostructures containing mainly high angle grain boundaries. This difference is fundamental, particularly in terms the unique properties that can be obtained. For example, in ultrafine-grained SPD alloys enhanced superplasticity can be observed, but does not occur in alloys with identically sized sub-grains.

There has been some debate whether SPD materials should be described as being nanocrystalline, since they often have a mean grain size of about 100–200 nm, somewhat above the 100 nm threshold that is often accepted as the upper bound of the nanocrystalline regime. However, the grain interior of SPD-processed materials usually contains some substructure due to a highly distorted crystal lattice. X-ray analysis demonstrates, as a rule, a crystallite size (coherent domains) of about 50 nm. This is why one may consider SPD materials as a class of bulk nanostructured materials [12], even though the measured grain size may be greater than 100 nm. Of course, some SPD methods, such as HPT, can produce sub-100 nm grain sizes.

Workshop Summary

This book consists of five chapters, each addressing key areas of the emerging field of severe plastic deformation, and corresponding to the five sessions of the workshop. These were: 1) SPD processing methods and process modelling, 2) characterization and modelling of microstructures produced by SPD, 3) microstructure evolution during SPD, 4) properties of SPD materials, and 5) applications and commercialization of SPD materials.

Chapter I, titled "Innovations in Severe Plastic Deformation Processing and Process Modelling," introduces the various methods of processing materials by SPD. The methods of SPD processing most frequently used are HPT and equal channel angular (ECA) pressing or extrusion. At the present time HPT is very attractive from the scientific point of view because it is capable of producing uniform nanostructures possessing smaller grain sizes than are attainable by any other SPD method. HPT also offers the advantage of being able to introduce continuously variable magnitudes of deformation. This is helpful in studies of microstructure evolution. The principal drawback of HPT is that the dimensions of the processed nanostructured disk-shape samples are small, usually not exceeding 20 mm in diameter and 1 mm in thickness. Also, the precise deformation conditions and constraints during HPT may vary since they depend on friction between the rotating punch and the sample.

ECA pressing has the advantage of allowing fabrication of large bulk ultrafine-grained ingots. Most of papers in this chapter address aspects of ECA pressing, including detailed studies of the influence of the number of passes, optimal choice of routes, temperature of deformation, tribological characteristics and others factors that influence the formation of microstructures.

Goforth *et al.* overview their extensive experience with ECA pressing of various metals and alloys and analyze the results of their research on development and application of the ECA pressing method. A highlight of this work is the favorable comparison of rigid-viscoplastic finite element analysis of ECA with upper bound analyses. They also show details of how the flow patterns and forces during ECAP depend upon material properties.

The following paper by Dobatkin examines ECAP and HPT of various steels, specifically exploring the strain, strain rate, and temperature dependence of dynamic recovery and recrystallization during ECA pressing. It provides a comprehensive perspective on the types of microstructures that can be obtained in steels and how the mechanisms of microstructure formation vary with temperature. This work is also distinct because it addresses concurrent phase transformations that occur during cold deformation, specifically the formation of martensites. Finally, Dobatkin discusses a variety of industrially viable processes beyond ECAP and HPT by which ultrafine grained steels can be produced.

However, the process of ECA pressing is complex and does not always result in the formation of uniform nanostructures and the associated enhancement of properties. In this connection, Kopylov in his paper "Application of ECAP-Technology for Producing Nano- and Microcrystalline Materials" introduces the term "clear" ECA pressing which identifies the macroscopic and microscopic conditions that must exist during each ECAP pass to ensure the formation of homogeneous nanostructures. By optimizing ECAP processing variables Kopylov is able to impart uniform properties to a variety of metals and alloys, increasing yield strengths up to a factor of six higher than conventional materials. In his paper he also describes his experience in ECA pressing of large bulk ingots (150 × 150 mm in cross section) out of hard-to-deform materials, namely Mo and its alloys.

The ECA pressing method allows fabrication of cylindrical or parallelepipeds shaped bulk ingots. However, industry often requires production of sheet materials. Ghosh and Huang introduce a new surface-shear based deformation technique for formation of microstructures in aluminum alloy plates and sheets which appears promising. Rolling of ECA pressed materials can also be used to significantly change their shape and their microstructure.

Among other SPD methods of structure refinement Salishchev *et al.* consider a method of hot working during which the deformation temperature is gradually decreased resulting in the formation of submicrocrystalline structure in bulk TiAl and Ti_3Al alloy billets via dynamic recrystallization.

Chapter II is devoted to the topic "Microstructural Characterization and Modeling of Severe Plastic Deformation Materials". This topic is particularly important because it provides the foundation for understanding the unusual physical and mechanical properties of SPD materials. Furthermore, the microstructures of SPD materials have been particularly difficult to characterize. SPD forms nanostructures with extremely small grain sizes, of several ten nanometers, and defect structures that cause a high level of crystal lattice microdistortions. Consequently, traditional microstructural observation methods (e.g. metallography) are not applicable at all, or cannot be applied without adaptation for finer scale structures.

For example, in his paper Zehetbauer has applied new diffraction methods like Electron-Back-Scatter-Diffraction and X-ray Bragg Peak Profile Analysis using synchrotron radiation with sufficient spatial resolution to examine the structure of SPD materials. To analyze the evolution of microstructure during large deformation this author applied and extended the concept of the classical stages of strain hardening and microstructures corresponding to them. Zehetbauer's conclusion that "by multiaxial deformation, stage IV can be markedly extended; the misoriented zones can reach sizes of a few 10 nm while the strength increases up to a factor of three" is particularly significant.

Papers by Ungar, Alexandrov, Ungar *et al.* are devoted to the development and application of methods of X-ray structural analysis for investigation of the structure of SPD materials. In particular, they demonstrate the application of the modified Williamson-Hall and Warren-Averbach methods of the X-ray analysis for precise determination of crystallite size distribution and analysis of lattice strains. They find that in copper subjected to ECA pressing, an increasing number of passes leads to a more homogeneous size distribution of grains or subgrains. The paper suggests the possibility of a fundamental lower limit to the grain/subgrain size that can develop during SPD.

The issue of long range elastic stress fields in nanostructured materials produced by SPD is studied in papers by Koneva *et al.* and Tyumentsev *et al.*

It is known that SPD can destroy short range order in intermetallics, changing their crystal structure, and causing their transformation to an amorphous state. These interesting processes are considered in papers by Greenberg *et al.* and Noskova *et al.*

Chapter III extends the topic of microstructure characterization in Chapter II and focuses in more depth on microstructural evolution. Langdon *et al.* introduce the topic, examining the different microstructures that develop during ECA. This work is similar to that of Prangnell *et al.* in Chapter I, but arrives at different conclusions about the relative effectiveness of different ECA processing routes on grain refinement. In the work of Langdon *et al.* aluminum is ECA pressed by all four standard routes. They find that a homogeneous equiaxed granular structure with high angle grain boundaries forms most quickly when the sample is rotated by an angle of $+90°$ in between each pass (by route B_c).

Prangnell *et al.* measure .orientation distributions by means of the EBSD method for different sections of aluminum alloy samples prepared by various routes of ECAP. In contrast to Langdon's work, they show that nanostructures with a maximum volume fraction of high angle grain boundaries are formed via route A (no rotation of billet between passes).

Thus, the results obtained in papers by Langdon *et al.* and Prangnell *et al.* appear to contradict each other, creating an important subject for further investigation.

Research examining the evolution of microstructure during SPD is very important since it will eventually lead to an understanding of what levels and modes of straining are required to form homogeneous equiaxed nanostructures. Details of microstructure evolution during SPD deformation, and post-deformation annealing are examined at scales ranging from a few nanometers by High Resolution Transmission Electron Microscopy (Horita *et al.*) to the macro-level (Panin). At the atomic level, the work of Horita *et al.* provides additional evidence that grain boundaries in SPD material exist in a highly energetic non-equilibrium state with a non-regular distribution of facets and steps. At the macro-level, the work by Panin shows how some mechanical properties of copper are better understood in terms of the formation of meso- or macro- deformation bands. However, the behavior of nanostructured materials is increasingly dominated by the grain boundaries, whose volume fraction grows rapidly with decreasing grain size to less than 100 nm. As noted in the paper by Horita *et al*, the grain boundaries of SPD materials contain very high defect densities.

SPD nanostructured materials exist in a metastable state. Thus, there is an urgent need to investigate their thermo-mechanical stability. The paper by Thiele *et al.* deals with this question, examining the evolution of microstructure of pure SPD nickel during annealing and cyclic loading.

The development of structural models of microstructure evolution during SPD on the basis of the experimental data is very important. Strain hardening theories have long served the field of large strain deformation in this manner, explaining the distinct stages of strain hardening in terms of the types of microstructures that evolve. Schafler *et al.* provide new experimental evidence for the transformation of cell wall character from polarized dipole dislocation arrangements to polarized tilt walls during the transition from Stage III to Stage IV hardening. This work complements the theoretical treatments of work hardening by Zehetbauer in Chapter II and by Naimark and Seefeldt in Chapter III. Zehetbauer examines Stage IV and Stage V hardening behavior as it occurs during SPD. Naimark examines the role of ensembles of grain boundary dislocations during the formation of a nanostructured states. Seefeldt *et al.* describe a work-hardening model "able to reproduce the saturation of the cell structure, a hyperbolic decrease of the mean fragment size and a linear increase of the misorientation between fragments."

Specific nanostructures produced by SPD lead to novel physical and mechanical properties. Research on these properties is contained in Chapter IV "Physical and Mechanical Properties of Severe Plastic Deformation Materials." Among the unusual mechanical properties of SPD materials, most notable are their high strength (Valiev, Provenzano *et al.*, Markushev *et al.*, Lapovok *et al.*) including at cryogenic temperatures, their superplasticity at low temperature and/or high strain rate (Mukherjee *et al.*, Smirnov *et al.*), and their significantly higher fracture stress (Glezer *et al.*). The question of increasing strength and ductility in nanostructured materials is closely connected with investigations of deformation localization and mechanisms of plastic deformation in SPD materials.

At present there exist rather contradictory data on fatigue behavior of materials in the nanostructured state. To a great extent this is connected with the variable quality of nanostructured samples, which sometimes do not possess the optimal structures that can produced by SPD, *i.e.* macroscopically homogeneous, with equiaxed grains and high angle grain boundaries. The potential of SPD to increase fatigue strength by structure refinement is analyzed in the papers by Mughrabi, Thiele *et al.*, and Kaneko, *et al.*

Nanostructured materials processed by SPD methods because of their metastable state are not thermostable. Therefore it is desirable to increase their thermostability via creation of nanocomposite materials (Buchgraber *et al.*, Sastry *et al.*). Studies of the kinetics and mechanisms of grain growth in nanostructured materials (Islamgaliev *et al.*, Kolobov *et al.*) show how enhanced diffusion behavior impacts stability and high temperature deformation.

The unusual physical and mechanical properties of SPD nanostructured materials have stimulated interest in their potential commercial applications. This topic is considered in Chapter V "Future Horizons for Severe Plastic Deformation Materials: Applications and Commercialization". The analysis provided by Lowe *et al.* shows the growth in the number of patents, indicating the increasing commercial importance of SPD methods. Lesuer *et al.* and Aernoudt show the properties of nanostructured ultrahigh-carbon steel wire that is already used in applications such as reinforcing wires in radial tires for vehicles. The possibility of using nanostructured materials in specialized applications is also being explored, such as using nanostructured titanium in traumatology and orthopaedy. Pilot samples of titanium implants, characterized by higher strength and fatigue properties, are already being produced for medical trials (Stolyarov *et al.*). Nanostructuring of commercial pure titanium by SPD eliminates some of the problems of bio-compatibility with a human tissue that exist for titanium alloy implants.

As SPD processing methods are developed for commercial scale production (*e.g.* as proposed by Dobatkin in Chapter I), other applications of bulk articles of nanostructured materials can be expected to emerge. The interest in these materials is caused not only by their high strength or superplastic properties but also their physical properties, such as enhanced magnetic properties [1].

1. R.Z. Valiev, (1996) ed., Ultrafine-grained materials processed by severe plastic deformation, *Special issue: Annales de Chimie. Science des Matériaux.* 21, 369.
2. H. Gleiter (1989), Nanocrystalline Materials, *Progr. Mater. Sci.* 33, 223.
3. J.R. Weertman (1993), Mechanical properties of nanocrystalline metals, *Mater. Sci. Eng* A166, 16.
4. C.C. Koch and Y. S. Cho (1992), Nanocrystals by high energy ball milling, *Nanostructured Mater.* 1, 207.
5. D.G. Morris (1998), Mechanical behaviour of nanostructured materials, *Trans Tech. Publ. Switzerland*, 85.
6. R.Z. Valiev, A.V. Korznikov and R.R. Mulyukov (1993), Structure and properties of ultrafine-grained materials, *Mater. Sci. Eng.* A168, 141.
7. V.M. Segal, V.I. Reznikov, A.E. Drobyshevskiy, A.E. Kopylov (1981) Metal working by simple shear,

Russian Metally **1**, 99.

8. V.A. Pavlov (1989), Amorphisation during intense rolling, *Phys. Met. Metallogr.* **67**, 924.
9. G. Langford, M. Cohen (1969) Microstructure of armco-iron subjected to severe plastic drawing, *Trans. ASM* **82**, 623.
10. V. V. Rybin (1987) Large plastic deformation and destructions of metals, *Metallurgia, Moscow.*
11. J. Gil Sevillano, P. Van Houtte and E.Aernoudt (1981), Large strain work hardening and textures, *Progr. Mat. Sci.* **25**, 2.
12. R.Z. Valiev, R.K. Islamgaliev, I.V. Alexandrov (2000) Bulk nanostructured materials from severe plastic deformation, *Progr. Mat. Sci.* **45**, 102.

I. INNOVATIONS IN SEVERE PLASTIC DEFORMATION PROCESSING AND PROCESS MODELING

SEVERE PLASTIC DEFORMATION OF MATERIALS BY EQUAL CHANNEL ANGULAR EXTRUSION (ECAE)

R.E. GOFORTH, K.T. HARTWIG, L.R. CORNWELL
Department of Mechanical Engineering
Texas A&M University
College Station, TX 77843-3123

1. Introduction

Equal Channel Angular Extrusion (ECAE) was invented in the former Soviet Union by Vladimir Segal [1]. It is an innovative process capable of producing relatively uniform intensive plastic deformation in a variety of material systems, without causing substantial change in geometric shape or cross section. Multiple extrusions of billets by ECAE permits severe plastic deformation in bulk materials. More importantly, by changing the orientation of the billet between successive extrusions, sophisticated microstructures and textures can be developed. According to the orientation chosen after each pass, four fundamental equal-channel angular extrusion routes are defined and used for obtaining different textures and microstructures [2]. Researchers in the Texas A&M University's (TAMU) Deformation Processing Laboratory in the Department of Mechanical Engineering have been conducting research on the ECAE process since 1992. The inventor of the process, Dr. V. Segal, was a research associate in the Lab from 1992 to 1995. Research has concentrated in two areas: (1) development of theoretical and practical knowledge of the mechanics of ECAE and the further expansion of that knowledge leading to the capability of effectively mathematically modeling the process, and (2) investigation of the various technological advantages of ECAE and possible industrial applications for a variety of materials. Process modeling has involved both geometric, analytical (upper-bound analysis, slip-line theory) methods and non-linear finite element analysis [2]. Many technological advantages were found for ECAE, including the creation of a variety of microstructures (*e.g.* submicron grains) as well as equiaxed, laminar, and fibrous textures for many of the materials that were investigated at TAMU. ECAE was also found to be an excellent method for powder consolidation.

The purpose of this paper is to summarize typical research and applications work done at TAMU since 1992. Publications documenting all the work done at TAMU in more detail can be found in specific publications referenced throughout this paper.

2. Process Modeling

As with other forming operations, the successful implementation of ECAE to process different materials and alloys to achieve the expected material properties, relies on the design, control and optimization of the forming process. To this end, knowledge

T.C. Lowe and R.Z. Valiev (eds.), Investigations and Applications of Severe Plastic Deformation, 3–12.
© 2000 *Kluwer Academic Publishers. Printed in the Netherlands.*

regarding the deformation patterns, stresses (loads), and heat transfer, as well as technological information associated with lubrication, material handling, die design and manufacture etc, is required. Generally speaking, compared with traditional time-consuming trial-and-error methods, modeling is an efficient way to acquire important information. The ECAE modeling research at TAMU has concentrated in the following areas: (1) geometric visualization, (2) analytical methods (slip line theory and upper bound analysis), and (3) finite element method (mixed finite element formulation).

2.1. VISUALIZATION OF ECAE

As mentioned before, one of the advantages of ECAE is that different microstructures and textures can be developed by changing the orientation of a billet of material between successive passes. According to the work-piece orientation chosen for each extrusion pass, there are four fundamentally different ECAE routes (A, B, C and C´) that can be defined. In route A, the billet orientation is unchanged for each pass, i.e. the billet is inserted into the extrusion die with no rotation about the extrusion axis. In Route B, the billet is deformed alternatively in two orthogonal directions by rotating about the extrusion axis +/- 90°. In Route C the billet is rotated 180° about its extrusion axis after each pass. In Route C´, the billet is rotated + 90° after each pass. Without a change of the shear plane, the deformation patterns of Routes A and C are two-dimensional in nature. On the other hand, the deformation processes of Routes B and C´ are three-dimensional due to the alternating changes of the shear planes. A unified mathematical description for the deformation processes of the four ECAE routes was established. The work shows that the deformation of any ECAE route within a typical cycle can be treated as a combination of two- dimensional deformation and a series of rigid body motions. A self-consistent velocity field characterized the two-dimensional deformation while a mathematical description of the rigid motion was derived for each route. Based on analytical derivations, a numerical scheme was developed to keep track of the movement of any material particle. With the aid of graphic software such as PATRAN™, a program for three-dimensional visualization of ECAE was developed.

Based on the equations developed by Cui [2], any material particle position during any ECAE deformation stage can be tracked, which allows the development of a visualization program to capture the billet's overall deformation pattern. Since this procedure is very similar to regular pre-processing and post-processing in finite element analysis, many software packages are available for visualizing and analyzing the deformation pattern. In this work, the numerical calculations are implemented by FORTRAN programming, and the visualization plot is realized by PATRAN™. Typical results are shown in Figures 1 and 2.

a   b   c

Figure 1. ECAE flow pattern after (a) the 1st pass, (b) the 2ⁿᵈ pass, and (c) 3ʳᵈ pass of Route A.

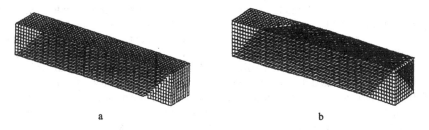

Figure 2. Flow pattern of Route B after (a) 1st pass and (b) 2nd pass.

2.2. ANALYTICAL METHODS

The exact calculations of the stress and strain distributions based on the controlling equations for a forming process with large deformation are often difficult, if not impossible. To circumvent such difficulties, some analytical methods such as upper bound analysis and slip-line field theory have been established by making a number of simplifying assumptions regarding the material properties and deformation modes. The following assumptions are common for both the upper bound analysis and the slip-line theory: (i) in most cases, deformation is two-dimensional (plane-strain assumption); (ii) the material is rigid-perfectly plastic, and the possible effects of temperature, strain hardening and strain rate are neglected.

Upper bound analysis ignores the stress equilibrium equations and predicts higher load levels than the exact solution. Slip line theory can provide a more accurate load distribution by satisfying the stress equilibrium equations, but there are inherent difficulties in the construction of slip-line fields, especially when the deformation areas are irregular or boundary conditions involve both velocity and traction components.

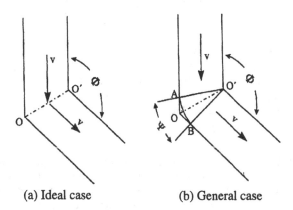

(a) Ideal case   (b) General case

Figure 3. Two deformation patterns of ECAE.

The intention of ECAE is to introduce large and uniform deformation to the material by simple shear. To this end, the billet movement during extrusion should be controlled precisely so that its direction of motion can be changed abruptly when the billet passes through the intersection of two channels as shown in Figure 3a. In reality, however, this ideal deformation pattern can hardly be achieved because the material deformation process is complicated by many factors such as the geometry of the die system, frictional characteristics between the die walls and the billet, and the properties of the billet material. The fan-like deformation area during ECAE depicted in Figure 3b is believed to be more realistic and adopted by some researchers in their work. Segal *et al.* [3] first used this model to analyze the effect of the friction force on ECAE by using slip-line field theory. In their work, the AOB area in Figure 3b was regarded as a dead zone caused by the friction force between the billet and die surfaces. As a result, the size of the fan area is uniquely determined by the friction factor m, which is predicted by the following equation [3]:

$$\Psi = \arccos m - \Phi = ar\cos\left(\frac{\tau}{k}\right) - \Phi \tag{1}$$

where τ and k are the interface shear strength and shear yield stress of the material, respectively. Note that m is τ/k. Iwahashi *et al.* [4] further pointed out that the fan-like deformation region could also be formed if the die system is fabricated with a rounded junction around corner O. With this consideration, AOB in Figure 3b can be regarded as a part of the die system and a formulation to calculate the total strain per ECAE pass based on pure geometric relationships is:

$$\Delta \varepsilon = \frac{(2 \, ctg \, \theta + \Phi)}{\sqrt{3}} \tag{2}$$

where $\theta = \dfrac{\phi + \psi}{2}$.

Using upper-bound analysis, Cui [2] has shown that the total strain per ECAE pass is:

$$\Delta \varepsilon = \Delta \varepsilon_a + \Delta \varepsilon_b = \frac{(2 \, ctg \, \theta + \psi)}{\sqrt{3}} \tag{3}$$

The difference between the two equations is due to an incorrect geometric relationship assumption in ref. [4]. Graphic comparisons between the two equations [2] show that the Iwahashi equation overestimates the influence of the fan angle on the effective true strain by only a small percent (3%) under certain conditions. The punch pressure can predicted by upper-bound analysis [2] and is:

$$\frac{p}{2k} = ctg\theta + \frac{\Psi}{2}. \tag{4}$$

The upper bound method is thus very useful in predicting the punch pressure (load) and flow patterns in forming processes, however, the core procedure of the upper bound approach is the construction of a geometric consistent (or self consistent) velocity field. This may not be an easy task under certain circumstances especially if the work-piece geometry is irregular and/or the flow pattern is complicated by factors such as material properties, boundary condition, and temperature effects. Moreover, since no attention is paid to satisfying the equilibrium equations, some important information such as stress and strain distributions are not available from upper bound analysis. The finite element method, which can satisfy all the controlling equations, at least in weak forms, provides a rational model by including more realistic factors. Also the effect of material properties on the deformation zone (fan angle) can be evaluated using FEM, as will be discussed in the next section.

2.3. FINITE ELEMENT METHOD

Currently, two kinds of finite element formulations, solid formulation and flow formulation, are widely used in the simulation of forming processes. The flow formulation was chosen for this study [2] because during large deformation processes such as ECAE, the amount of elastic deformation is much less that that of plastic deformation. Therefore, elastic deformation can be safely neglected and the material behavior can be described by rigid-viscoplasticity. In the flow formulation, Zienkiewicz [5] was the first to suggest mixed finite element technology. Generally speaking, mixed finite element technology can provide greater accuracy in stress calculations, since the formulation is based upon both velocity (strain-rate) and stress. Based upon this mixed finite element formulation, a code named ECAE-FEA (ECAE finite element analysis) was developed [2]. The engine code was written in FORTRAN 90 and was implemented in UNIX, VAX/VMS and Windows environments. By outputting the numerical results in a neutral format, results can be post-processed using a graphic software package such as MSC/PATRAN™. Results are shown in Figures 4 through 6.

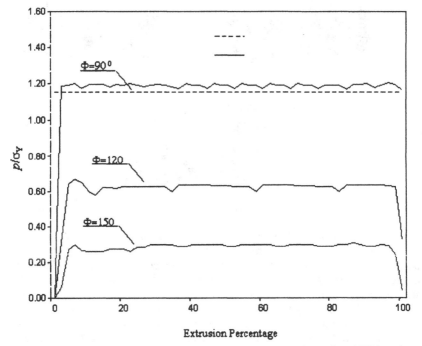

Figure 4. The comparison of the punch pressures between the upper bound analysis and FEM result.

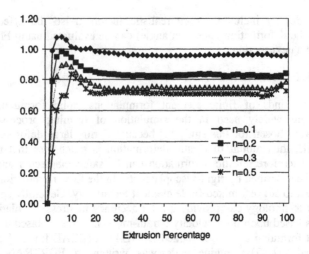

Figure 5. The influence of hardening exponent (n) on the normalized punch pressure.

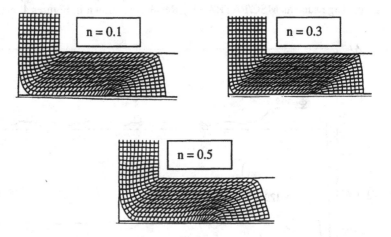

Figure 6. Deformation patterns for work hardening materials with different n values.

A comparison of punch pressures between upper bound analysis and FEM for various ECAE angles (non-strain hardenable material) is shown in Figure 4. Extrusion percentage represents percent of total stroke. They compare favorably. The influence of the strain hardening exponent (n) on normalized punch pressure using the FEM code is shown in Figure 5. These results, of course, cannot be predicted using upper bound analysis. It is interesting to note that a higher *n* value results in a lower punch pressure level. In Figure 6 it can be seen that as n increases, the elements in contact with the back leg wall of the die experience less deformation. As n increases, the fan angle, (angle) increases. From Equation (3) it is seen that as the fan angle increases, the total strain decreases. Complete details of this work can be found in ref. [2].

3. Materials Processed by ECAE

An innovative die system was designed, built, tested, and patented. The design is based on sliding components in order to reduce friction forces significantly. Both 1 inch x 1 inch and 2 inch by 2 inch die systems were constructed. Some of the materials that were investigated using the ECAE die systems include: aluminum alloys [6,7,8], aluminum powders, pure copper [9], Cu-Nb metal matrix composites [10], nickel-based super-alloys, intermetallic compounds [11], titanium and high carbon steels [12]. A substantial portion of the work was done in order to develop materials with stable sub-micron grain structures capable of superplastic deformation at lower temperatures and at higher strain rates than possible for conventional superplastic sheet materials. The combination of ECAE and sheet rolling offers potential for this type of processing. Also the production of ECAE processed fine grained structure in bulk material offers the potential for superplastic extrusion of intricate configurations for the fabrication of, e.g. integrally stiffened structures.

In the following sections, some of the above listed investigations will be summarized. Details of each of the investigations are given in the references listed above.

Al-Cu-Li base alloys offer an attractive property combination of low density, high specific strength and modulus and exceptional cryogenic properties which makes them excellent candidates for a variety of aerospace applications (e.g. space launch vehicle hardware). They are currently manufactured either by ingot metallurgy (IM), powder metallurgy (PM), and/or mechanical alloying (MA). The superplastic behavior and deformation mechanisms of aluminum alloys have been found to be very much dependent on the type of material and its processing conditions. Severe plastic deformation has been found to be effective in producing an ultra-fine grained sub-micron grained (SMG) structure, which may permit lower superplastic forming temperatures and/or higher strain rates, thus more cost effective (higher production rates). ECAE, which has the capability of producing superfine sub-micron grained structure by intense plastic deformation, may improve the superplastic behavior of these alloys.

It has been well established that superplastic deformation occurs by grain boundary sliding accommodated either by diffusional flow or by some kind of dislocation activity. Grain boundary sliding can be facilitated by grain refinement. In this investigation one objective was to determine if ECAE processing is capable of reducing the grain size to the sub-micron level in a 2090 (Al-Cu-Li) alloy. The effect of warm rolling on the superplastic behavior of the ECAE processed alloys was also investigated. Results were compared to conventionally processed 2095 superplastic sheet (rolling).

The 2095 as-cast (direct chill) plates (average grain size 105 μm) were solution heat-treated at 510°C for 70 minutes followed by peak aging at 1600°C for 20 hours. The 1 in.by 1 in. by 5 in. billets were subjected to multiple ECAE passes (routes A and C) up to effective true strains ranging between 4.68 (four passes) and 9.36 (eight passes). Some billets were extruded (ECAE) 4 passes at 350°C followed by 4 passes at 200°C via routes A and C followed by warm rolling (WR), to a strain of 1.8, at 300°C.

10

These thermomechanical schedules were selected based upon the results of other researchers (with similar materials) and preliminary experimental work at TAMU. Samples were selected from the processed billets and statically annealed at the selected superplastic deforming temperatures (450°C and 490°C) for 20 minutes (corresponding to the temperature equilibration time prior to step-strain rate and constant strain-rate tests used to evaluate the superplastic characteristics of the material. Characterization involved the experimental determination of (1) activation energy, (2) strain-rate sensitivity, (3) thresh-hold stress, and (4) percent elongation to failure.

The rate-controlling deformation mechanisms were investigated by comparing the experimental creep rates with those predicted by the Ashby-Verrall [1], Ball and Hutchinson [14], Mukherjee [15], and Sherby [16] models. The Sherby model creep rate prediction was found to best describe the experimental results for the processing conditions investigated. This could possibly be predicted since the Sherby model was developed specifically for SMG structures and the others were for conventionally processed alloys with grain sizes in the range of 2-12 μm. Typical results are given in Figure 7. More complete details are given in ref. [8]. The ECAE +WR processed material exhibited percent elongation to failure greater that 400% with strain rate sensitivities on the order of 0.38 at 490°C and 0.44 at 450°C respectively and at strain rates in the range of $2x10^{-3}$ s^{-1}. The ECAE processed alloys and those conventionally rolled had activation energies corresponding to lattice diffusion, while the ECAE + WR processed alloys had activation energy corresponding to grain boundary diffusion.

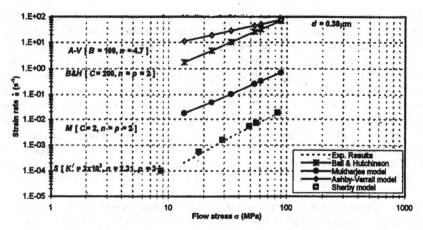

Figure 7. Comparison between the different models for 2095 billets processed via 4A, 350°C + WR at 300°C and SPD at 450°C.

TEM examination was used to evaluate sub-micron grain sizes and grain misorientation angles. In Figure 8 a and b, a well defined homogeneous structure (0.38 μm subgrains) is observed at different magnifications for 2095 billets that were ECAE processed (4 passes, route A, 350°C) plus WR processed at 300°C after annealing at 450°C for 20 minutes. A grain boundary angle of 15-18° was measured. In Figure 8 c and d, different magnifications of 2095 billets processed by ECAE (4 passes, Route A) at 350°C plus 4 ECAE passes (Route A) at 200°C are shown. A heterogeneous structure is observed consisting of roughly equiaxed subgrains (0.85 μm) plus elongated subgrains within which a recrystallized structure of 0.64 μm subgrains is observed. Helical dislocations are observed within the subgrains and at the subgrain boundaries.

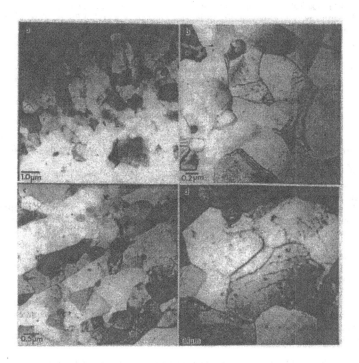

Figure 8. Bright field TEM photomicrographs for 2095 billets processed via ECAE and ECAE+WR, annealed for 20 minutes at 450°C (a) and (b) billets ECAE+WR processed via 4A, 350°C + WR at 300°C and (c) and (d) billets ECAE processed via 4A, 350°C = 4A, 200°C

4. Conclusions

The modeling efforts at TAMU have led to a better understanding of the mechanics of ECAE. A geometrically based computer program allows a clear visualization of flow patterns in multi-pass and processed material. Results indicate that multi-pass processed material may contain large volume regions of differing strain. An analysis using the upper bound method is found to be useful in predicting total strain and punch pressure. A key result of FEM analysis is incorporation of the strain hardening exponent in determining deformation patterns (fan angle) and punch pressure. A host of experiments on a variety of materials show that multi-pass ECAE produces submicron sized grains in bulk material with potential for superplastic forming.

12

5. References

1. Segal, V.M., Invention Certificate of the USSR, No. 575892, 1977.
2. Cui, H. (1996), Computational modeling of equal channel angular extrusion, Ph.D. Dissertation, Department of Mechanical Engineering, Texas A&M University, College Station, Texas.
3. Segal, V.M., Reznikov, V.I., Drobyshevkiy, A.E., and Kopylov, V.I. (1981) Plastic working of metals by simple shear, *Russia, Metall.* (English translation) **1**, 99-105.
4. Iwahashi, Y., Wang, J.T., Horita, Z., Nemoto, M., and Langdon, T.G. (1996) Principle of equal-channel angular pressing for the processing of ultra-fine grained materials, *Scripta Mater.* **35**, 143-146.
5. Zienkiewicz, O.C. (1984) Flow formulation for numerical solution of forming processes, in J.F.T. Pittman, O.C. Zienkiewicz, R.D. Wood and J.M. Alexander (eds.), *Numerical Analysis of Forming Processes*, Wiley, New York, 1.
6. Ferrasse, S., Segal, V.M., K.T. Hartwig, and Goforth, R.E. (1997) Development of a submicrometer-grained microstructure in aluminum 6061 using equal channel angular extrusion, *J. Mater. Res.* **12, No. 5**, 1-9.
7. Salem, H.G., Goforth, R.E., and Hartwig, K.T. (1998) Superplastic characterization of 2095 al-li alloy processed by equal channel angular extrusion, *Superplasticity and Superplastic Forming* 1998, Proceedings, TMS, ed. by A. Ghosh and T. Bieler., 165-178
8. Salem, H.G. (1997) Influence of equal channel angular extrusion processing on the physical, mechanical, and microstructural properties of Al-Cu-Li base alloy, Ph.D. Dissertation, Department of Mechanical Engineering, Texas A&M University, College Station, Texas.
9. Ferrasse, S., Segal, V.M., Hartwig, K.T., and Goforth, R.E. (1997) Microstructure and properties of copper and aluminum alloy 3003 heavily worked by equal channel angular extrusion, *Metallurgical and Materials Transactions A* **28A**, 1-11.
10. Segal, V.M., Hartwig, K.T., and Goforth, R.E. (1997) In situ composites processed by simple shear, *Materials Science and Engineering A* **A224**, 107-115.
11. Semiatin, S.L., Segal, V.M., Goetz, R.L., Goforth, R.E., and Hartwig, K.T. (1995) Workability of a gamma titanium aluminide alloy during equal channel angular extrusion, *Scripta Metallurgica and Materialia* **33**, 535-540.
12. Semiatin, S.L., Segal, V.M., Goforth, R.E., Frey, N.D., and DeLo, D.P. (1999) Workability of commercial-purity titanium and 4340 steel during equal channel angular extrusion at cold-working temperatures, *Metallurgical and Materials Transactions A* **30A**, 1425-1435.
13. Ashby, M. and Verrall, R., (1973) Diffusion-accommodated flow and superplasticity *Acta Metall.* **21**, 149-163.
14. Ball, A. and Hutchinson, M., (1969) Superplasticity in the Al-Zn eutectoid, *Met. Sci, J.* **3**, 1.
15. Mukherjee, A. (1971) The rate controlling mechanism in superplasticity, *Mater. Sci. Eng* **8**, 83-89.
16. Kim, W., Taleff, E., and Sherby, O.D. (1995) A proposed deformation mechanism for high strain-rate superplasticity, *Scripta, Metall.* **32**, 1625-1632.

SEVERE PLASTIC DEFORMATION OF STEELS: STRUCTURE, PROPERTIES AND TECHNIQUES

S.V. DOBATKIN
Moscow State Steel and Alloys Institute, Leninsky prospekt 4, 117936 Moscow, Russia

Abstract

This article focuses on the results from recent studies of severe plastic deformation of various types of steels, mainly, austenitic, ferritic and ferritic-pearlitic, in temperature ranges of hot, warm and cold deformation. The principles of formation of various types of structures are considered: 1) dynamically recrystallized and dynamically recovered structures during severe hot deformation. 2) dynamically recovered and submicrostructures during severe warm deformation. 3) submicro- and nanostructures during severe cold deformation. The favorable effect of phase transformations and microalloying in steels on the formation of nanostructures is discussed. The methods and techniques of severe plastic deformation for processing of steels and favorable factors for attaining large plastic strains are considered.

1. Introduction

Grain refinement by dynamic recrystallization during hot deformation is well known for producing final grain sizes larger than one micron [1, 2]. Recently it has been shown that severe plastic deformation at low temperatures promotes the formation of ultrafine-grained structure [3–6]. The specific nano- and submicrocrystalline structures are characterized by large numbers of grain boundaries whose non-equilibrium condition causes singularity of both mechanical and physical properties [4–6].

The majority of research work on microstructures and properties induced by severe plastic deformation has been done on copper, nickel, iron, titanium, as well as on aluminum and magnesium alloys [3, 4, 6]. Steels have been investigated insufficiently. The purpose of the present work is to study and compare structures and properties of steels after large strain in hot, warm and cold deformation conditions.

2. Experimental Procedure

Hot deformation was carried out on low and medium carbon structural steels, high carbon tool steels, high silicon (3.0–6.5%) ferritic electric steels, austenitic (such as 0.08%C–18%Cr–10%Ni–Ti), ferritic (such as 0.08%C–18%Cr–Ti) and austenitic-ferritic stainless steels by compression, torsion, tension (including superplastic tension),

T.C. Lowe and R.Z. Valiev (eds.), Investigations and Applications of Severe Plastic Deformation, 13–22.
© *2000 Kluwer Academic Publishers. Printed in the Netherlands.*

hydrodynamic extrusion, as well as flat, strip, section, tube rolling in the range of temperature 600–1300 °C, strain rates 0.001–180 s^{-1}, and strains up to 20 [7–10].

For warm deformation conditions, equal channel angular (ECA) pressing was used to achieve large strains [5, 11]. Plain carbon steels with 0.17 % C, 0.32% Mn and 0.18% Si were chosen for this processing because of their rather low yield stress [12]. Both normalized and hot rolled initial states of steels were used.

ECA pressing was carried out using the samples of 20 mm in diameter and 120 mm in length at the temperature 500° C. Samples were pressed through a die for 2 and 4 passes, with 180° turn around the axis of pressing after each pass, producing alternating deformation. The die had an angle of intersection of the channels of 90° In our experiments, four passes of ECA pressing corresponded to the maximal strain the samples could sustain before destruction.

Cold deformation was performed by torsion under 6 GPa pressure on the samples of 10 mm in diameter and 0.5 mm in thickness. Deformation was continued up to 9 revolutions, producing strains of $\varepsilon = 6.4$. The 0.08% C–18% Cr–10% Ni–0.6% Ti and 0.05% C–15% Cr–10% Ni–V–N austenitic steels having fcc lattices and 0.08% C–18% Cr–0.6% Ti ferritic steel as well as armco iron with bcc lattices were chosen as materials of research [13]. The average size of an initial grain was 40–45 microns for austenitic steels and 100–110 μm for ferritic steel and armco-iron, respectively.

3. Results and Discussion

3.1. HOT SEVERE DEFORMATION OF STEELS

Depending on the deformation temperature, strain and strain rate, one can obtain random dislocation substructures, dynamically recovered (polygonized) substructures, or dynamical recrystallized structures [1, 2, 7–10]. The task is then to select the desired structural state and to prescribe the deformation conditions necessary to attain this state.

To determine temperature and strain rate ranges for dynamic recovery and recrystallization under hot deformation conditions and to quantitatively evaluate the resulting structure, structure state maps (SSM) are plotted in coordinates "temperature - strain rate" for given strain and initial grain size (Figure 1) [8, 10].

The T_1 line corresponds to the dynamic recrystallization start temperature, while the T_2 line corresponds to 100% dynamic recrystallization. Major lines quantitatively describing the microstructure are: D_R for equal dynamically recrystallized grain sizes, d_{sg} for equal subgrain sizes and V_R for equal recrystallization volume fractions.

In bcc polycrystals with higher stacking fault energy, maps of structural state differ in that the dynamic recrystallization threshold more strongly depends on strain rate than in fcc polycrystals, and the region of dynamic recrystallization is limited by high temperatures and strain rates (Figure 1b).

It was found (in bcc polycrystals especially), that subgrain coalescence leads to

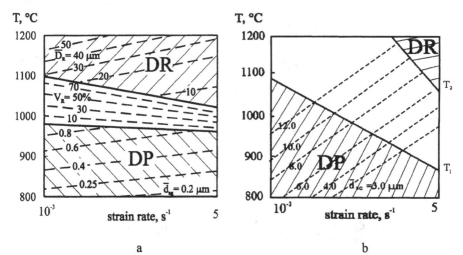

a   b

Figure 1. Structure state maps during high temperature compression (ε=0.5)
(a) 0.08% C - 18% Cr - 12% Ni - Ti austenitic steel (initial grain size - 30 μm)
(b) 0.08% C - 18 % Cr - Ti - ferritic steel (initial grain size - 150 μm)
DR - dynamic recrystallization, DP - dynamic polygonization (recovery).

formation of larger subgrains with higher misorientation of subboundaries. Thus, continuous increase in subgrain misorientation due to coalescence can be attributed to dynamic recrystallization *in situ*, or to continuous recrystallization. However, we believe that this process can be more correctly characterized as dynamic continuous recovery.

Structure state maps (SSM) vary with strain (Figure 2). At a given strain rate, dynamic recrystallization threshold shifts to lower temperatures with increasing strain at steady state. For example, steel with 0.08% C, 18% Cr, 1% Ti may be subject to hot torsion to very large strains, up to ε = 20 at 1100° C, yet the threshold strain for dynamic recrystallization is so low that fully dynamically recovered structure could not be obtained.

The systematic behaviors of low-, medium- and high carbon structural and tool steels, high silicon ferritic electric steels, ferritic, austenitic and austenitic-ferritic stainless steels deformed in compression, torsion, tension (including superplastic tension), hydrodynamic extrusion, flat, strip, section and tube rolling in the ranges of T = 600–1300°C, $\dot{\varepsilon}$ = 0.001–180 s^{-1}, ε = 0–20 allowed us to determine a general sequence of stages of structure evolution: random hot working (RHW) – unsteady dynamic recovery (UDR) - steady state dynamic recovery (SSDR) – dynamic continuous recovery (DCR) – dynamic recrystallization (DRX) – dynamic normal grain growth (DNGG) [7, 8].

All of these stages were observed only in torsion at moderate strain rate and T and in tension under the conditions of superplasticity. In real hot working (decreased temperatures, high strain rates, fine initial grain size) most of the stages shorten and even degenerate. Decrease in strain rate, initial grain coarsening, alloying and increase in stacking fault energy are favorable for developing dynamic recovery.

16

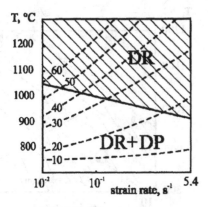

DR - dynamic recrystallization
DP - dynamic polygonization

Figure 2. Structure state map during high temperature torsion of 0.08% C - 18 % Cr -Ti ferritic steel (ε= strain of destruction, initial grain size - 150 μm) (recovery).

In steels, the recovered structure of the high temperature phase results in room temperature tensile strength increase by 10–80% as compared to statically and dynamically recrystallized structures [8, 9].

This is accompanied by:
– decreasing toughness and plasticity in austenitic stainless steels;
– decreasing toughness but growing plasticity in austenitic-ferritic stainless steels with 30–50% of ferrite;
– increasing toughness, especially at low temperatures but decreasing plasticity in ferritic stainless steels and in ferritic-pearlitic controlled rolled steels;

– simultaneously increasing plasticity and deformation and fracture resistance in martensitic quenched steels; in high carbon steels red-shortness reduces due to precipitation of carbides.

The recovered structure does not deteriorate and in some cases (especially in a weak acid environment) even enhances pitting and intercrystalline corrosion resistance of austenitic chromium-nickel steels due to homogeneous distribution of both alloying elements Cr and Ni, and impurities, particularly phosphorus.

Thus, during hot severe deformation we could obtain both dynamic recovered and dynamic recrystallized structures. The grain size of steels obtained after severe hot deformation was 5–10 μm, although theoretically it could be 1 μm [14].

3.2. WARM SEVERE DEFORMATION OF STEELS

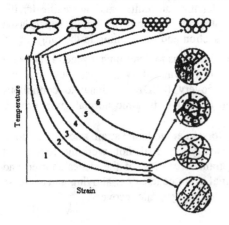

1 - random hot working
2 - unsteady dynamic recovery
3 - steady state dynamic recovery
4 - dynamic continuous recovery
5 - dynamic recrystallization
6 - dynamic grain growth

Figure 3. Sequence of structure formation processes during hot deformation.

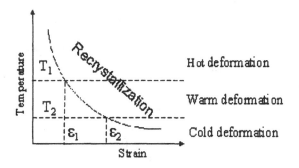

 Let's propose that the scheme in Figure 3 is applicable not only for hot deformation conditions, but for the whole temperature range (Figure 4). To achieve start temperature for dynamic recrystallization at warm deformation conditions we need strains higher that we usually use. This prospect is addressed for ECA pressing by V.M. Segal [5, 11]. The ferritic-pearlitic steel with 0.17% C, 0.32% Mn, 0.18% Si was chosen as having a rather low yield stress[12].

Figure 4. Strain dependence of dynamic recrystallization start temperature.

Electron microscopy reveals the formation of two types of structure during deformation: 1) dynamically recovered – with low – and middle angle subgrain misorientation, and 2) submicrocrystalline– with high angle grain boundaries (Figure 5). Both elongated and rather equiaxed subgrains were observed in the recovered structure of ferrite. There is also substructure inside pearlite between plates of cementite. Submicrocrystalline grains exist in the substructure of the basic ferrite as well as in the ferritic substructure of pearlite. (Figure 5). The sizes of grains in these cases are defined by a distance between subboundaries (0.3–0.4 μm) and between plates of cementite (0.1–0.2 μm), that causes large dispersion of the distribution of average dimensions. Average distances between boundaries both in submicrocrystalline – and other substructures are 0.25–0.45 μm. The fragmentation and spheroidization of cementite plates is observed (Figure 5b). Average sizes of structural elements in steel with

0.5 μm

0.3 μm

Figure 5. Structure of 0.17% C steel after ECA pressing with T = 500^{0}C, φ = 90^0; N = 4.

18

different initial states, measured on foils cut out both in longitudinal and in cross directions are the same.

The microhardness of 0.17% C – steel after ECA pressing are by 30–50 % higher in comparison with the normalized condition. Already with N = 2 the high meanings of microhardness are reached which with increase of shear strain up to N = 4 raise insignificantly (Figure 6).

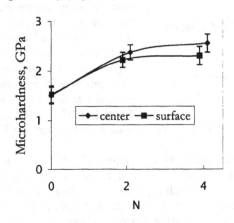

Figure 6. Microhardness of 0,17 %C steel after ECA pressing (T=500 ^{0}C; φ=90^0).

The shape of flow curves after tension as well as the close meanings of YS and UTS (Table 1) testify to a high strength state of 0.17% C steel after ECA pressing.

Already with N = 2 a level of strength achieves high meanings with satisfactory plasticity and changes a little with increase of shear strain.

The strengthening of 0.17% C–steel after ECA pressing with initially hot rolled state is higher, than in initially normalized one. If take into account similarity of substructure of ferrite in both cases, the increase of properties in hot-rolled state, apparently, is explained by different states of pearlite.

The level of mechanical properties (YS = 700–800 MPa with EL = 10–15%) of 0.17% C–0.32% Mn–0.18% Si steel after ECA pressing with T = 500° C corresponds to a level of high-strength low alloyed steels used now in building designs.

Thus, the opportunity of formation both dynamically recovered and submicrocrystalline (0.1 μm < D < 1 μm) structures in steels during severe warm deformation is shown.

TABLE 1. Mechanical properties of 0.17% C steel after
normalization and ECA-pressing (T = 500° C and φ = 90^0)

Treatment	UTS, MPa	YS, MPa	δ, %	ψ,%
Normalization	495	295	30	75
N = 2 (origin state-normalized)	725	715	15	53
N = 4 (origin state-normalized)	740	715	13	53
N = 4 (origin state-hot rolled)	850	840	9.5	52

3.3.   COLD SEVERE DEFORMATION OF STEELS

The use of equal channel angular pressing for steels at low temperature is too difficult and depends upon the techniques applied. Usually torsion under high pressure is used for steels.

Figure 7. Structure of 18 % Cr-Ti ferritic steel after SPD, x 50000.

TEM analysis of the investigated materials after torsion with ε = 4.3 (1 revolution) showed that separate structural elements with high angle boundaries are formed [13]. The process of high angle boundaries formation in ferritic steel and, especially, in armco-iron, occurs faster, than in austenitic steels, probably, due to a greater stacking fault energy. In ferritic steel already after ε = 5.3 (3 revolutions) a large volume fraction is occupied by structural elements with high-angle misorientation and with size of 100 – 400 nm. For ε =5.3–5.8 (3–5 revolutions) mostly oriented submicrocrystalline structures were formed, gradually becoming more equiaxed with increasing strain. In all materials the size of structural elements becomes 50–200 nm with ε = 6.4 (9 revolutions) (Figure 7). The character of diffraction pattern indicates the presence of both middle – and high-angle boundaries between structural elements. The greatest size of structural elements (100–200 nm) is revealed in armco-iron and the least (50–100 nm) – in austenitic steels.

Thus, in austenitic steels the nanocrystalline structure (the size of grains should be less than 100 nm) is only formed during severe plastic deformation by torsion. In other steels it is possible to form partially nanocrystalline structures. Therefore, henceforth we shall characterize the structures obtained as submicrocrystalline (SMC).

Martensitic $\gamma \rightarrow \alpha$ and $\gamma \rightarrow \varepsilon \rightarrow \alpha$ transformations in austenitic steels during cold deformation were observed by X-ray diffraction analysis [13]. There was approximately $\approx$ 50% of α-martensite, $\approx$ 35% of austenite, $\approx$ 10% of δ-ferrite and $\approx$ 5% of ε-martensite in the final structure of 18% Cr – 10% Ni – Ti steel after ε = 6.4 (9 revolutions).

In nitrogen containing 15% Cr – 10% Ni–V steel, $\approx$ 35% of α-martensite and $\approx$ 10% of ε-martensite was revealed after torsion with ε = 6.4. Thus, it is possible to speak already about two-phase austenitic-martensitic steels.

Severe plastic deformation results in significant increase of microhardness in all materials (Figure 8).

The high level of microhardness for all investigated steels after torsion with ε = 5.8 – 6.4 (5–9 revolutions), and also smaller microhardness for armco-iron indicates that submicrocrystalline strengthening, first of all, is due to growth of the concentration of grain boundaries, while the influence of alloying on hardening through its affect on phase composition (austenite, ferrite, martensite) is insignificant.

Thermal stability of SMC-structures obtained by torsion was compared to stability of cellular structures after cold compression (Figure 8). After compression the difference in hardening both at the expense of different alloying for the same lattices and due to distinction of lattices for close alloying are kept.

20

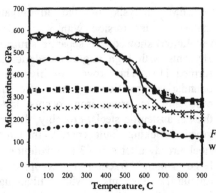

Legend:
——— - torsion (5 turns)
------ - compression (ε=0.5)
• - 18 Cr - Ti steel
◆ - Armco-iron
X - 15 Cr-10 Ni - N - steel
▲ - 18 Cr-10 Ni - Ti - steel

Figure 8. Thermostability of materials studied with submicrocrystalline and cellular structure.

The microhardness of deformed materials with SMC structures is 1.8–2.5 times higher, than in materials with cellular structures (Figure 8). The microhardness in materials with SMC-structure begins falling already at 400°C, which is lower by 300° C than in materials with cellular structure. But up to 650°C the microhardness of SMC structure remains higher.

The recrystallization start (grain growth) temperatures on heating in austenitic steels (especially with nitrogen) are higher than in ferritic steels due to presence of martensitic phase.

The process of recrystallization usually includes the nucleation and growth of new grains in the deformed matrix. During severe cold deformation the appearance of new grains with high-angle boundaries appear with the formation of submicrocrystalline structure. There is a complete absence of grain growth because of low deformataion temperature.

This low temperature process may be characterized as dynamic fragmentation "in situ" or more precisely as dynamic nucleation of new grains [13]. So this is the first stage of recrystallization - nucleation.

With heating submicrocrystalline grains grow, *i.e.* the second stage of recrystallization takes place. This process is reminiscent of meta- or postdynamic recrystallization due to the preexisting nuclei in the deformed matrix.

Thus, during cold severe deformation of steels it is possible to obtain and submicro- and nanocrystalline structures.

3.4. SPD TECHNIQUES

For attaining of large plastic strains the following factors are favorable:

1. Deformation mode approaches simple shear.
2. Deformation occurs without final shaping.
3. Alternating deformation.

3.4.1. Alternating Bending

Prof. N.I. Krylov *et al.* [15, 16] proposed a number of ways of hot deformation based on single and bi-plane bending. They constructed two types of bending machines: roller and rotor. Single plane roller machines were used for deformation sheets and tubes from stainless austenitic steels and low-carbon low-alloying steels. Bi-plane roller machines – for deformation of tubes from plain carbon steels by alternating bending in mutually perpendicular planes.

Rotor machines deform tubes in longitudinal and transversal directions by alternating bending. These machines have four cassettes with rollers tilted from the longitudinal axis. Four cassettes rotate in pairs in opposite directions to prevent rotation of the tube.

Using such deformation, the thermomechanical treatment of electric weld tubes was carried out. It should be mentioned that the bending schemes proposed by Prof. Krylov were used under hot deformation conditions.

3.4.2. All-Round Multiple Forging

All-round forging utilizes multiple free-forging operations of upsetting, with changing loading axes [17, 18]. In this scheme the deformation is less homogeneous than in other schemes of intensive plastic deformation but this scheme allows one to deform highly brittle materials and to obtain UFG states in them. The temperature range of all-round forging is 0.1–0.5 T_m.

3.4.3. Equal Channel Angular Pressing

ECA pressing is used now for steels mainly at high temperatures [5]. So the prospects to use ECA pressing for steels depends on two challenges: 1) utilization of ECA pressing for steels at room temperature to obtain bulk steels with nanocrystalline structure; and 2) developing continuous schedules of ECA pressing. The first task should be feasible by using back pressure and decreasing the friction between die and sample due to three moving walls [5] or active forces of friction. The feasbility of the second task may be apparent in ECA pressing with continuous processes of direct pressing such as "conform", "linex" or "extrolling" [5].

3.4.4. Extrusion with Torsion and Bending

This scheme allows for large shear strains [19]. It resembles the equal channel angular pressing process but with a rotating exit channel. By varying the inclination angle of channels the geometry of their longitudinal and cross-section, compression and direct pressing can be added to torsion and bending. This scheme is currently used mainly for consolidation of powders at room temperature.

4. Conclusions

1. During severe plastic deformation the following structures form:
   - dynamiclly recrystallized ($D > 1$ μm) and dynamically recovered in hot deformation conditions;
   - dynamically recovered and submicrocrystalline (0.1 μm $< D < 1$ μm) in warm deformation conditions;
   - submicro- and nanocrystalline ($D < 0.1$ μm) in cold deformation conditions.

2. For obtaining nanocrystalline structures in steels during severe deformation microalloying and the presence of a second phase are favorable.

3. For attaining large plastic strains the following factors are favorable:
   - deformation mode approaches simple shear;
   - deformation without final shaping;
   - alternating deformation.

5. References

1. Jonas, J.J., Sellars, C.M., and Tegart, W.J. McG. (1969) Strength and structure under hot-working conditions, *Metallurgical Reviews* **130**, 1-33.

2. Sakai, T. and Jonas, J.J. (1984) Dynamic recrystallization — mechanical and microstructural considerations, *Acta Metall.* **32**, 189—209.

3. Smirnova, N.A., Levit, V.I., Pilyugin, V.P., Kuznetsov, R.I., and Degtyarev, M.V. (1986) Low-temperature recrystallization peculiarities of nickel and copper, *Fizika metallov i metallovedenie* **62**, 3, 566—570 (in Russian).

4. Valiev, R.Z., Korznikov, A.V., and Mulyukov, R.R. (1993) Structure and properties of ultrafine-grained materials produced by severe plastic deformation, *Mater. Sci. Eng.* **168**, 141—148.

5. Segal, V.M., Reznikov, V.I., Kopylov, V.I. *et al.* (1994) *Processes of plastic structure formation in metals*, Science and Engineering Publ. House, Minsk, Belorus (in Russian).

6. Ultrafine-Grained Materials Produced by Severe Plastic Deformation. A thematical issue. Ed. by R.Z. Valiev, *Ann. Chim. Fr.* (1996} **21**, 369—480.

7. Bernshtein, M.L., Kaputkina, L.M., Prokoshkin, S.D., and Dobatkin, S.V. (1985) Structural changes during hot deformation of austenite in alloy steels, *Acta Metall.* **33**, 247—254.

8. Bernshtein, M.L., Dobatkin, S.V., Kaputkina, L.M., and Prokoshkin, S.D. (1989) *Hot stress-strain curves, structure and properties of steels*, Publishing house "Metallurgiya", Moscow (in Russian).

9. Kaputkina, L.M. and Dobatkin., S.V. (1997) Efficiency of polygonized structure obtained by thermomechanical processing of steels, *Proc. of the Inter. Conf. on Thernonechanical Processing of steels and other materials (THERMEC-97), Wollongong, Australia* **1**, 587—593.

10. Dobatkin, S.V. and Kaputkina, L.M. (1997) Application of hot deformation structure state maps to sheduling thermomechanical processing of steels, *ibid*, 907—910.

11. Segal, V.M., Ganago, O.A., and Pavlik, D.A. (1980) Processing of cast samples by simple shear, *Kuznechno-shtampovochnoe proizvodstvo* **2**, 7—9 (in Russian).

12. Dobatkin, S.V., Valiev, R.Z., Krasilnikov, N.A., and Konenkova, V.N. (1999) Structural inhomogeneity of plain carbon steel during equal channel angular pressing at low temperature, *Proc. of the Fourth Inter. Conf. on Recrystallization and Related Phenomena (REX'99), Tsukuba City, Japan*, 913—918.

13. Dobatkin, S.V., Valiev, R.Z., Kaputkina, L.M. et al. (1999) Recrystallization of steels with submicrocrystalline structure, *ibid*, 907—912.

14. Shtremel, M.A. (1999) *Strength of alloys. Part 1. Defects of lattice*, Publishing house of Moscow State Steel and Alloys Institute, Moscow (in Russian).

15. Kitaysky, V.E., Krylov, N.I., and Slonim, A.Z. (1978) Thermoplastic processing of low-carbon steel tubes using roller deformation machines, *Zbornik trudov "Termicheskoe i termomechanicheskoe uprochnenie metallov", Moscow*, 21—23 (in Russian).

16. Krylov, N.I. et al. (1981) New methods and machines for strengthening of weld tubes, *Zbornik trudov "Novye materialy i metody ih obrabotki", Kiev, IES*, 40—41 (in Russian).

17. Salishev, G.A., Valiakhmetov, O.R. and Galeyev, R.M. (1993) *J. Mater. Sci.*, **28**, 2898.

18. Kaibyshev, O.A., Kaibyshev, R.O., and Salishev, G.A. (1993) *Mater. Sci. Forum*, **113-115**, 423.

19. Perelman, V.E. (1999) The usage of combine influence of large plastic strain for obtaining ultradense and ultrafirm blanks from powder materials, *Abstracts of NATO ARW "Investigations and applications of severe plastic deformation", Moscow*, 100.

APPLICATION OF ECAP – TECHNOLOGY FOR PRODUCING NANO- AND MICROCRYSTALLINE MATERIALS

V.I. KOPYLOV
Physical-Technical Institute of National Academy of Science of Belarus
Kuprevicha str., 10, Minsk, Belarus

Abstract

Conditions for producing nano- and microcrystalline structures (NMC) by ECAP technology are described. The concept of "clear ECAP" for describing uniform ECAP deformation is introduced. A model of grain refinement and temperature-strain rate conditions of effective refinement is described.

1. Introduction

Equal channel angular pressing (ECAP) is an effective method for controlling metal microstructures [1–2]. One of the most promising applications of ECAP technology is production of nano- and microcrystalline (NMC) materials [3]. Recently, significant successes were demonstrated in this field (see overview [4]). The characteristics of some NMC materials produced by ECAP at the Physical-Technical Institute of National Academy of Sciences of Belarus are described in Table 1. Many fundamental problems of using ECAP to produce NMC materials remain unsolved. In particular, the problem of producing the desired NMC structure requires discussion. The present work is devoted to the consideration of this problem.

2. Conditions to Produce NMC Structures using ECAP Technology

It is important to consider deformation processes at the different structural levels in order to specify the special conditions necessary to produce NMC structures by the ECAP method. These conditions can be very complicated.

On the macro-level, for NMC structures, it is necessary to create the conditions for produce severe plastic deformation. Here there is the problem of determining the deformation magnitude $\varepsilon_0(d_0)$ that is required to obtain the desired grain size d_0. For the ECAP-process, as a rule it is necessary impose multi-pass deformation $\varepsilon_0(d_0) = N_1 \varepsilon_1$ (N_1 – the pass number) to obtain sufficiently severe plastic strains. It is further necessary to ensure that the deformation is uniform during every ECAP pass. The condition of uniformity is the major macroscopic constraint for obtaining nanostructured material by severe plastic deformation. This condition is ensured with the right choice of ECAP temperature and strain-rate and the right choice of ECAP technological regimes.

23

T.C. Lowe and R.Z. Valiev (eds.), Investigations and Applications of Severe Plastic Deformation, 23–27.
© 2000 *Kluwer Academic Publishers. Printed in the Netherlands.*

TABLE 1. Mechanical properties of materials after standard and ECAP treatments

Material	Method of treatment	Vickers hardness, HV	Yield Stress, $\sigma_{0,2}$ MPa	Ultimate strength, σ_b MPa	Elongation, $\delta\%$	Reduction of Area, $\psi\%$
α-Fe	ECAP	300	850	980	6	55
	Standard	90	200	350	21	70
Cu (99,98%)	ECAP	155	400	480	11	67
	Standard	55	100	240	19	87
Ni (99,86%)	ECAP	310	600	920	8	62
	Standard	70	170	400	22	90
Austenitic steel	ECAP	640	1340	1550	27	41
	Standard	200	210	560	70	69
Fe-36% Ni INVAR	ECAP	300	820	850	16	74
	Standard	120	280	460	30	88
Fe-45% Ni ELINVAR	ECAP	360	1540	1650	10	30
	Standard	120	760	930	15	44

The magnitude of the optimum deformation ε_1 per pass is limited by the processes taking place on the microlevel. This magnitude should satisfy two contradictory conditions. On one hand, ε_1 should be rather large to supply the effective grain refinement during deformation. On the other hand, the magnitude ε_1 should not be so large as to cause failure of the material. The complexity of determining the optimum magnitude ε_1 is associated with the dependence of ε_1 on the physical nature and structural parameters of materials. Macroscopically optimizing the magnitude ε_1 involves imposing limits on the temperatures and strain-rate during ECAP deformation. Calculation of the optimum ε_1 is carried out using the theory of severe plastic deformation [5] and the theory of the non-equilibrium grain boundaries [6].

It is important to note that the temperature during ECAP is also limited by the requirement for stability of the NMC microstructure during deformation. The ECAP temperature should not exceed the temperature of recrystallization for NMC metals which, because of the grain boundary non-equilibrity, is much less than the usual value and depends on strain-rate and the deformation value ε_1.

3. Macro-Level Conditions: Clear ECAP

The condition of ECAP deformation macro-uniformity must be provided by special static boundary conditions on every deformation pass. These conditions depend upon the friction between the sample and press equipment, the contact area, and the special back pressure. In this case the uniform localized deformation of simple shear concentrates on the line of channel intersection. In the case when the special boundary conditions are not observed, different versions of non-uniform plastic flow occur.

Figure 1 illustrates the important role of the requirement for uniform deformation. As shown in Figure 1b, c, after one pass of plastic deformation during incorrect ECAP, the plastic flow is non-uniform. Such non-uniform flow leads to non-uniformity of both structure and properties. On the next pass, the structure becomes more non-uniform, leading to fracture of the sample. A different situation is observed during the correct realization of ECAP deformation (Figure 1a). In this case, which we have named "clear ECAP", the deformation is uniform, and it is possible to realize the multi-pass uniform deformation required for producing NMC structures.

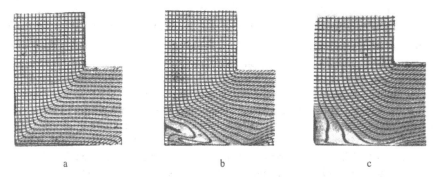

a   b   c

Figure 1. Experimental pictures of nets marked on samples deformed for different ECAP conditions: (a) uniform deformation (clear ECAP); (b), (c) - non-uniform deformation.

It is important to emphasize again that not any realization of pressing the billet through the intersecting channels provides the formation of NMC structures. For the NMC structure formation it is necessary to ensure uniform ECAP deformation which is named "clear ECAP".

4. Micro-Level Conditions: Limit of Grain Refinement

The fundamentals of the modern theoretical conceptions of the plastic grain refinement mechanisms are based on the theory of severe plastic deformation [5] and theory of the non-equilibrium grain boundaries [6]. As demonstrated in [7], during intragranular deformation orientational misfit dislocations appear in the grain boundary, and after dislocation spreading, complicated systems of defects occur in the grain boundary joints. As a first approximation, such defect systems can be described as the joint disclination system. The power of such disclinations ω increases with intragranular deformation $\omega = \psi \cdot \varepsilon_v$ (where ψ – the parameter characterizing the level of plastic deformation uniformity, ε_v – the magnitude of intragranular deformation). The joint disclinations cause the internal stress $\sigma_i \sim A G \omega$ ($A \sim 1$ – the numerical coefficient, G – shear modulus) which also increases with deformation. The various accommodation processes are activated by internal stress σ_i and lead to decrease of the internal stress.

Depending on the temperatures and strain-rate conditions and the structure parameters of a material (the grain size, the size and volume fraction of the second phase particles *etc.*), the single accommodation process usually becomes dominating and determines the mechanical behavior of the material.

Under ECAP deformation conditions, one of the mechanisms of joint disclination stress relaxation can be the local rotation of the crystalline lattice. This process can be described as the joint disclination generation of the partial disclination or the generation of the gapped boundaries. The motion of these defects through a grain leads to the grain fragmentation. These processes are possible under conditions of nonconservative motion of dislocations. Hence, for the generation of defects mentioned above and, therefore, for the fragmentation of material, the stress σ_i generated by the joint disclinations must exceed the threshold stress σ_0 which is necessary for the activation of accommodation slip systems near the joints: $\sigma_i > \sigma_0$. In the elementary case when $\sigma_i \sim \omega \sim \varepsilon_v^c$, this condition can be described as the condition for the critical intragranular deformation ε_v^c. If the value ε_v^c exceeding value $\varepsilon_v^{(1)} > \varepsilon_v^c = \sigma_0/[A\psi G]$ the grains fragment. The magnitude ε_v^c evaluation proposed above is equitable only in the case when the power of joint disclinations continuously increases with deformation: $\omega \sim \psi\varepsilon_v$. However, such dependence is realized under the specific conditions of plastic deformation when diffusion accommodation processes are absent. Under real conditions, the processes of diffusion accommodation of joint disclinations always contribute in the kinetics of accumulation ω. According to [6], at room and elevated temperatures of ECAP, such contribution is rather large and caused by the non-equilibrium grain boundary state and the low values of activation energy for grain boundary self-diffusion.

Under the condition of the diffusion accommodation, the joint disclination power ω can be determined by the kinetic equation: $\dot{\omega} = \psi\dot{\varepsilon}_v - \omega/t_r$ where the first addend describes the power ω increasing as a consequence of intragranular plastic deformation, and the second addend describes the power ω decreasing as a consequence of diffusion mass transfer [8]. Under the steady-state deformation conditions it can be written: $\dot{\omega}$ =0 and $\omega^s = \psi\dot{\varepsilon}_v t_r$. Characterizing time t_r of diffusion accommodation of a disclination dipole with the rotation vector ω_0 and the dipole arm $1 = d$ can be obtained from the expression [8]: $t_r = Md^3kT/\delta D_b^* G\Omega$. Here D_b^* is the non-equilibrium grain boundary self-diffusion coefficient [6], M – the numerical factor, δ – thickness of boundary, Ω – atomic volume.

On the basis of the above relationships, it is possible to evaluate the characteristic grain size d_0. If the magnitude of grain size reaches the value d_0, the diffusion accommodation of joint disclinations occurs so fast that the internal stresses σ_i have no time to increase up to the level σ_0, which is required for activation of the accommodation slip systems and the initiation of the grain fragmentation. The value of d_0 is determined by the following expression $d_0 = (\sigma_0 \Omega D_b^* \delta/MA\psi kT_v)^{1/3}$. This expression allows one to evaluate the minimum grain size, or sometimes-called the "limit of grain refinement," which can be reached during ECAP deformation.

5. Temperature and Strain Rate Conditions of ECAP Grain Refinement

The grain refinement mechanisms stated above allow one to evaluate the optimum parameters of ECAP processing. In particular, from the proposed model, the simple expression for the optimum strain rate of ECAP deformation $\dot{\varepsilon}_\gamma(d_o)$ to obtain a given grain size d_0 is   . This value determinates the strain rate which is required for the desired grain size d_o. It is important to underline that this expression describes the local strain rate rather than the average of intragranular deformation.

The optimum ECAP deformation temperature can be similarly calculated too.

A limitation on the ECAP deformation temperature is associated with the necessity to hinder recrystallization during both the ECAP process and the heat treatment of material between deformation passes. As is shown in [9], the recrystallization process in NMC materials produced by the ECAP methods has some specific singularities. In particular, the recrystallization temperatures of NMC materials produced by ECAP technology are much lower than the recrystallization temperatures of conventional metals. This is due to the special non-equilibrium state of the grain boundaries and the special distribution of defects in the grain boundaries of NMC metals. For pure metals this recrystallization temperature is defined by expression [9]: $T_r = Q_b^*/k ln$ $(D_{b0}^* \delta G \Omega^*/d^3 kTA_I)$, where t^* is annealing time. As it is obvious from this expression, the magnitude T_r mainly depends on activation energy of non-equilibrium grain boundary diffusion and, consequently, the grain boundary state.

6. References

1. Segal, V.M., Reznikov, V.I., Kopylov, V.I., Pavlik, D.A., Malyshev, V.F. (1994) *Processes of Plastic Structural Formation of Metals*, Nauka i Technika Publishers, Minsk (in Russian).
2. Segal, V.M., Reznikov, V.I., Drobyshevski, F.E., Kopylov, V.I. (1981) Plastic treatment of metals by simple shear, *Izv. Akad. Nauk SSSR, Metally* 1, 115–118.
3. Akhmadeev, N.A., Valiev, R.Z., Kopylov, V.I., Mulyukov, R.R. (1992) Producing of ultrafine grain size structure in copper and nickel with using intensive shear deformation, *Metally* 5, 96–101.
4. Valiev, R.Z. (1996) Ultrafine grained materials produced by severe plastic deformation: an introduction, *Annales de Chimie - Science des Matériaux* 21, 369–378.
5. Rybin, V.V. (1986) *Severe Plastic Deformation and Fracture of Metals*, Metalurgia Publishers, Moscow (in Russian).
6. Chuvil'deev, V.N. (1996) Micromechanism of deformation-stimulated grain-boundary self-diffusion, Part I-III, *The Physics of Metals and Metallography* 81, 463–468; **81**, 583–588; **82**, 71–77.
7. Perevezentsev, V.N., Rybin, V.V., Chuvil'deev, V.N. (1992) The theory of structurual superplastisity, Part I-IV, *Acta Metall. Mater.* 40, 5, 887–923.
8. Perevezentsev, V.N., Pirozhnikova, O.Ed., Chuvil'deev, V.N. (1991) Grain growth during superplastic deformation of microduplex alloys, *The Physics of Metals and Metallography* 71, 4, 33–41.
9. Kopylov, V.I., Malashenko, L.M., Kukareko, V.A.. Makarov, I.M., Chuvil'deev, V.N. (1999) Recovery and recristallization in microcrystalline metals prodused by ECAP, *Part I-II. Submitted to The Physics of Metals and Metallography.*

SEVERE DEFORMATION BASED PROCESS FOR GRAIN SUBDIVISION AND RESULTING MICROSTRUCTURES

A.K. GHOSH AND W. HUANG
Department of Materials Science and Engineering
The University of Michigan, Ann Arbor, MI 48109

Abstract

Significant interest exists within the materials science community to create nanocrystalline microstructures in metallic alloys in an effort to obtain improved strength and processability of these alloys. Repeated deformation methods, such as 3-axis forging, ECAE (Equal Channel Angular Extrusion), pressurized torsion *etc.* have been used to create bulk materials with 50 – 500 nm size microstructures. Stability of these microstructures to further deformation and to thermal exposure is however a major problem which needs to be addressed in the future years. Another issue for process scale-up is die-wall sticking with the moving workpiece, which make scale-up of billet size a nontrivial problem. To minimize these structure-altering aspects of rolling deformation, a surface-shear based deformation technique has been developed and applied to plates and sheets of aluminum alloys to create desired fine-scale microstructures. Initial studies using this and other processes on several Al alloys with and without dispersoid particles have been analyzed by optical, and electron microscopy. Preliminary results on mechanical behavior and texture are reported.

1. Introduction

Nanocrystalline and ultrafine grain microstructures (50-500 nm grain size), characterized by a large volume fraction of grain boundary regions, can provide unique mechanical properties [1,2]. Often such structures are synthesized by vapor deposition and thermal treatments, electrodeposition, *etc.* [1,3-6]. For production of bulk materials however, deposition based processes are generally too slow and impractical. Alternative methods such as repeated metal working approaches involving severe plastic straining have been of interest for subdividing grains into fine scale microstructures [7-9]. In these methods, shear localization in internal shear zones is generally the precursor mechanism. When the process is repeated to produce extensive distribution of these zones throughout the bulk with attendant recovery mechanisms, extensive grain subdivision and grain refinement is observed.

Development of shear zones can be due to externally imposed shape changes as well as internal obstacles for flow, *e.g.* hard second phase particles [10,11]. It has been well known in studies on thermomechanical processing that rolling and extrusion are preferred processes for developing a product form which can also introduce extensive

29

T.C. Lowe and R.Z. Valiev (eds.), Investigations and Applications of Severe Plastic Deformation, 29–36.
© *2000 Kluwer Academic Publishers. Printed in the Netherlands.*

redundant shear and subsequent continuous recovery or discontinuous recrystallization processes to produce ultrafine grain microstructures [12-14]. The use of internal particles has been a major focus for not only initiating "particle-stimulated nucleation", but for pinning and stabilizing grain boundaries to prevent coarsening [11-16]. Grain sizes well below 10 μm, going down to 200-1000 nm have been achieved by these conventional processing techniques when suitable alloy chemistries and particle distributions are used.

Present efforts on multipass deformation methods take advantage of continuous recovery of the high dislocation density aided by changes in deformation path between passes while introducing additional dislocations during the new passes. Among these processes, ECAE (Equal Channel Angular Extrusion), Repeated Pressurized Torsion and Three-Axis Forging processes have been utilized for fabricating small bulk samples [7-9] with grain sizes below 500 nm [process illustration shown in Figure 1(a) and (b)].

These multipass deformation methods preserve the initial shape of the billet, so the resulting fine grain material is amenable to net shape forging at low temperature. There are 3 main problems with these processes. The first problem results from the lower flow stress of the fine grain regions at the process temperature (~150 - 250°C for Al alloys), which makes it difficult to further deform the larger grain regions homogeneously. The inhomogeneity of the microstructure can lead to inhomogeneous mechanical behavior, *e.g.* multiple necking in a tensile test. Secondly, large frictional effects and high pressure in ECAE and Pressurized Torsion processes create difficulties for scale-up. Thirdly, when the desired product is sheet or plate, further rolling of the billet is necessary, which can partially destabilize the billet microstructure, resulting in somewhat coarser and elongated grains [9]. Thus for fabrication of nanograin sheet material, a preferred approach is to process a plate geometry rather than roll a nanograin block causing a drastic change in strain path. With these goals in mind, a new process for sheet fabrication has been developed, hereafter referred to as MCF (Multipass Coin-Forge) process, as described below.

1.1. NEW PROCESSING METHOD

Figure 1(c) schematically shows the first of the two main steps of the MCF material processing method. A plate of material is isothermally forged partly between die platens containing "sinewave" grooves (actual cross-section: circular). This is primarily a plane strain coining operation which causes indentation and extensive surface shear to push material upward between the indenters, producing a wavy surface with no more than 0.1 - 0.2% overall thinning. The workpiece is then rotated through a small angle (15°), and forged again between the sinewave platens. Now the pressure on the contacting portions of sine wave crests creates compressive deformation in those areas (previously under tension), and previously compressed (indented) regions undergo tension now. With many angular rotations between each coining step, all parts of the surface is exposed to repeated tension-compression cycles. Figure 2 shows a schematic deformation zone geometry under each indenter which can produce (in each coining step) an average surface strain of 0.24 by overlapping strain fields. The microphotograph showing the much finer structure in this heavily deformed surface region (size ~500 μm) is contrasted against the less deformed center of the plate. The dimension of this surface region is clearly a function of die and workpiece geometry. Combined tension-compression cycles can lead to dynamic recovery and grain refinement as is well known from fatigue research, but during cyclic deformation persistent slip bands can also form. The repeated workpiece rotation in the MCF

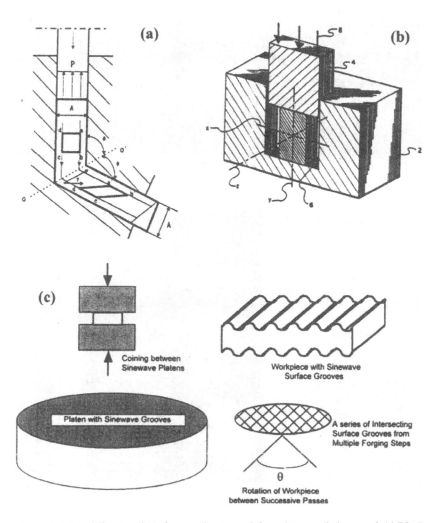

Figure 1. Three different methods for imparting severe deformation to workpiece metal: (a) ECAE process schematic showing billet shear[7], (b) Plane strain die and punch for forging repeatedly in 3 orthogonal directions[9], (c) Multipass Coin-Forge Process, in which metal plate surface is coined between two sinewave dies, with successive rotation of workpiece, followed by flat forging or rolling step. Process is repeated until deformed surface zones meet.

32

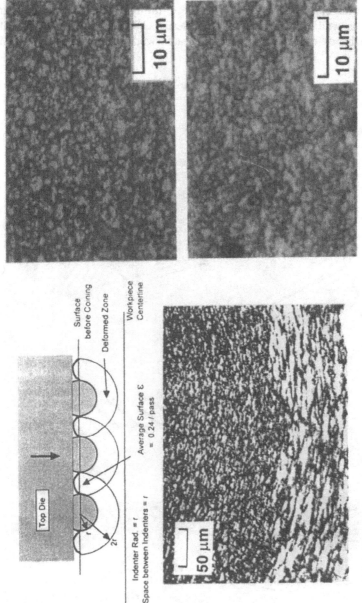

Figure 3. Optical Micrographs of MCF processed 5083 Alloy showing recovered ("recrystallized") grains - after 40 passes (top), and after 80 passes (bottom)

Figure 2. Schematic illustration of coining operation showing the heavily deformed region under the indentors, and corresponding micrograph near the workpiece surface shows finer microstructure at the surface than within the interior.

process precludes slip bands from successive tension and compression paths to become overlapping, thereby continuing to add more residual shear. By applying 40 repeated steps of alternate tension and compression, it is possible to introduce an effective strain of 7-10 near the surface, depending on the degree of dynamic recovery at the process temperature.

The second step in this method is to forge the mutipass-coined plate between flat platens to remove the surface waves and displace the heavily deformed surface layers close to each other. The average compressive deformation during this step is approximately 15%. Depending on the thickness of the plate at this stage, repetition of the above-described steps is carried out, by taking a 15% reduction each time until the intensely deformed surface layers contact each other. Finally, the plate is either stepwise forged or lightly rolled to final gauge. For this last step, a temperature between room temperature and the coining temperature may be maintained depending on the remaining ductility of the workpiece.

2. Results and discussion

Processing of several Al alloys have been carried out by the MCF method, as well as 3-axis forging method. These include an Al-Mg alloy containing Al_6Mn dispersoids (5083 Al), and an Al-Zn-Mg-Cu-Sc-Zr alloy (see Table 1), and two modified PM 7000 alloy samples processed by 3-axis forging. Alloys 1 and 2 are from Ingot Metallurgy source.

TABLE 1. Chemical composition (wt%), Processing and Grain Size of Al Alloys

Alloy #	Mg	Zn	Cu	Mn	Cr	Zr	Sc	Al	Process	Grain Size(μm)
1	4.6	-	-	0.7	0.2	-	-	bal.	IM/MCF	0.4 - 1.5
2	3.5	4.2	1.	0.8	0.4	0.3	0.4	bal.	IM-RST/MC	0.3 - 1.5
3-1	2.5	7.2	2.0	-	0.3	0.3	-	bal.	PM/3axisFO RGE	0.6 - 2.5
3-2	2.5	7.2	2.0	-	0.3	0.3	-	bal.	PM/TMT	0.7 - 3.0
4	2.6	7.3	2.1	0.9	-	0.3	-	bal.	PM/3axisFO RGE	0.6 - 2.0

IM = Ingot metallurgy source, PM = atomized powder source, TMT = Thermomechanical processing involving overaging, rolling reduction > 90% and recrystallization steps [12], MCF = present process

Alloy 1 refers to conventional 5083 commercial alloy in hot rolled condition. Alloy 2 has been produced by depositing a rapidly solidified liquid by spin casting [16]. Alloys 3-1, 3-2 and 4 refer to consolidated and extruded (17:1) powder alloys. TMT processing refers to a processing similar to the Rockwell process [12], but mainly using a large cold rolling reduction [13]. All alloys in the processed condition were crystalline, possessed nano- or micrograin structure of similar sizes. Figure 3 shows microstructures of the 5083 alloy at various stages of MCF processing. Although the multipass deformation produces recovered and "recrystallized-like" structures, produced by a continuous recovery process, loose dislocations and unrecovered grains remain within the material. These photos also indicate that grain size in the 0.7 - 0.8 μm is readily achievable, but overall grain sizes in the range of 400 nm to 1.5 μm were seen, while the 3-axis forged materials were slightly coarser.

2.1. TEXTURE AND MECHANICAL BEHAVIOR

Crystal texture changes were examined by developing x-ray pole figures of the processed materials. (111) pole figures are shown in Figure 4 for Alloy 1. As hot rolled material showed normal pole distribution for partially recrystallized f.c.c. materials. The 90% cold rolled alloy showed intense (111) poles approximately 30^0 away along the RD from the sheet normal, which agrees with normal f.c.c. cold rolled textures, although the remaining details of minor intensities are different. The difference may be due to the alloying and dispersoid content in this alloy.

It is clear that cold rolling increases pole intensities from 7x random to 17x random. MCF processing on the other hand tends to bring the (111) poles to within 15° from sheet normal and with a high intensity of 48x random. It appears that this high (111) pole intensity distribution close to sheet normal should render a higher in-plane strength in the sheet. A corresponding Transmission Electron Microscopy photo of MCF processed material (Figure 4b) shows grain sizes less than 1 μm, the average being around 0.8 μm, but loose dislocations are also present. The room temperature yield strength of this alloy is approximately 400 MPa, as compared with 150 MPa for the hot rolled alloy. This significant rise in the yield strength is believed to be partly due to the finer grain structure, remnant dislocations and the intense (111) orientation closer to the sheet normal, which alters the Schmid factor.

Figure 4c shows a compilation of high temperature and room temperature data for various processing conditions and microstructures of 5083 Al. Strength is plotted against the inverse of absolute temperature starting from the melting temperature. The data on threshold stress for superplastic flow is from Ref. [17] for 8 μm and 31 μm grain size materials. At lower temperatures, Handbook yield strength data is used, in addition to the present test results. It is clear that grain size has an influence on yield or threshold strength which changes with temperature. When grain size becomes greater than 1 μm, these curves become more linear since low temperature strength is elevated while the high temperature strength is reduced. Thus the grain size effect on strength in the different temperature regimes are opposite of each other, with least effects seen around the so-called recrystallization temperature ($T_m/2$). As discussed in Refs. [17,18], elevated temperature deformation is glide-climb controlled, and with reduced fraction of high angle grain boundaries in the MCF material, m value does not exceed 0.5 causing a somewhat nonuniform strain distribution in superplastic tensile specimens.

3. Conclusions

A new multipass surface deformation/forging process (MCF) has been developed to fabricate sheet materials with grain sizes less than 1μm. The microstructure appears to be recovered and partially recrystallized with intense texture and high room temperature strength.

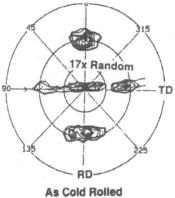

As Hot Rolled **(a)** **As Cold Rolled**

As MCF Processed

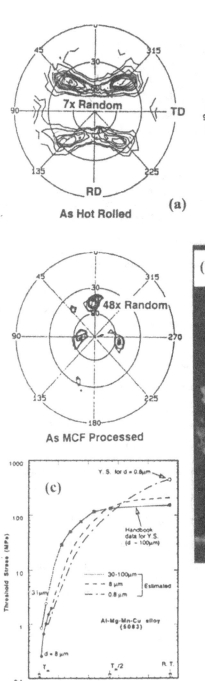

(b)

Figure 4a. (111) poie figures of 5083 Al in different processing conditions,

b. TEM photo in MCF condition showing fine grain structure,

c. Threshold (yield) strength against inverse temperature for different grain sizes

36

4. Acknowledgements

This work was performed under support from US Dept of Energy under grant FG02-96ER45608-A000, and Century Aluminum Co. Acknowledgements are also due to Air Force Office of Scientific Research for partial support of A. Ghosh through contract F33615-94-C-5800. Discussions and help from Dr. S.L. Semiatin of the U.S.Air Force Laboratory and Mr. D.H. Bae are gratefully acknowledged.

5. References

1. Birringer, R., Gleiter, H., Klein, H.P., and Marquart, P. (1984) Nanocrystalline materials - an approach to a novel solid structure with gas-like disorder? *Phys. Lett.* **102A**, 365.
2. Averback, R.S. and Hofler, H.J. (1991) Processing and properties of nanophase materials, in D.C.Van Aken, G.S. Was, and.A.K. Ghosh, (Eds.), *Microcomposites and Nanophase Materials*, TMS Publication, pp. 27.
3. Tench, D.M.and White, J.T. (1991) Electroforming of nanostructured metal composites" *ibid*, 1991, p. 53.
4. Ramani, A.S. and Howell, P.R. (1991) A preliminary study of a mechanically alloyed niobium-yttria microcomposite, *ibid*, pp. 97
5. Vyletel, B., Ghosh, A.K., and Van Aken, D.C. (1991) Microlaminates from the Mo-Si System Developed by Vapor Phase Synthesis, *ibid*, pp. 137
6. McCormick, P.G., Wharton, V.N., Reyhani, M.M., and Schaffer, G.B. (1991) Synthesis of Nanophase Titanium Alloys by Chemical Reduction during Mechanical Alloying, *ibid*, pp. 65
7. Segal, V.M. (1993) Working of metals by simple shear deformation process, *Proc. 5th Intl. Aluminum Extrusion Tech Seminar* **3**, 403.
8. Valiev, R.Z., (1993) Structure and properties of ultrafine grained materials produced by severe plastic deformation, *Mat. Sci. and Eng.* **A168**, 141.
9. Ghosh, A.K., Method for Producing a fine grain aluminum alloy using three axes deformation, U.S.Patent 4,721,537, Jan.26, 1988
10. Ashby, M.F. (1966) Work hardening of dispersion-hardened crystals, *Phil. Mag.* **14**, 1157.
11. F. J. Humphries and M. Hatherly, (1995) *Recrystallization and Related Annealing Phenomena*, Pergamon Press, pp. 262.
12. Paton, N.E., Hamilton, C.H., Wert, J.A., and Mahoney, M.W. (1982) Characterization of fine-grained superplastic aluminum alloys, *J. Metals* **34**, (8), 21.
13. Ghosh, A.K.and Gandhi, C.Grain, refinement and superplastic forming of an aluminum base alloy, U.S. Patent 4,770,848, Sep. 13, 1988
14. T.R. Bieler, (1988) Superplastic-like behavior at high strain rates in mechanically alloyed aluminum, *Scripta. Metall.and Mater.* **22**, p.81
15. Higashi, K., (1991) Positive exponent strain rate superplasticity in mechanically alloyed aluminum, *Scripta Metall. Mater.* **25**, 2053.
16. Ghosh A. K., Method and apparatus for rapidly solidified ingot production, U. S. Patent No. 5,346,184, Sep. 13, 1994.
17. Bae D.H. and Ghosh, A.K. (1999) Grain size and temperature dependence of superplastic deformation in an Al-Mg Alloy under isostructural condition, to appear in *Acta Mater*.
18. Ghosh, A.K. (1994) A physically based model for superplastic flow, *Materials Science Forum* **170-172**, 39, Trans Tech Publications, Switzerland.

MODELING OF CONTINUAL FLOWS IN ANGULAR DOMAINS

B.V. KOUTCHERYAEV
Moscow State Steel and Alloys Institute
Leninsky prospekt 4, 117936 Moscow, Russia

Abstract

The continuous kinematically admissible velocity fields for domains with arbitrary angles between two channels are constructed by the K. Shvarts-A. Christoffel's (S–C) integral method. In complex form such important kinematical parameters as vector velocity, strain rate and shear strain degree are obtained. These parameters are used as a basis for determining corrected parameters by the standard procedure of simulating the actual metal motion in the domains.

1. Introduction

In some approximations the angular domains D (Figure 1a) can be used to represent metal motion in treatment processes such as cutting [1], extrusion [2] etc. In the general case there are several basic parameters which determine metal motion in such domains. Among them are the geometrical parameters: cross-section ratio coefficient $\mu = H/h$ and the angle $\alpha\pi$, characterizing the intersection of the two channels. In Mechanics of Continua (M of C) one of the most common variants of the kinematical boundary problem formulation in such domains may be based on the set of kinematically admissible (KA) velocity fields $\bar{V}$ from which, according to that or this criteria, we can choose one nearly the same as the real (R) velocity field or coinciding with it. For example in M of C this may be done by the J. Lagrange functional minimization [3, 4, 5]. There are several methods, allowing as the first approximation, to construct a KA velocity field $\bar{v}$, taking into account the geometric parameters of the domain only. According to M.M. Philonenko-Boroditch [3] let us call such KA velocity field $\bar{v}$ a primary velocity field or a basic solution of the boundary problem and the set of KA velocity fields $\bar{V}$ constructed on it, a corrected solution or a correction.

2. Principles of Using Complex Functions

From M of C it is known that the incompressibility condition is satisfied identically if the vector $\bar{V}$ construction is fulfilled with the help of basic ψ and corrected Ψ stream functions. The last function is connected with the first [4, 5] $\Psi = \psi + \Phi$, where the correcting function Φ and its necessary co-ordinates derivatives satisfy the

37

T.C. Lowe and R.Z. Valiev (eds.), Investigations and Applications of Severe Plastic Deformation, 37–42.
© 2000 *Kluwer Academic Publishers. Printed in the Netherlands.*

homogeneous boundary conditions. Then components V_k of the vector velocity $\vec{V}$ are determined by components v_j of the vector velocity $\vec{v}$: $V_1 = v_1(1 + + \partial\Phi/\partial\psi) + \partial\Phi/\partial x_2$; $V_2 = v_2(1 + \partial\Phi/\partial\psi) - \partial\Phi/\partial x_1$. The construction of the continuous KA-velocity field $\overset{1}{v} = v_1 + iv_2$ may be fulfilled by Complex Functions Theory (CFT) methods. For polygonal domains such fields are obtained simply with the help of the S-C integral [6]. The considerable domain D (Figure 1a) in the physical plane Z is represented as a tetragon $A_1A_2A_3A_4$. In CFT the kinematic parameter construction problem is reduced to the complex potential construction

$$w(z) = \varphi + i\psi, \tag{1}$$

which conformally maps the domain D in the physical plane Z on the strip E (Figure 2a) in the complex potential plane W. In this case the complex velocity

$$w' = \frac{dw}{dz} = v_1 - iv_2 \tag{2}$$

is complexly conjugated with the vector velocity

$$\vec{v} = \overline{w'} = v_1 + iv_2 \tag{3}$$

in the domain D and the complex strain rate

$$w'' = \frac{d^2 w}{dz^2} = \xi_{11} - i\xi_{12}, \tag{4}$$

where ξ_{ik} - are components of the strain rate tensor [4]. From CFT it is known [6] that the S-C integral carries out the conformal map on the upper semi-plane Δ of the auxiliary plane ζ. By fulfilling such a mapping for domain E we obtain a function $w = w(\zeta)$ and for domain D a function $z = z(\zeta)$. Thus in the parametric form with the help of the auxiliary parameter ζ, the function $w = w(z)$ can be obtained.

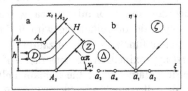

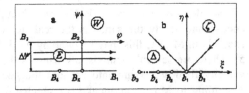

Figure 1. Conformal mapping D onto Δ
a - the flow in the physical plane Z;
b - the image D in the auxiliary plane ζ.

Figure 2. Conformal mapping of E onto Δ
a - the strip E in the plane W;
b - the image E in the auxiliary plane ζ.

3. The Conformal Mapping of *D* onto *Δ*

In the general case the mapping of domain D of the physical plane Z (Figure 1a) onto upper semi-plane Δ of the auxiliary plane ζ (Figure 1b) with the help of the S-C integral is written as follows:

$$z(\zeta) = c_1 \int_{\zeta_0}^{\zeta} \prod_{k=1}^{n} (\zeta - a_k)^{\alpha_k - 1} d\zeta + c_2, \tag{5}$$

where α_k are the inner angles (expressed in part of π) of the n-angled domain D ($\sum\limits_{k=1}^{n} \alpha_k = n-2$) and a_k are the S-C integral constants. According to the S-C integral properties any three constants a_k (the S-C integral normalization) can be chosen arbitrarily (Table 1). If one of these constants is equal to infinity, then the corresponding multiplier in the integrand in (5) is equal to unity.

Let us carry out the velocity field construction for the common configuration of a domain D with an arbitrary angle α ($0 \le \alpha \le 1/2$) and for different sizes of inlet H and outlet h channels. According to the normalization in Table. 1, the point $a_2 = 1$ in the

TABLE 1. Normalization of the S-C integral for the mapping of D onto Δ

k	A_k	a_k	α_k
1	$-\infty$	0	0
2	0	1	$1-\alpha$
3	∞	∞	0
4	A_4	a	$1+\alpha$

plane ζ corresponds to the point $A_2 = 0$ in the plane Z. That is why if the down limit ζ_0 of integral (5) is located at point a_2, than for $\zeta_0 = 1$ we have $z(1) = c_2 = 0$. After rewriting (5) using the normalization presented in Table 1, the following is obtained:

$$z(\zeta) = c_1 \int_{1}^{\zeta} \left(\frac{\zeta - a}{\zeta - 1}\right)^{\alpha} \frac{d\zeta}{\zeta}. \tag{6}$$

In order to find constants c_1 and a let us use two increments Δz in the plane Z and the corresponding bypasses in the semi-plane Δ along the semi-circles C_R and C_r of infinitely large radius R and infinitesimal radius r.

The first bypass along C_R in the plane Z corresponds to the transition from the beam A_2A_3 to the beam A_3A_4 for the increment $\Delta z = H (-\sin\alpha\pi + i\cos\alpha\pi) + 0 (1/R)$, where 0 (1/R) is an infinitesimal value of the 1/R order smallness. Just for the same bypass in the plane ζ ($\zeta > \|a\|$ and $\zeta > 1$) from (6) we obtain: $c_1 = H (\cos\alpha\pi + i\sin\alpha\pi)/\pi = H e^{i\alpha\pi}/\pi$. The second bypass along C_r in the plane Z corresponds to the transfer from the beam A_2A_1 to the beam A_1A_4 for the increment $\Delta z = ih + 0(r)$, where $0(r)$ is infinitesimal value of the r order smallness.

Just for the same bypass in the plane ζ ($\zeta \ll \|a\|$ and $\zeta \ll 1$) from (6) we obtain: $a = -\left(h/H\right)^{1/\alpha}$. Finally, the integral (6) with respect to c_1 and a after substitution

$(\zeta-a)/(\zeta-1) = \omega^{\gamma}$, where $\gamma = 1/\alpha$, is transformed to the sum of two table integrals

$$z(\zeta) = \frac{H\gamma}{\pi} e^{i\alpha\pi} \left(\int_\omega \frac{\omega^{\gamma-2} d\omega}{1-\omega^\gamma} - a^{-\alpha} \int_s \frac{s^{\gamma-2} ds}{1-s^\gamma} \right), \tag{7}$$

where $s = \omega_a^{-\alpha}$. The integrals in (9) are identical.

4. The Conformal Mapping of E onto $\varDelta$

In the same manner as D-Δ mapping the S-C integral

$$w(\zeta) = c_3 \int_{\zeta_0}^{\zeta} \prod_{k=1}^{n} (\zeta - b_k)^{\beta_k - 1} d\zeta + c_4 \tag{8}$$

maps the strip E (Figuire 2a) in the complex potential plane W onto the semi-plane Δ of the auxiliary plane ζ (Figure 2b).

For the holomorphic function (1), realizing the conformal mapping D in Z onto E in W, the conservative function φ with the stream function ψ satisfies the J. D'Alamber-L. Euler equation [6] and isolines of constant values φ and ψ make up the orthogonal net in the Z and W planes. Bounds $A_1A_2A_3$ and $A_3A_4A_1$ of physical flow in the domain D (Figure 1a) belong to the set of stream lines with constant values ψ (let $\psi = \psi^-$ for $A_1A_2A_3$ and $\psi = \psi^+$ for $A_3A_4A_1$). The same lines ψ^+ and ψ^- in the domain E are straight lines. They are boundaries of the bi-angular domain E ($\psi^- \le \psi \le \psi^+$) with vertices B_1 and B_3 (Figure 2a).

Let us choose the constants b_k according to the S–C integral (8) normalization which is shown in Table 2.

TABLE 2. Normalization of the S-C integral
for the mapping of E onto Δ

k	B_k	b_k	β_k
1	∞	0	0
2	0	1	1
3	$-\infty$	∞	0
4	B_4	b	1

For the upper limit $\zeta = \zeta_0$ in (11) we have $w(\zeta_0) = c_4$. If to place ζ_0 onto the point b_2, then (see Table 2) we have $w(1) = 0$ and $c_4 = 0$. In order to determine the constant multiplier c_3 in (8) let us introduce onto the plane ζ the an auxiliary point $b_5 = -1$, which is symmetrical to the point $b_2 = 1$ (Figure 2b). The image B_5 of the new point in the plane W is located on the intersection of the abscissa axis ψ and the down branch $B_5 = -i\psi^-$ of the strip E (Figure 2a). Then $-i\psi^- = ic_3\pi$ and $c_3 = -\psi^-/\pi$. So the S–C integral (11) gets the following view $w(\zeta) = -(\psi^-/\pi)\ln\zeta$.

Thus the complex potential (1) $w = w(z)$ is constructed in parametric form: $z = z(\zeta)$ (10); $w = w(\zeta)$ (10), taking into account the previous results.

5. Kinematic Parameters

Now we can determine the kinematic parameters in the angular domain D. Using (7) and (8) let us find the complex velocity (2) in the parametrical form. First of all from (7) we obtain $dz/d\zeta = H_e^{i\alpha\pi}[(\zeta-a)/(\zeta-1)]^{\alpha}/\pi\zeta$ and from (9) find $dw/d\zeta = -\psi^{-}/\pi\zeta$. The complex velocity (2) then becomes

$$w' = \frac{dw}{d\zeta}\frac{d\zeta}{dz} = -\frac{\overline{\psi}}{H} \cdot e^{-i\alpha\pi}\left(\frac{\zeta-1}{\zeta-a}\right)^{\alpha}. \tag{9}$$

In order to check formula (11) let us use the known vector velocity values at points A_k of the physical domain D in the plane Z. If the initial velocity module at point A_3 is equal to v_0 ($v_1 = -v_0\cos\pi\alpha$; $v_2 = -v_0\sin\pi\alpha$;), then substituting in (9) the value of the image of that point $a_3 = \infty$ in the plane ζ (Table 1) we obtain $\psi^{-}/H = v_0$. For the point A_1 ($a_1 = 0$) from (9) we have $v_1 = -v_0H/h$, $v_2 = 0$. For point A_2 ($a_2 = 1$) from (9) it is obtained that $v_1 = 0$, $v_2 = 0$; for the point A_4 ($a_4 = a$) from (9) we have $\bar{v} = \infty$. Finally, the complex velocity (2) can expressed as

$$w' = -v_0 e^{-i\alpha\pi}\left(\frac{\zeta-1}{\zeta-a}\right)^{\alpha}. \tag{10}$$

Then we can derive the complex strain rate (4):

$$w'' = \frac{dw'}{d\zeta}\frac{d\zeta}{dz} = v_0\left[\frac{\pi\alpha\zeta(a-1)}{H(\zeta-1)(\zeta-a)}\right]e^{-i2\alpha\pi}\left(\frac{\zeta-1}{\zeta-a}\right)^{2\alpha}. \tag{11}$$

Among all deformation characteristics of shaping, only the shear strain degree $\Lambda = \int H dt$ (and others connected with it) allows one to calculate the additional values as a result of discrete shaping processes. For the two-dimensional motion of incompressible continua, the shear strain rate intensity is $H = 2\sqrt{\xi_{11}^2 + \xi_{12}^2}$, where ξ_{jk} are components (4) of the shear strain rate tensor. Some works [7, 8] note the correlation between the non-uniformity of the Λ-distribution with the texture of deformation and properties of deformed metals. The basic solution has been used for the construction of the corrected KA-velocity fields set. From this set the best approximation to the R-velocity field is chosen by the minimization of J. Lagruge's functional [5]. Strain rates Ξ_{jk} of the corrected solution allow one to calculate H and estimate Λ. As an example the distribution of these kinematical parameters in the equal-channel domain D is shown in Fig. 3.

In this example, the flow of the ideal rigid-plastic media when angle α equals to 1/12 (as a particular case) is considered. The dependency of the non-homogeneity Λ coefficient $K_\Lambda = \Lambda_{max}/\Lambda_{min}$ from the angle α inside the domain is shown in Figure 4. It is seen that an increasing of the angle α leads to a reduction of the non-homo geneity of

42

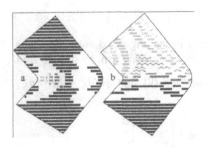 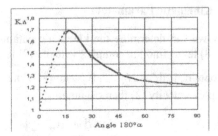

Figure 3. Isolines: a-H=const; b-Λ=const.　　　　*Figure 4*. Dependence of K_A on anlgle α.

the shear strain degree Λ. The numerical simulation on the computer shows that the value of Λ in the whole domain D increases as α increases.

6. Conclusion

1. The formulae for the kinematic parameters of the basic solution in the angular domain D with arbitrary sizes of channels (H; h) and an angle α between them are obtained with the help of CFT.
2. Correction of the basic solution with the help of M of C variational principles allows the simulation of metal deformation in angular domains.
3. The flow of an ideal rigid-plastic media in an equal-channel domain is considered. The use of the model is helpful in investigations and for an understanding of the different effects associated with metal deformation in angular domains.

7. References

1. Goun, R.E. (1982) The construction of the basic solution for extrusion, rolling and cutting of metals. In book *Plastic deformation of metals and alloys*, Moscow, Metallurgy, (MS&AI, Scient. Trans. N 140), pp. 54–56.
2. Segal, V.M., Reznikov, V.I., Kopylov, V.I. *et al.* (1994) *Processes of Plastic Structure Forming of Metals*, Minsk, Science and Technique, 232.
3. Pobedria, B.E. (1981) *Numerical Methods in Elasticity-Plasticity Theory*, Moscow, MSU, 344.
4. Goun, G.Ya. (1980) *Theoretical Foundations of Metal Shaping*, Moscow, Metallurgy, 446.
5. Koutcheryav, B.V. (1999) *Mechanics of Continua. (Theoretical Foundations of the Composite Metals Shaping)*, Moscow, MISIS, 320.
6. Lavrentiev, M.A., Shabat, B.V. (1965) *Methods of Complex Functions Theory*, Moscow : Science, 716.
7. Siebel, E. (1941) *The theory of rolling processes between unequally driven rolls*, Archiv fur Eisenhuttenwesen, **15**, No. 9, 125–128.
8. Koutcheryav B.V., Koutcheryav V.V., Fedoseev A.I. (1987) The strain heterogeneity control in asymmetrical rolling, *Russ. Non-Ferrous Metals* **8**, 69–70.

SYNTHESIS AND CHARACTERIZATION OF NANOCRYSTALLINE TIAL BASED ALLOYS

O.N. SENKOV AND F.H. (SAM) FROES
Institute for Materials and Advanced Processes
University of Idaho, Moscow, ID 83844-3026

Abstract

Gamma TiAl based alloys and a TiAl/Ti$_5$Si$_3$ composite with nanocrystalline and submicrocrystalline structures were produced by mechanical alloying (MA'ing) and hot isostatic pressing (HIP'ing). MA'ing in a high-energy mixer mill led to severe plastic deformation and amorphization of the powder particles. The γ-TiAl alloys produced from the amorphous powders had a nanocrystalline structure. The grain size increased when the HIP temperature and time increased. Grain growth in the TiAl alloys during annealing for up to 800 hours in the temperature range of 725°C to 1200°C was studied. No effect of the Ti$_5$Si$_3$ phase particles on grain growth kinetics was found.

1. Introduction

Gamma TiAl-based alloys are being considered as potential candidates to replace presently used Ni-based superalloys in thermal protection systems and engine components working at temperatures up to 900°C, as they have half the density and similar high temperature properties to the superalloys [1-3]. However, there is a limitation in the forming methods which can be used with TiAl alloys because of their brittleness at temperatures below 600°C. Attempts to improve ductility of the alloys by chemistry modifications or microstructural control have shown limited success [3, 4].

It has been shown [5] that generally very brittle ceramics exhibit a fair degree of ductility at room temperature if they have a nanocrystalline structure. A critical grain size has been theoretically predicted [6, 7] below which ceramic and intermetallic materials can become ductile. This has been supported experimentally on NiAl [8] and Ti$_3$Al [9]. Grain refinements to a nanoscale range also open up possibilities of a considerable decrease in temperatures for compaction or superplastic forming, as the rates of these diffusion-related processes increase dramatically when the grain or particle size decreases [10-12].

The main problem of a nanocrystalline structure is its instability at high temperatures. Because of the large excess free energy, significant grain growth has been observed in several nanocrystalline materials [13, 14]. On the other hand, a stabilization of the nanocrystalline grain structure was detected in many materials after continuous annealing [14]. Second phase particles, chemically stable at high temperatures, can stabilize the grain structure.

T.C. Lowe and R.Z. Valiev (eds.), Investigations and Applications of Severe Plastic Deformation, 43–48.
© 2000 *Kluwer Academic Publishers. Printed in the Netherlands.*

The present work has been directed at analyzing the stability of the nanocrystalline structure in TiAl-based materials at different annealing temperatures and times. The effect of second phase Ti_5Si_3 particles on grain growth was also studied. The Ti_5Si_3 phase has the closest coefficient of thermal expansion to that of TiAl, which, together with its high stiffness and strength, makes Ti_5Si_3 the favored reinforcing phase in TiAl [15]. Fully dense nanocrystalline compacts were produced by HIP'ing of MA'd amorphous powders.

2. Experimental Procedures

Gas-atomized (GA) powders of nominal alloy compositions Ti-47Al-3Cr and Ti-48Al-2Cr-2Nb (at. %) were supplied by Crucible Research Center, Pittsburgh, PA, and TiH_2 and Si powders were supplied by Johnson Matthey Electronics, Ward Hill, MA. To produce a $TiAl/Ti_5Si_3$ composite, the Ti-47Al-3Cr powder was blended with 22.5wt.% TiH_2 and 7.5 wt%Si. The GA and blended powders were MA'd in a SPEX 8000 mill under an argon atmosphere for 15 hours. A 10:1 ball-to-powder weight ratio was used. The MA'd powders were then HIP'd in a 20-cm diameter unit in Bodicote IMT, Inc., London, OH. The HIP conditions are shown in Table 1.

Microstructural and phase characterizations of HIP'd and annealed samples were carried out using a transmission electron microscope JEOL JEM 2010 operating at 200 kV and X-ray diffraction (XRD, Cu-K_α radiation). The sizes of individual grains were directly measured from the TEM photomicrographs.

TABLE 1. Powders and Processes Used to Produce Nanocrystalline Materials.

Goal Composition	Powders Used	MA Duration (Hours)	HIP Conditions	Final Grain Size (nm)
Ti-47Al-3Cr	Ti-47Al-3Cr	15	725°C, 207MPa, 2h	42
			850°C, 207MPa, 2h	88
			975°C, 105MPa, 2h	180
Ti-48Al-2Nb-2Cr	Ti-48Al-2Nb-2Cr	15	850°C, 207MPa, 2h	92
Ti-47Al-3Cr / 30vol.% Ti_5Si_3	Blend of Ti-47Al-3Cr, TiH_2 and Si	15	850°C, 207MPa, 4h	140
			1050°C, 105MPa, 4h	580

3. Results and Discussion

3.1. NANOCRYSTALLINE TI-47AL-3CR AND TI-48AL-2NB-2CR ALLOYS

The HIP'd materials had a very fine grain structure, Figure 1. The grain size decreased when the HIP'ing temperature was decreased, Table 1, and it was 42 nm in the Ti-47Al-3Cr after HIP'ing at 725°C. The two materials, Ti-47Al-3Cr and Ti-48Al-2Nb-2Cr, had nearly the same grain size (around 90 nm) after HIP'ing at the same temperature, 850°C.

Considerable grain growth occurred during the initial stages of annealing of the HIP'd samples, Figure 2, with a tendency to reach a saturation stage where grains grew very slowly. After annealing for 500 hours at 850°C and 975°C, for example, the average grain size in Ti-47Al-3Cr was 170 and 415 nm, respectively. The same tendency for grain growth was observed in the nanocrystalline Ti-48Al-2Nb-2Cr alloy [16], Figure 2.

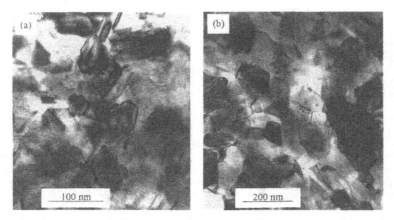

Figure 1. Microstructure of as-HIP'd materials: (a) Ti-47Al-3Cr, 725°C; (b) Ti-48Al-2Cr-2Nb, 850°C.

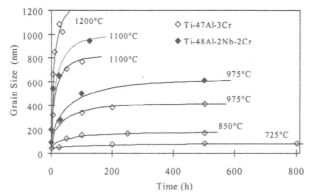

Figure 2. Dependence of the mean grain size on the annealing time and temperature in the two nanocrystalline TiAl-based alloys.

3.2. SUBMICROCRYSTALLINE TI-47AL-3CR/TI5SI3 PARTICULATE COMPOSITE

Figure 3 shows TEM micrographs of the TiAl/Ti$_5$Si$_3$ compacts after HIP'ing at 850°C and 1050°C. The microstructure of the compacts consists of equiaxed grains with an average grain size of 140 nm and 580 nm after HIP'ing at 850°C and 1050°C. The γ-TiAl and Ti$_5$Si$_3$ grains were almost the same size and were homogeneously distributed.

Figures 4a and 4b show the effects of temperature and annealing time on grain size,respectively. At a constant (100 hours) annealing time, the grain size exponentially increased from 170 nm to 730 nm when the temperature increased from 850°C to 1100°C. Annealing at 1100°C for up to 500 hours led to a continuous increase in the grain size so that after 500-hour annealing it was 1300 nm.

EDS chemical analysis of individual grains in the HIP'd and annealed specimens showed that from 30 to 35% of silicon atoms in the Ti_5Si_3 ordered hexagonal phase were substituted by aluminum atoms. On the other hand, the solubility of silicon in the γ-TiAl phase did not exceed 2 at.%.

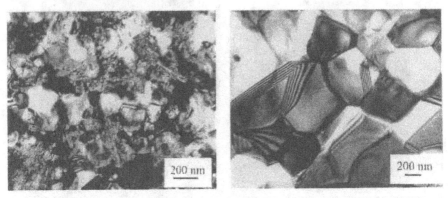

Figure 3. TEM photomicrographs of the compacts after HIP'ing at (a) 850°C and (b) 1050°C.

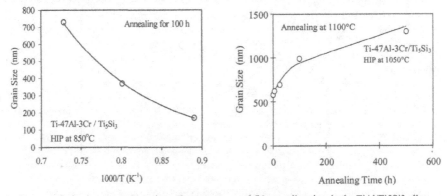

Figure 4. Grain size versus (a) reciprocal temperature and (b) annealing time in the TiAl/Ti5Si3 alloy produced by MA'ing for 15 hours followed by HIP'ing at (a) 850°C and (b) 1050°C.

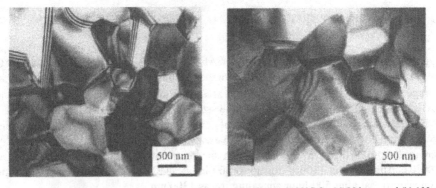

Figure 5. TEM photomicrographs showing grains after annealing at 1100°C for (a) 25 hours and (b) 100 hours in the TiAl/Ti_5Si_3 composite HIP'd at 1050°C.

3.3. KINETICS OF GRAIN GROWTH

Grain growth was described by the equation [17]:

$$D^n - D_o^n = K_o \exp(-Q/RT)t \tag{1}$$

where D_o and D are the initial and current grain sizes, t is the annealing time, T is the absolute temperature, R is a gas constant, Q is the activation energy for grain growth, and n and K_o are material constants. Fitting this equation to the experimental data, Figures 2 and 4, gave the values $n = 4$, $K_o = 1.2\text{-}2.4 \times 10^{22}$ nm^4/h ($= 3.3\text{-}6.5 \times 10^{-18}$ m^4/s), and $Q = 330$ kJ/mol to be the same for all three alloys studied. The same temperature dependencies of the kinetic parameter $K=(D^4 - D_o^4)/t$ were obtained for the alloys and composite, Figure 6. These results indicate that the Ti$_5$Si$_3$ phase has a little effect on grain growth. The value of the grain growth exponent $n=4$ is twice that predicted for normal grain growth in high purity metals when no dragging or pinning forces are present [17]. However this value ($n=4$) is frequently observed in two-phase alloys when the grain growth is limited by a permanent pinning force on grain boundaries [18], which is probably the case for the alloys studied in the present work.

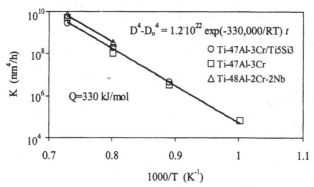

Figure 6. Temperature dependence of the kinetic parameter K = $K_o$exp(-Q/RT), Equation (1), for the submicrocrystalline TiAl/Ti$_5$Si$_3$ composite and nanocrystalline TiAl alloys.

4. Conclusions

1. Fully dense TiAl-based alloys and composite with nanocrystalline (grain size <100 nm) and submicrocrystalline (grain size 100-600 nm) structures were produced by hot isostatic pressing of mechanically alloyed amorphous powders. The grain size of the HIP'd alloys decreased when the HIP'ing temperature decreased.

2. The Ti$_5$Si$_3$ phase was homogeneously distributed in the TiAl matrix. Both phases had almost the same grain size. A high solubility of aluminum in the Ti$_5$Si$_3$ phase was detected. About 30 at.% Si was replaced by Al, so that the phase can be described by formula Ti$_5$Si$_2$Al. About 1.5 at.% silicon were soluble in the γ- TiAl phase.

3. The grain growth in the TiAl-based alloys and composite in the temperature range 850°C to 1100°C was described by the equation $D^4 - D_o^4 = K_o \exp(-Q/RT)$

t where $Q = 330$ kJ/mol and $K_o = 3.3\text{-}6.5 \times 10^{-18}$ m^4/s. The kinetics of grain growth in these materials was almost the same indicating that the Ti$_5$Si$_3$ phase had little effect on the grain growth in the gamma titanium aluminide.

5. Acknowledgements

This work was supported by the US Army Research Office, Grant No. DAAG55-98-1-0008 and the Idaho State Board of Education, Grant No. S96-016.

6. References

1. Froes, F.H., Suryanarayana, C., and Eliezer, D. (1992) Synthesis, properties and applications of titanium aluminides, *J. Mater. Sci.* **27**, 5113-5140.
2. Bartolotta, P., Barrett, J., Kelly, T. and Smashey, R. (1997) The use of cast Ti-48Al-2Cr-2Nb in jet engines, *JOM* **49 (5)**, 48-50, 76.
3. Kim, Y-W. and Dimiduk, D.M. (1997) Designing gamma TiAl alloys: fundamentals, strategy and production, in M.V. Nathal, R. Darolia, C.T. Liu, P.L. Martin, D.B. Miracle, R. Wagner, and M. Yamaguchi (eds.), *Structural Intermetallics 1997*, TMS, Warrendale, PA, pp. 531-543.
4. Imayev, V.M., Salishchev, G.A., Imayev, R.M., Shagiev, M.R., Gabdullin, N.K. and Kuznetsov, A.V. (1997) An approach to ductility improvement of TiAl and Ti3Al titanium aluminides based on microstructure control, in M.V. Nathal, R. Darolia, C.T. Liu, P.L. Martin, D.B. Miracle, R. Wagner, and M. Yamaguchi (eds.), *Structural Intermetallics 1997*, TMS, Warrendale, PA, pp. 505-514.
5. Karch, J., Birringer R. and Gleiter, H. (1987) Ceramics ductile at low temperature, *Nature* **330**, 556-558
6. Schulson, E.M. (1981) Comments on the brittle to ductile transition of long-range ordered alloys, *Res. Mech. Letters.* **1**, 111-114.
7. Chan, K.S. (1990) Theoretical analysis of grain size effects on tensile ductility, *Scripta Metall.* **24**, 1725-1730.
8. Schulson, E.M. and Barker, D.R. (1983) Brittle to ductile transition in NiAl of a critical grain size, *Scripta Metall.* **17**, 519-524.
9. Imayev, R.M., Gabdullin, N.K., Salishchev, G.A., Senkov, O.N., Imayev, V.M. and Froes, F.H. (1999) Effect of grain size and partial disordering on ductility of Ti$_3$Al in the temperature range of 20-600°C, *Acta Mater.* **47**, 1809-1821.
10. Chang, H., Aktstetter, C.J. and Averback, R.S. (1992) Characterization of nanophase TiAl produced by inert gas condensation, *J. Mater. Res.* **7**, 2962-2970.
11. Imayev, V.M., Salshchev, G.A., Shagiev, M.R., Kuznetsov, A.V., Imayev, R.M., Senkov, O.N. and Froes, F.H. (1999) Low-temperature superplasticity of submicrocrystalline Ti-48Al-2Nb-2Cr alloy produced by multistep forging, *Scripta Materialia* **40**, 183-190.
12. Senkov, O.N., Srisukhumbowornchai, N. and Froes, F.H. (1998) Synthesis of titanium aluminide based compounds and composites with nanocrystalline and bimodal structures, in W.D. Cho and H.Y. Sohn (eds.), *Value-Addition Metallurgy*, TMS, Warrendale, PA, 61-69.
13. Birringer, R. (1989) Nanocrystalline materials, *Mater. Sci. Eng.* **A117**, 33-43.
14. Malow, T.R. and Koch, C.C. (1996) Thermal stability of nanocrystalline materials, *Mater. Science Forum* **225-227**, 595-604.
15. Powell, P.W. and Reynolds, G.H. (1993) Gamma titanium aluminide composite development activities, in F.H. Froes (ed.), *P/M in Aerospace, Defense and Demanding Applications 1993*, MPIF, Princeton, NJ, 291-297.
16. Senkov, O.N., Froes, F.H., Srisukhumbowornchai, N. and Ovecoglu, M.L. (1998) Grain growth in titanium-aluminide-based alloys with nanocrystalline and microcrystalline structures, in H. Weiland, B.L. Adams and A.D. Rolett (eds.), *Grain Growth in Polycrystalline Materials III*, TMS, Warrendale, PA, 701-706.
17. Atkinson, H.V. (1988) Theories of normal grain growth in pure single phase systems, *Acta Metall.* **36**, 469-491.
18. Higgins, G.T., Wiryolukito, S. and Nash, P. (1992) *Grain Growth in Polycrystalline Materials*, Trans. Tech. Publ., Aedermannsdorf, Switzerland.

FORMATION OF SUBMICROCRYSTALLINE STRUCTURE IN TIAL AND TI₃AL INTERMETALLICS VIA HOT WORKING

G. SALISHCHEV, R. IMAYEV, V. IMAYEV, N. GABDULLIN, M. SHAGIEV, A. KUZNETSOV, O.N. SENKOV*, F.H. (SAM) FROES*
Institute for Metals Superplasticity Problems
Russian Academy of Sciences, Ufa 450001, Russia
*Institute for Materials and Advanced Processes
University of Idaho, Moscow, Idaho 83844-3026, USA

Abstract

A method based on initiation of dynamic recrystallization (DRX) during hot working has been developed to produce a submicrocrystalline (SMC) structure (d < 1 μm) in massive work-pieces of hard-to-deform materials, like titanium aluminides, The method involves continuous grain refinement due to dynamic recrystallization at a decreasing temperature. A microstructure with a grain size of 0.1 to 0.4 μm and no porosity was produced in different TiAl and Ti₃Al based alloys. Partial disordering was detected in a Ti₃Al alloy with the SMC structure. The grain refinement hardened the intermetallic alloys at room temperature (RT). In a fully ordered Ti₃Al alloy RT ductility increased when the grain size decreased, while the ductility of a partially disordered SMC Ti₃Al and TiAl alloys was close to zero.

1. Introduction

Intermetallic TiAl and Ti₃Al based alloys are hard-to-deform materials [1]. Chemistry and microstructure control can improve ductility of these materials [1–3]. In particular, by grain refinement to a SMC level, ductility can be considerably increased and superplasticity (SP) can be achieved at 600°C–900°C, *i.e.* 200–400°C below the temperature range for SP in the materials with micron-sized grains [2, 4, 5]. Methods of powder metallurgy (P/M) and severe plastic deformation are commonly used to produce the SMC structure [6, 7]. Dynamic recrystallization (DRX) occurring during hot working of titanium aluminides results in grain refinement [2, 8]. The grain size produced by DRX decreases when the deformation temperature decreases or strain rate increases [9]. SMC structures were produced by DRX in titanium alloys and steels [10], as well as in a TiAl-based alloy [2], by a step-by-step decrease in the deformation temperature. Because the critical strains required to initiate DRX and to obtain a fully recrystallized structure increase considerably when the deformation temperature decreases, material must be ductile enough to get the SMC structure. It is therefore not clear yet how effectively the DRX approach can be applied to low ductility intermetallic alloys of various compositions, what minimal grain size can be achieved,

T.C. Lowe and R.Z. Valiev (eds.), Investigations and Applications of Severe Plastic Deformation, 49–55.
© *2000 Kluwer Academic Publishers. Printed in the Netherlands.*

and what mechanical properties can be obtained. The present work is undertaken to answer these questions.

2. Experimental Procedure

The DRX approach to produce a SMC structure was initially applied to a two-phase Ti-50 at.% Al (Ti–50Al) P/M alloy. Then it was scaled up to a series of two-phase γ-TiAl based alloys such as Ti–50Al, Ti–48Al–2Cr–2Nb, and Ti–46Al–2Cr–2Nb–1Ta cast alloys and a Ti–46Al P/M alloy, and to a single phase α_2–Ti$_3$Al cast alloy (Ti–25Al). To initiate DRX and refine microstructure, multi-directional isothermal forging at decreasing temperature was used. After forging, all billets were annealed for 2 hours at a temperature which was 50°C below the temperature of the final forging step. Details of forging, mechanical tensile tests and microstructure investigations can be found in earlier publications [2, 4, 5].

3. Results and Discussion

Figure 1a shows deformation curves of the P/M Ti–50Al alloy at various temperatures. The initial microstructure of this alloy with the average grain size of 18 μm is shown in Figure 2a. During deformation, DRX occurred leading to formation smaller grain sizes at lower deformation temperatures, although the volume fraction of the recrystallized regions decreased, when the temperature decreased, because of strain localization (Table 1). The strain localization led to crack formation in specimens deformed at 950°C to ε = 60%. The smallest grain size produced in this material was 1.3 μm.

The grain size, d, produced by DRX determines the steady state flow stress, σ_s, through the empirical equation $\sigma_s = kd^{-n}$, where k is a constant and the grain size exponent $n \cong 0.5$ [9]. The dependence of σ_s on d for the Ti–50Al alloy is given in Figure 2, with n = 0.49. By extrapolation, σ_s = 860 MPa can be estimated to obtain the DRX grain size of 0.1 μm. This value of stress can be achieved by a decrease in temperature, but the material must be ductile enough to initiate DRX. Ductility of the Ti–50Al alloy was considerably increased by decreasing the initial grain size to 5 μm so that even at 700°C the specimens were deformed to a strain of 80% without fracture, Figure 1b. DRX developed during deformation (Table 2) and the grain size of ~0.1 μm was produced by DRX at 700°C (Figure 2b) at a level of the flow stress σ_s = 665 MPa. The dependence of σ_s on d in the Ti–50Al alloy specimens with the initial grain size d_{in} = 5 μm is given in Figure 3, with n = 0.49, similar to the specimens with d_{in} = 18 μm, however lower steady state stresses were achieved in the specimens with smaller initial grain size. The latter is probably due to the effect of the initial grain size on pct of the recrystallized volume (Table 1, 2). It is also necessary to point out that grain refinement at the temperature-strain rate conditions studied may lead to SP behavior with a linear grain size dependence of the flow stress [11]. The contribution of this factor is apparently increased with an increase in the DRX volume.

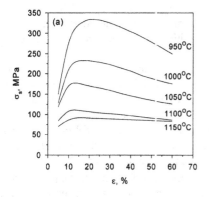
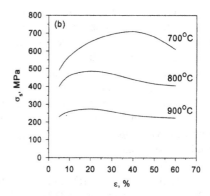

Figure 1. True stress-engineering strain curves for the P/M Ti-50Al alloy with different initial grain sizes: (a) 18 μm, (b) 5 μm (ε=8.3×10⁻⁴ s⁻¹).

Figure 2. Microstructure of a compacted Ti-50Al alloy:
a – initial condition, b – after isothermal deformation at 700° C.

TABLE 1. Temperature dependence of the average size of recrystallized grains, d_{rec}, and specific recrystallized volume, V_{rec}, at $\varepsilon = 8.3 \times^{-4}$ s⁻¹ in the P/M Ti-50Al alloy with an initial grain size of 18 μm.

Microstructure Parameters	Temperature of deformation, °C			
	950	1000	1050	1100
d_{rec}, μm	1-2	4.5	10.5	12.5
V_{rec}, %	80	87	90	93

TABLE 2. Temperature dependence of the size of the dynamically recrystallized grains, d_{rec}, and specific recrystallized volume, V_{rec}, at ε=8.3×10⁻⁴ s⁻¹ in the P/M TiAl alloy with an initial grain size of 5 μm.

Microstructure Parameters	Temperature of deformation, °C					
	700	750	800	850	900	950
d_{rec} (μm)	0.13	0.25	0.45	0.75	1.2	1.9
V_{rec} (%)	56	70	78	84	90	96

52

Based on the above-mentioned results, a method of production of the SMC structure was suggested which consisted of isothermal deformation of material at a decreasing temperature after each step of deformation. The initial temperature and strain rate are determined by the initial structure of the material to achieve 100% DRX. The fraction of the recrystallized volume can be increased by an increase in strain using soft modes of deformation such as multi-axial deformation. As the grain refinement during DRX leads to SP behavior, the microstructural homogeneity can be improved because of extensive diffusion and grain boundary sliding peculiar to SP. On the other hand, SP conditions prevent achieving DRX and, therefore, grain refinement [11]. As the SP conditions are reached at particular temperature/strain rate conditions due to DRX-initiated grain refinement, the conditions must be changed (*e.g.* by a temperature decrease) to impede SP flow and facilitate further DRX and grain refinement. Use of the empirical equation $\sigma_s = kd^n$ allows determination of the DRX conditions for production of a required grain size. These conditions can be achieved not only by the step-by-step temperature decrease (as described above) but also by a discrete increase in strain rate at a constant temperature. In this case, however, post-dynamical grain growth should be prevented by rapid cooling or by use of second phase particles suppressing grain growth considerably.

A SMC structure was produced in a number of gamma TiAl–based alloys and a Ti₃Al alloy by multi-step isothermal forging (Table 3). The ordering energy and, therefore, the specific energy of the antiphase interface in the Ti₃Al intermetallic are much lower as compared to TiAl [12, 13]. Therefore the rate of recovery in the Ti₃Al intermetallic is lower and the tendency for DRX is higher as compared to TiAl.

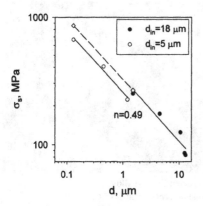

Figure 3. Size of dynamically recrystallized grains vs. steady state stress in the Ti-50Al alloy with the initial grain size of 18 μm and 5 μm. Dashed line is an extrapolation to smaller grains.

It can be seen from Table 3, that different processing parameters are required for different γ-alloys to produce almost the same SMC structure. For example, it was very difficult to obtain a submicron grain size in the Ti–48Al–2Cr–2Nb and Ti–46Al–2Cr–2Nb–1Ta cast alloys with an initial coarse-grained structure. However, in the P/M Ti–46Al alloy with a finer initial grain structure, the SMC structure was easily formed even in large-scale billets, for example, by 2-stage a-b-forging. After the same number of forging stages, the grain size of 0.3 μm was produced in a single phase α₂ Ti–25Al cast alloy while the grain size of only 0.4 μm was produced in the Ti–50Al alloy.

The lowest grain size produced by the method described is determined by the lowest forging temperature as well as by the rate of grain growth between forging cycles. Dispersion of second phase particles hindering grain growth can therefore lead to a finer final grain size. Smaller grain size produced in the Ti$_3$Al as compared to TiAl alloys can be explained by weaker recovery and more intense DRX.

TABLE 3 Alloy compositions, processing conditions, density and grain sizes of workpieces.

Composition, Type of Microstructure, Grain/Colony Size, μm	Processing	Grain Size, μm	Density, Initial / After Processing g/cm³
Ti-50Al, duplex, r$_\gamma$=20%, d=100/1000-2000	Cast, 3-5-stage a-b-c-forging at T=1000-800°C	0.4	-
Ti-46Al, near γ D=20-30	PM, 2-stage a-b-forging: 1) at T=1000°C, 2) at T=750°C	0.2	-
Ti-48Al-2Cr-2Nb, lamellar, d=500	Cast+HIP, 20-stage a-b-c-forging at T=1000-750°C	0.3	3.974 / 3.972
Ti-46Al-2Cr-2Nb-Ta, lamellar, d=300-400	Cast+HIP, 15-stage a-b-c-forging at T=1000-740°C	0.3	-
Ti-25Al, lamellar d=200-300	Cast, 6-8-stage a-b-c-forging at T=1050-700°C	0.3	4.206 / 4.205
Ti-25Al, lamellar D=200-300	Cast, 8-10-stage a-b-c-forging at T=1050-650°C	0.1	-

An increase in the number of forging stages did not eliminate the chemical non-homogeneity of the TiAl cast alloys, which was manifested by a non-homogeneous distribution of the α_2 phase [14]. At the same time, the P/M alloys with SMC structures were characterized by an increased homogeneity of the second phase distribution. The specific density of the alloys was determined before and after processing to be nearly the same, indicating the absence of porosity and cracks in the forged material (Table 3).

The effect of hot working on the degree of long-range order, η, was found to be different in the TiAl and Ti$_3$Al alloys. This parameter was not changed after hot working in the two-phase γ alloys. Contrary to this, η decreased in the SMC Ti–25Al alloy after forging at different temperatures 650 and 700°C to produce the grain sizes of 0.1 and 0.3 μm, respectively. Specimens with d = 0.8 μm, processed at higher temperatures (750°C) of forging were fully ordered. Specimens with d = 0.1 and 0.3 μm were partially disordered (η = 0.80 and 0.85). There is probably a relationship between the value of ordering energy and tendency to partial disordering during forging. Therefore, the Ti$_3$Al alloy with lower ordering energy partially disordered during forging while the TiAl alloys were fully ordered. The degree of long-range order must be taken into account together with the grain size in formation of final mechanical property combinations.

Both in TiAl and Ti$_3$Al stoichiometric alloys a substantial increase in strength at room temperature was observed when a submicron-sized grain structure was formed. The TiAl alloy with the grain sizes of d = 0.4 μm and had the yield stress of 990 MPa, whereas the yield stress at the grain size of 14 μm was 270 MPa. The yield stress of the Ti$_3$Al alloy with d = 0.3 μm and d = 0.8 μm was 1360 MPa and 960 MPa, respectively. Ductility of the TiAl alloy with submicron grain size was close to zero, whereas ductility of the fully ordered Ti$_3$Al at d = 0.3 μm increased up to 4.8%. In the TiAl alloy a RT peak ductility (~ 6%) was observed at a grain size of ~8 µm due to the operation of twinning that facilitated the stress relaxation [3]. Probably the difference in behavior of TiAl and Ti₃Al alloys is caused by the difference in their superlattices and deformation mechanisms [3]. Partial disordering of the Ti₃Al led to embrittlement because of the lower dislocation mobility [5].

4. Conclusions

1. A universal method of obtaining a SMC structure in semi-finished products made of hard-to-deform TiAl and Ti₃Al intermetallics was developed by employing DRX. The grain refinement was achieved with the use of multi-step isothermal forging at decreasing temperature. During each forging step performed at a constant temperature, DRX led to a grain size decrease allowing deformation at a lower temperature with a further decrease in grain size.
2. A fully recrystallized structure with a grain size of 0.1 to 0.4 µm was produced in TiAl and Ti₃Al based alloys. Initial chemical non-homogeneity of the alloys was not removed by the extensive hot forging. Partial disordering occurred in Ti₃Al during deformation at temperatures 600–700°C.
3. Room temperature strength increased considerably when grain size decreased to a submicron value. Ductility of fully ordered Ti₃Al increased while ductility of TiAl-based alloys and partially disordered Ti₃Al decreased when the grain size decreased below 1 µm.

5. Acknowledgement

This work was supported by the NATO Linkage Grant #HTECH.LG 961178.

6. References

1. Kim, Y-W. (1994) Ordered intermetallic alloys, Part III: gamma titanium aluminides, *JOM* **46** (7), 30-40.
2. Imayev, R.M., Imayev, V.M., and Salishchev, G.A. (1992) Formation of submicrocrystalline structure in TiAl intermetallic compound, *J. Mat. Sci.* **27**, 4465-4471.
3. Imayev, V.M., Salishchev, G.A., Imayev, R.M., Shagiev, M.R., Gabdullin, N.K., and Kuznetsov, A.V. (1997) An approach to ductility improvement of TiAl and Ti₃Al titanium aluminides based on microstructure control, in *Structural Intermetallics 1997*, eds. M.V. Nathal, R. Darolia, C.T. Liu, P.L. Martin, D.B. Miracle, R. Wagner, M. Yamaguchi, The Minerals, Metals and Mater. Soc., Warrendale, PA, 505-514.
4. Imayev, R.M., Salishchev, G.A., Imayev, V.M., Shagiev, M.R., Gabdullin, N.K., Kuznetsov, A.V., Senkov, O.N., Froes, F.H. (1998) Low-temperature superplasticity of submicrocrystalline intermetallics, *Mater. Sci. Forum* **304-306**, 195-200.
5. Imayev, R.M., Gabdullin, N.K., Salishchev, G.A., Senkov, O.N., Imayev, V.M., and Froes, F.H. (1999) Effect of grain size and partial disordering on ductility of Ti₃Al at temperatures of 20°C to 600°C, *Acta Mater.* **47** (6), 1809-1821.
6. Koch, C.C. and Cho, Y.S. (1992) Nanocrystals by high energy ball milling, *Nanostruct. Mater.* **1**, 207-212.
7. Valiev, R.Z., Korznikov, A.V., and Mulyukov, R.R. (1993) Structure and properties of ultrafine-grained materials produced by severe plastic deformation, *Mater. Sci. Eng.* **A168**, 141-148.

8.   Kaibyshev, O.A., Glazunov, S.G., Salishchev, G.A., Imayev, R.M., and Ivanov, V.I. (1987) Effect of hot deformation on the structure of cast Ti-36wt.%Al alloy, *Phys. Met. Metall.* **64(5)**, 1005-1010.

9.   Glovers, G. and Sellars, C.M. (1973) Recovery and recrystallization during high temperature deformation of a-iron, *Met. Trans.* **A4**, 765-774.

10.   Salishchev, G., Zaripova, R., Galeev, R., and Valiakhmetov, O. (1995) Nanocrystalline structure formation in materials and their deformation behaviour, *Nanostruct. Mater.* **6**, 913-916.

11.   Kaibyshev, O.A. (1992) *Superplasticity of Alloys, Intermetallics, and Ceramics*, Springer-Verlag Berlin Heidelberg, Berlin, 1-317.

12.   Blackburn, M.J. and Williams, J.C. (1969) Strength deformation modes and fracture in TiAl alloys, *Trans. Am. Soc. Metals* **62 (2)**, 398-409.

13.   Shechtman, D., Blackburn, M.J., Lipsitt, H.A. (1974) The plastic deformation of TiAl, *Met.Trans.* **A23**, 1373-1381.

14.   Imayev, V.M., Imayev, R.M., Salishchev, G.A., Shagiev, M.R., and Kuznetsov, A.V. (1997) Effect of strain rate on twinning and room temperature ductility of TiAl with fine equiaxed microstructure, *Scripta Mater.* **36 (8)**, 891-897.

SEVERE PLASTIC DEFORMATION PROCESSES: MODELING AND WORKABILITY

S.L. SEMIATIN, D.P. DELO, T.C. LOWE*
Air Force Research Laboratory
Materials and Manufacturing Directorate
AFRL/MLLM, Wright-Patterson Air Force Base, Ohio USA
**Los Alamos National Laboratory*
Los Alamos, New Mexico USA

Abstract

Research efforts to model severe plastic deformation (SPD) processes and to quantify the workability of materials during such operations are summarized. The majority of research on the modeling of the equal channel angular extrusion (ECAE) process invented by Segal has focused on the kinematics of metal flow under nominally frictionless conditions. Physical (visioplasticity) models suggest that such assumptions yield reasonable estimates of imposed strain over the bulk of the sample. Recent work on the stress state developed during ECAE has shown that a high degree of compression exists in the deformation zone except near the inner corner of the tooling. At this location, tensile stresses which may lead to fracture are generated. Thus, ductile metals can often be deformed through multiple passes in ECAE unless they exhibit flow softening which can lead to flow localization in the form of shear bands. Research on the modeling of the second major SPD process, that of torsion under superimposed compression, a method developed by Bridgman, has been investigated to a less extent primarily because of the mixed displacement-load boundary conditions and the statically indeterminate nature of the operation.

1. Introduction

Despite the recent surge in interest in severe plastic deformation (SPD), relatively little work has been conducted in order to model the processes, let alone to establish or understand those factors which limit the working of metals during these operations. Hence, the majority of research has focused on relatively ductile metals whose processing conditions (*e.g.*, strain rate and temperature) have been determined largely by trial-and-error methods. In this paper, research related to the modeling of two SPD operations, equal channel angular extrusion (ECAE) and torsion under superimposed compression, is summarized. In addition, workability considerations for ECAE, particularly for difficult-to-work materials, are reviewed.

T.C. Lowe and R.Z. Valiev (eds.), Investigations and Applications of Severe Plastic Deformation, 57–64.
© *2000 Kluwer Academic Publishers. Printed in the Netherlands.*

2. Modeling/Analysis of SPD Processes

A number of approaches have been taken to model the deformation that occurs during severe plastic deformation processes such as ECAE and torsional straining. These include descriptions based on slab, slip-line field, finite-element, and physical models, among others. Predictions of deformation patterns, loads, and the effect of tooling design on metal flow have been developed from such analyses.

2.1. MODELING OF ECAE

Segal, the inventor of ECAE, performed the first analysis of the kinematics and loads involved in the process using the slip line field method [1]. Denoting the channel angle as 2ϕ, the imposed shear strain per pass for the frictionless case in which the channels intersect sharply and deformation occurs abruptly when material elements cross the shear plane was shown to be $\gamma = 2 \cot \phi$. Hence, a channel angle of $2\phi = 90°$ produces a shear strain of 2 and a von Mises effective strain $\bar{\varepsilon} = 2 / \sqrt{3} = 1.155$ per pass. Similar slip-line field analysis of ECAE was performed by Cui, et al. [2]. Segal, et al. [1] also found that the ram pressure P for frictionless ECAE in a sharp-cornered channel is given by:

$$P = 2k \cot\phi + \sigma_2, \tag{1}$$

in which k denotes the flow stress in shear, and σ_2 is the backpressure applied in the exit channel. For frictionless ECAE with zero backpressure, the ram pressure is thus approximately equal to the flow stress.

Iwahashi, Furukawa, and their coworkers [3, 4] extended the work of Segal, et al. [1] for the frictionless case in which the channels intersect gradually such that deformation occurs as material elements pass through a centered-fan of angle ψ. The effective strain per pass is then the following:

$$\bar{\varepsilon} = \left\{ 2 \cot \left(\phi + \psi / 2\right) + \psi \csc \left(\phi + \psi / 2\right) \right\} / \sqrt{3}. \tag{2}$$

Insight into the possible effects of *multipass* deformation on the formation of mechanical and crystallographic texture was also provided by Furukawa, et al. [4] and Ferrasse, et al. [5] who defined three possible scenarios. These were denoted as routes A, B, and C, which comprised rotations of 0°, 90°, or 180°, respectively, about the billet axis between passes. Because the shear plane is fixed in eulerian (tooling/laboratory) coordinates and the microstructure/substructure rotates upon passing through the shear plane, the principal axes of strain do not coincide with the principal axes of the microstructure except at the end of every other pass in route C and every fourth pass in route B. Thus, the kinetics of microstructure refinement during ECAE are a strong function of processing route. Routes B and C generally lead to refinement via continuous dynamic recrystallization more quickly than route A [6, 7].

The validity of the expressions for shear strain and the general characteristics of the flow pattern during ECAE have been evaluated using both physical modeling and finite element simulation techniques. Wu and Baker [8] and Semiatin and DeLo [9] conducted ECAE experiments using plasticene and gridded AISI 4340 samples, respectively. In the former work, samples were extruded through plexiglas tooling with no sliding walls/floor, i.e. a simple, solid die. The tooling had a channel angle of $2\phi =$ 90° or 120° with or without a sharp corner; the rounded-corner setups had $\psi = 28°$ (2ϕ = 90°) or $\psi = 60°$ ($2\phi = 120°$). Measurements of the shear strain from single and multipass extrusions showed that the center of the billets (away from the die walls) did indeed undergo deformations which were well predicted by expressions such as Equation (2). However, the billet regions near the die wall underwent substantially lower strains due to sticking friction. By contrast, when a 'complex' tooling design with front and back walls and a bottom "floor" that moved with the billet [10] was used to conduct ECAE on 4340 steel, uniform deformation across the billet section was obtained [9].

The finite-element-modeling (FEM) work of Prangnell, *et al.* [11] and DeLo and Semiatin [12] provided further insight into the effect of die-geometry and friction on metal flow. Prangnell, *et al.* [11] focused on a solid die geometry with 2ϕ = 90° or 100°. Frictionless conditions with the larger die angle tended to promote uniform flow. On the other hand, ECAE through the channel with 2ϕ= 90° under realistic friction conditions (friction shear factor = 0.25) yielded predictions of the formation of dead metal zones in the die corner. A similar, nonuniform metal flow pattern through a solid die was found in the FEM simulations with 2ϕ = 90° of DeLo and Semiatin [12]. Subsequent work by Semiatin and DeLo [13] revealed that metal flow nonuniformities developed when using a solid die dampen out once a length equal to the breadth of the channel has been extruded. FEM simulations for the solid die with 2ϕ = 90° also revealed that the billet separated from the die in the bottom corner, a phenomenon replicated by the physical models of Wu and Baker [8]. When the FEM model incorporated sliding tooling members while maintaining realistic friction conditions [12], uniform flow and no separation in the corner was predicted, results in line with the physical modeling observations for such tooling.

Unfortunately, the focus on the final imposed strain after material elements exit the die has diverted attention from the details of the stresses and strains within the deformation zone during ECAE. For example, substantial compressive stresses are generated across the shear zone due to the punch pressure and the reaction forces set up by the tooling. In fact, the stress state is almost totally compressive in nature over the majority of the deformation zone. Furthermore, FEM simulations have shown that tensile stresses (of the order of 25 percent the flow stress) are generated in those regions of the workpiece which flow near and around the inner-channel corner [13]. These low levels of tensile stress are quite beneficial in reducing the tendency for failure by ductile fracture processes. FEM simulations have also shown that the strain within the deformation zone is slightly more complex than a state of simple shear along a shear plane [13].

DeLo and Semiatin [12] also conducted an extensive FEM investigation of equal channel angular extrusion under *nonisothermal* conditions. Here, the efficacy of 3-D simulations in predicting the magnitudes of the temperature gradients due to die chill was demonstrated. However, it was found to be more difficult to predict the details of the deformation in the 3-D FEM model because of problems associated with mesh density and element behavior. Therefore, an approach involving 3-D heat transfer simulations and 2-D deformation simulations was developed.

2.2. MODELING OF SEVERE TORSIONAL DEFORMATION

Two forms of severe torsional deformation of thin disks have been described in the literature. The first, due to Bridgman [14], comprises simultaneous compression and torsion of a disk which is not constrained laterally; its diameter is thus free to expand beyond that of the tooling anvils. The second approach, developed by Valiev and his colleagues at Ufa State Aviation Technical University in the former Soviet Union, appears to comprise compression/torsional deformation of a disk situated between a tight fitting cylindrical plunger and die, a geometry which prevents lateral expansion of the workpiece [15]. A number of approximate analyses have been conducted to describe various aspects of the Bridgman test, but apparently there are none in the western literature for that of Valiev, *et al.* Therefore, attention will be focused on the former deformation mode.

Kuhlmann-Wilsdorf, *et al.* [16] have summarized and discussed some of the key features of the Bridgman opposed-anvil test. These include descriptions of sample shearing and thinning, slippage between the sample and anvils, contact/rubbing of the anvils when the sample becomes very thin, and "turbulent flow." Unfortunately, most of the analyses of the test appear to make unrealistic assumptions. For example, the variation of shear strain and shear stress across the sample diameter is often neglected, and shear stresses and torques are associated with the shear strain at the outer diameter of the sample (or the anvils when the sample overhangs the tooling). By this means, the occurrence of anvil-sample slippage is determined by comparing the torque associated with the applied pressure and coefficient of friction to that required to induce shear in pure torsion, neglecting the influence of imposed normal (thickness) stresses and friction generated (radial, circumferential) stresses on plastic flow [17].

Unfortunately, as reported by Kuhlmann-Wilsdorf, *et al.* [16], most of the attempts to model the pressure distribution between the sample and the anvils have been based on simple friction hill analyses of the compression of a cylindrical disk [18,19] without due regard to the stresses associated with the torsional deformation. Needless to say, such shear stresses would modify the yield criterion used in conjunction with the load equilibrium equation to estimate the applied pressure and internally generated stresses due to friction. Proper analysis of the stresses would allow the determination of the pressure required to prevent anvil-sample slippage while minimizing radial material flow and sample thinning. Because of the complexity of the boundary conditions in the Bridgman opposed-anvil test, it appears that an FEM approach would be required to obtain solutions for this deformation process.

3. Workability During SPD Processes

Factors limiting the deformation that can be imposed without failure during SPD processes have received limited attention. However, such considerations can be quite important in selecting processing conditions. For ECAE, investigations conducted at the Air Force Research Laboratory (AFRL) have provided useful insights for conditions involving both cold working as well as hot working.

In the cold-working regime, isothermal single-pass ECAE deformation was conducted at temperatures between room temperature and 325°C in a die with $2\phi = 90°$ for commercial-purity titanium (CP Ti) and AISI 4340 steel [20]. Both alloys exhibited a transition in behavior as a function of temperature and strain rate. For CP Ti, nominally uniform shear flow occurred only at high temperatures and low strain rates; under other deformation conditions, failure leading to segmentation of the workpiece was observed. By contrast, the 4340 steel exhibited uniform flow at all strain rates at low-to-moderate test temperatures. Metallographic examination of sections of failed ECAE samples revealed that all of the failures were a result of flow localization in the form of shear bands preceding fracture. The flow localization trends were successfully interpreted in terms of the flow localization parameter in simple shear, or the ratio of the nondimensional hardening/softening rate to the strain-rate sensitivity value m. At cold-working temperatures, values of m were low for both CP Ti and 4340 steel. Thus, the development of nonuniform flow in shear was found to be controlled principally by the occurrence or absence of flow softening at the large strains typical of ECAE.

Because flow softening tends to occur at large strains at cold-working temperatures, the selection of die angle (and hence imposed strain) and number of passes is particularly important with respect to process design at these temperatures. The specific processing route during multipass operations (*e.g.*, A, B, or C) can have a substantial influence on flow behavior through the effect of strain path on strain hardening behavior and/or fracture strain. For example, the absence or occurrence of shear localization during ECAE of samples of CP Ti at 325°C which were put through routes A and C, respectively, was demonstrated by Semiatin and DeLo [21].

The occurrence of nonuniform flow and fracture during ECAE of Ti-6Al-4V and TiAl under conventional hot working conditions has also been investigated [22-24]. At hot-working temperatures, flow softening is very common and is typically greatest in magnitude at *low* strains. Estimates of the flow localization parameter in shear as well as detailed FEM calculations were used to understand the occurrence of shear banding and shear failure for the conventional and intermetallic titanium alloys. This understanding led to the design of a modified process consisting of an initial increment of upset deformation preceding the shear flow in the ECAE deformation zone [22, 23]. Because of geometric hardening, the compression mode of deformation of a flow-softening material is inherently more stable than simple shear. Therefore, imposing an initial increment of upset deformation at low strains (at which the flow-softening rate is high) can place the material in a condition which is less likely to lead to subsequent localization when undergoing deformation in the ECAE shear zone.

Semiatin and DeLo [13] have also shed light on the propensity for fracture during ECAE under those conditions in which flow localization and shear fracture do not occur. Such failures usually occur in regions that flow near and around the inner corner radius at which a tensile component of stress σ_t is generated. The 'damage' C generated by these tensile stresses was quantified using the maximum tensile work integral proposed by Cockcroft and Latham [25]:

$$C = \left(\sigma_t / \bar{\sigma} \right) d\dot{\varepsilon}, \tag{3}$$

in which $\bar{\sigma}$ denotes the effective stress. FEM calculations [13] have shown that the maximum values of C (at the top surface of the ECAE workpiece) lie generally in the range between 0.15 to 0.40 for a single pass through a die with $2\phi = 90°$. Generous inner corner radii, high slider friction (which gives rise to a high backpressure in the complex tooling design), and material behavior characterized by strain hardening tend to produce the lowest values of tensile damage. To a first approximation, the maximum number of passes n_c that can be imposed (using route A) is $\dot{\varepsilon}_f / C$, where $\dot{\varepsilon}_f$ denotes the true fracture strain corresponding to the reduction in area in uniaxial tension. Semiatin and DeLo [13] validated this broad approach to predict fracture using observations for cases in which localization was avoided - 4340 steel deformed at ambient temperatures and a near-gamma titanium aluminide alloy deformed at high hot working temperatures.

4. Research Issues/Future Challenges

A number of aspects regarding modeling of and workability during SPD are yet unclear. Some of the areas which may be fruitful from a research standpoint include the following:

- Evolution of microstructure, substructure, and texture during ECAE. Because the microstructure rotates relative to the principal axes of stress and strain during ECAE, correlation of microstructure/substructure/texture to the imposed deformation is a complex undertaking. Crystal plasticity FEM analysis of single and multipass ECAE may provide useful insights.
- Strain path effects for ECAE. Strain path affects both strain hardening and cavitation, processes which impact flow localization and fracture modes of failure during ECAE. Thus, work to obtain an understanding of such effects under both cold and hot working conditions would be useful.
- Analysis of ECAE. Although FEM analyses have provided useful insights into the deformation during ECAE, the details of metal flow around the inner corner, where deformation tends to be most severe, are not well described. Alternate FEM approaches (e.g. Reference 26) may provide useful information regarding deformation and microstructure evolution in such regions.
- Analysis of severe torsional deformation. There is a great need to develop a complete model of this process. Such a model should include plastic deformation of disk samples due to torsionally-induced shear stresses and pressure-induced normal (axial, radial, circumferential) stresses. Such an analysis should include appropriate descriptions of anvil-sample interface friction and tooling deflection.

A number of future challenges for SPD processes relate to process scaleup including the following:

- Tooling design/container design for ECAE. Large hydrostatic stresses are generated during ECAE, thus requiring properly designed shrink fit tooling in order to avoid tooling distortion, flash formation between sliding members, etc.

- Billet conditioning between passes for ECAE. Tooling designs/process concepts to eliminate billet conditioning between passes are needed from an economic standpoint. The elimination of billet conditioning is especially important for hot working in which intermediate cooldown and reheating may eliminate the beneficial effects of previous ECAE steps.
- Product yield in ECAE. Because of the nature of the less highly deformed ends in ECAE-processed billets, product yield and process economics are an important challenge.
- Scaleup potential of severe torsional deformation. Because of the need for relatively high aspect ratio disks/pancakes, methods to apply the large torques/normal forces on a production scale must be developed.
- Competing processes. As with any new metals technology, the technical and economic factors associated with other methods of manufacturing nanocrystalline metals should be considered.

5. Acknowledgments

The writing of this paper and related efforts on equal channel angular extrusion were made possible by support from the Air Force Research Laboratory's Materials and Manufacturing Directorate and the Air Force Office of Scientific Research. One of the authors (DPD) was supported by Air Force contract F33615-94-C-5804. The assistance of L.A. Farmer in preparing the manuscript is also much appreciated.

6. References

1. Segal, V.M., Reznikov, V.I., Drobyshevskiy, A.E., and Kopylov, V.I. (1981) Plastic working of metals by simple shear, *Russ. Metall* 1, 99-105.
2. Cui, H.J., Goforth, R.E., and Hartwig, K.T. (1998) The three dimensional simulation of flow pattern in equal-channel angular extrusion, *JOM-e* 50, no. 8, 1-5.
3. Iwahashi, Y., Wang, J., Horita, Z., Nemoto, M., and Langdon, T.G. (1996) Principle of equal channel angular pressing of ultra-fine grained materials, *Scripta Mater.* 35, 143-146.
4. Furukawa, M., Iwahashi, Y., Horita, Z., Nemoto, M., and Langdon, T.G. (1998) The shearing characteristics associated with equal-channel angular pressing, *Mater. Sci. Eng. A* A257, 328-332.
5. Ferrasse, S., Segal, V.M., Hartwig, K.T., and Goforth, R.E. (1997) Microstructure and properties of copper and aluminum alloy 3003 heavily worked by equal channel angular extrusion, *Metall. Mater. Trans. A* 28A, 1047-1057.
6. Iwahashi, Y., Horita, Z., Nemoto, M., and Langdon, T.G. (1997) An investigation of microstructural evolution during equal-channel angular pressing, *Acta Mater.* 45, 4733-4741.
7. Iwahashi, Y., Horita, Z., Nemoto, M., and Langdon, T.G. (1998) The process of grain refinement in equal-channel angular pressing, *Acta Mater.* 46, 3317-3331.
8. Wu, Y. and Baker, I. (1997) An experimental study of equal channel angular extrusion, *Scripta Mater.* 37, 437-442.
9. Semiatin, S.L. and DeLo, D.P. (1998) Unpublished research, Air Force Research Laboratory, Wright-Patterson AFB, OH.
10. Segal, V.M., Goforth, R.E., and Hartwig, K.T. (1995) Apparatus and method for deformation processing of metals, ceramics, plastics, and other materials, U.S. Patent 5,400,633.
11. Prangnell, P.B., Harris, C., and Roberts, S.M. (1997) Finite element modeling of equal channel angular extrusion, *Scripta Mater.* 37, 983-989.
12. DeLo, D.P. and Semiatin, S.L. (1999) Finite-element modeling of nonisothermal equal-channel angular extrusion, *Metall. Mater. Trans. A* 30A, 1391-1402.

64

13. Semiatin, S.L. and DeLo, D.P. (1999) Unpublished research, Air Force Research Laboratory, Wright-Patterson AFB, OH.
14. Bridgman, P.W. (1935) Effects of high shearing stress combined with high hydrostatic pressure, *Physical Review* **48**, 825-847.
15. Valiev, R.Z. (1997) Structure and mechanical properties of ultrafine-grained metals, *Mater. Sci. Eng. A* **A234-236**, 59-66.
16. Kuhlmann-Wilsdorf, D., Cai, B.C., and Nelson, R.B. (1991) Plastic flow between Bridgman anvils under high pressures, *J. Mater. Res.* **6**, 2547-2564.
17. Jackson, J.W. and Waxman, M. (1963) in A.A. Giardini and E.C. Lloyd (eds.), *High Pressure Measurement*, Butterworth's, Washington, pp. 39-58.
18. Chan, K.S., Huang, T.L., Grzybowski, T.A., Whetten, T.J., and Ruoff, A.L. (1982) Pressure concentrations due to plastic deformation of thin films or gaskets between anvils, *J. Appl. Phys.* **53**, 6607-6612.
19. Riecker, R.E. and Towle, L.C. (1967) Shear strength of grossly deformed Cu, Ag, and Au at high pressures and temperatures, *J. Appl. Phys.* **38**, 5189-5194.
20. Semiatin, S.L., Segal, V.M., Goforth, R.E., Frey, N.D, and DeLo, D.P. (1999) Workability of commercial-purity titanium and 4340 steel during equal channel angular extrusion at cold-working temperatures, *Metall. Mater. Trans. A* **30A**, 1425-1435.
21. Semiatin, S.L. and DeLo, D.P. (1999) Equal channel angular extrusion of difficult-to-work alloys, *Materials and Design*, in press.
22. DeLo, D.P. and Semiatin, S.L. (1999) Hot working of Ti-6Al-4V via equal channel angular extrusion, *Metall. Mater. Trans. A* **30A**, in press.
23. Semiatin, S.L. and DeLo, D.P. (1999) Equal channel angular extrusion of difficult-to-work alloys, U.S. Patent 5,904, 062.
24. Semiatin, S.L., Segal, V.M., Goetz, R.L., Goforth, R.E., and Hartwig, T. (1995) Workability of a gamma titanium aluminide alloy during equal channel angular extrusion, *Scripta Metall. et Mater.* **33**, 535-540.
25. Cockcroft, M.G. and Latham, D.J. (1966) A simple criterion of fracture for ductile metals, NEL Report 240, National Engineering Laboratory, Glasgow, Scotland.
26. Mori, K., Osakada, K., and Fukuda, M. (1983) Simulation of severe plastic deformation by fine element method with spatially fixed elements, *Inter. J. Mech. Sci.* **25**, 775-783.

THE EFFECT OF STRAIN PATH ON THE RATE OF FORMATION OF HIGH ANGLE GRAIN BOUNDARIES DURING ECAE

P.B. PRANGNELL, A. GHOLINIA, M.V. MARKUSHEV*
Manchester Materials Science Centre, UMIST
Grosvenor St. Manchester, M1 7HS, UK
Institute of Metals Superplasticity Problems
450001, 39, St. Khalturina St. Ufa, Russia

Abstract

There is currently a great deal of interest in the use of severe deformation for producing alloys with submicron grain structures. High-resolution electron back scattered diffraction (EBSD) analysis has been employed to measure the boundary misorientations within the deformation structures of Al-alloys processed by equal channel angular extrusion (ECAE), using different strain paths. The strain path was varied by rotating the billet through 0, 90 and 180 degrees, between each extrusion cycle. This has highlighted great differences in the deformed state, as a function of the processing route. It has been found that the most effective method of forming a submicron grain structure is to maintain a constant strain path. Cyclic redundant strains lead to a far lower density of new high angle grain boundaries being formed.

1. Introduction

There are many claimed property advantages of submicron grained alloys produced by severe deformation [1]. In recent years several novel methods have been proposed for deforming bulk aluminium alloys to very high true strains. These include reciprocating extrusion [2], repeated roll bonding [3] and equal channel angular extrusion (ECAE) [4]. The common novel feature of these methods is that there is no reduction in the overall thickness of the billet and there is therefore no geometric restriction to the strain that can be achieved. In comparison, with conventional metal working operations, like rolling and extrusion, strains greater than 5 can only be obtained in filaments, or foils, which have little structural application. Because of the requirement of maintaining a constant billet geometry, several of the severe deformation processes are cyclic in nature and involve non-conventional strain paths. For example, in reciprocating extrusion the total strain is redundant. With ECAE it is relatively easy to systematically vary the strain path by rotating the billet between each extrusion cycle. Several processing schemes have been investigated, including rotating the billet by 90°, and 180°, around its axis [5, 6]. Of these processing routes, it has been proposed that a constant rotation of 90° is the most effective for producing an isotropic submicron grain structure [6]. However, such schemes of rotation can lead to the reversal of shear strains developed during earlier extrusions. These results thus appear to be in conflict

65

T.C. Lowe and R.Z. Valiev (eds.), Investigations and Applications of Severe Plastic Deformation, 65–71.
© 2000 *Kluwer Academic Publishers. Printed in the Netherlands.*

with previous studies that have compared the effects of redundant deformation to deformation with a constant strain path and have shown that deformation structures and textures are to a large extent reversible (*e.g.* [7, 8]). In the context of submicron grained alloys it is therefore important that the effects of the strain path during severe deformation, and redundant strains in particular, are more rigorously investigated. In this paper the deformation structures in alloys severely deformed by ECAE are compared for a range of different strain paths. Efforts have been made to obtain statistically significant, quantitative, data by measuring the density and distribution of low and high angle boundaries in the deformed state, using orientation mapping with a high resolution EBSD system over large sample areas.

2. Experimental

A DC cast Al-3%Mg-0.2%Zr-0.2%Fe (Al-Mg) alloy was homogenised prior to being machined into 90mm long by 15mm diameter rods. The rods were processed using an ECAE press with a die angle of 120°, through 16 extrusion cycles, giving a total effective strain of 10.7 (γ=1.15 per pass). Four different strain paths were used.

1) *No rotation*   Constant strain path: no rotation of the billet.
2) *180° rotation*   The billet was rotated through 180° clockwise between each cycle.
3) *±90° rotation* The billet was rotated through 90° clockwise and then 90° anticlockwise each alternate cycle.
4) *+90° rotation*   The billet was rotated through 90° clockwise, between each cycle.

The rotation was carried out around the rod's axis. In the literature routes 1, 2 and 4 are often referred to as route A, C, and B, respectively [5, 6]. The deformed samples were examined in a field emission Philips XL30 microscope and orientation mapping was carried out over several representative areas from each sample by using electron back scattered diffraction (EBSD). The spatial resolution of the EBSD system used was 20 nm. Because the angular accuracy of the EBSD system is around 1°, misorientations of less than 1.5° were excluded from the data. Low and high angle boundaries were conventionally defined as being less, or greater, than 15° misorientation, respectively.

3. Results

Using the high resolution EBSD technique, orientation maps were obtained from vertical sections along the center plane of the processed Al-Mg rods for the four different processing routes. In figure 1 examples of reconstructed grain structures are shown taken from 20x30 μm orientation maps measured with a 0.05 μm step size. In figure 2 the boundary misorientation distributions obtained from the maps in figure 1 are depicted and in figure 3 the average diameters of the grains and fractions of high angle boundaries are given for the different processing routes. (The grain sizes are the diameters of a circle of equivilant area). From figure 1 it can be seen that the deformation structures are generally inhomogeneous and with some of the processing routes very large grains are still present. The map area used to measure the average grain diameters thus tends to under represent the differences apparent in the images in figure 1, because many of the larger grains border the maps, or in the case of 180° rotation, are greater than the map area and are not included in the statistics. However, it is still evident that even after severe plastic deformation there are very significant differences in the deformation structures formed by the different strain paths.

3.1. DEFORMATION WITH A CONSTANT STRAIN PATH

The samples deformed with no rotation had the smallest "average" grain size of 0.5 µm, and the most homogeneous grain structure, as well as the highest fraction of high angle grain boundaries (~60%). Its deformation structure was largely comprised of submicron grains that were elongated in the direction of shear, with an aspect ratio of ~2.5. However, the grain structures still contained some much larger fibrous grains and there were low angle boundaries within many of the submicron grains.

3.2. 180° ROTATION

Despite being subjected to a total effective strain of ~10, with 180° rotation of the billet the map area in figure 1b is smaller than the largest grains present and much larger area lower resolution maps showed the presence of grains of similar dimensions to the initial

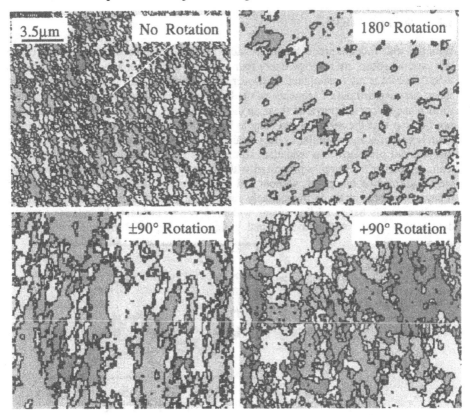

Figure 1. Reconstructed grain structures from the Al-Mg billets after 16 extrusion cycles for each of the four, processing routes.

68

grain size (200 μm). These large grains contained a few fine grains appearing in bands and concentrated in the region of the original grain boundaries. The sample thus consisted of very large equiaxed 100-200 μm grains containing an inhomogeneous distribution of a few new 0.3-2.0 μm grains. Because the large original grains are far bigger than the areas mapped in figure 3b this has resulted in a misleading average grain size in fig. 3a. Of the materials deformed by the four different processing routes, the sample processed by 180° rotation contained the lowest fraction of high angle grain boundaries (~ 0.3) and its microstructure was comprised largely of subgrains.

3.3. ±90° ROTATION

With the ±90° rotation processing route, shear is built up in two mutually orthogonal planes and this appears to result in less grain refinement than deforming to the same effective strain on one shear plane (*i.e.* with no rotation). In figure 3 it can be seen that many bigger elongated grains are still present, and significant proportions of the material were comprised of subgrains. The microstructure appears similar to that seen for samples deformed without rotation to lower strains.

3.4. +90° ROTATION

With the +90° rotation processing route the billet is again sheared in two planes, but in this case the shears are reversed every four extrusion cycles, *i.e.* after the sixteen cycles used the strain is fully redundant. However, unlike with the 180° rotation route, the two shears are reversed out of sequence as the billet is first sheared further on the alternate shear plane before the previous shear on the first plane is reversed. From figures 1-3 it can be seen that the +90° rotated sample has a slightly higher fraction of high angle grain boundaries and a similar average grain size compared to the ± 90° rotated sample that was not subjected to a redundant strain. This is the opposite result to that found from comparing the samples processed without rotation, and with 180° rotation, which were deformed on one plane, without and with redundant strain. However, with 180° rotation the shear strain is immediately reversed after each cycle. The +90° rotated sample contains bands of fine grains which appear to delineate boundaries between deformation bands, within the original grains, resulting in them being split into a

Figure 2. Boundary misorientation distributions after 16 extrusion cycles for the Al-Mg billets produced by the four different processing routes.

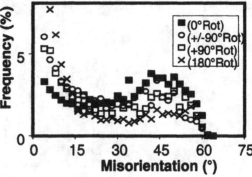

bimodal distribution comprised of submicron grains and grains of 4-6 μm in size. It is important to note that with both the 90° rotation routes the level of grain refinement, fraction of high angle boundaries, and homogeneity of the distribution of submicron grains, was markedly less than for the sample deformed with a constant strain path, but was far better than for 180° rotation.

4. Discussion

During plastic deformation at ambient temperatures, with a constant strain path, it is known that additional high angle boundary area can be created by two main processes; (1) grain subdivision and (2) distortion of the original grain shapes in proportion to the applied strain [9]. The material of interest here had initial grain size of ~200 μm and the majority of new high angle boundary area would thus be expected to have been created by grain subdivision [9]. A fully constrained grain must deform with five independent slip systems [10]. However, it is usually energetically easier for a grain to deform by splitting into cell blocks, which individually deform on a lower number of slip systems but collectively fulfil this condition. Each cell block is defined as a region within a grain deforming on the same set of slip systems and contains incidental cell boundaries, with a low average misorientation (< 2°) [11]. At low strains very coarse deformation bands are formed. As the strain increases cell blocks bordered by dense dislocation walls develop within the coarse bands and rotate progressively towards stable end orientations that make up the deformation texture [11]. On average the misorientation between adjacent cell blocks increases with strain and some dense dislocation walls will evolve into new high angle boundaries. Certain grain orientations are particularly unstable during deformation and split, such that different parts of the grain rotate towards very different end orientations, and this can form highly misorientated boundaries.

When the billet is rotated by 180° between each extrusion cycle this results in the shear strain being reversed every alternate cycle. Research using torsion has found that on reversing the sense of the shear, not only are the original grain shapes reconstructed, but the initial texture is also recreated [9]. The results obtained here are in agreement with these findings, as the original grains were still visible in the EBSD maps. An effective strain of 0.7 each extrusion cycle is sufficient for a grain to develop deformation bands and to start to subdivide into cell blocks. However on reversal of the strain, because the constraint is unaltered, the deformation pattern within the grain will remain the same and the cell blocks will rotate back towards their starting orientations, recreating the initial texture, and the misorientations across cell block boundaries will reduce towards zero. It is therefore not surprising that with a cyclic redundant strain the formation of additional high angle grain boundaries occurs at a very slow rate. The formation of new high angle boundaries, when a redundant strain process is used, can thus only be attributed to imperfections in the reversibility of the local deformation pattern within a grain. This is consistent with the few new fine grains being observed to be arranged in bands, from which it can be inferred that they are formed at the boarders of deformation bands, and near the original grain boundaries, where strain incompatibility occurs between adjacent grains leading to a more complex local deformation pattern.

With the ±90° rotation processing route, the billet was deformed in two shear planes and the rate of formation of high angle boundaries and grain refinement, was slower than when the same total strain was accumulated in one shear plane. This behavior can also be readily explained in terms of the known mechanisms of high angle boundary formation. With shear taking place alternately on two planes, the rate of rotation of cell blocks towards stable end orientations will be slower, as the intensity of each shear is less, compared to deformation with a constant strain path, and this leads to a lower rate of high angle boundary formation. +90° rotation involves the non-sequential reversal of shear strains developed on two mutually orthogonal shear planes. Despite the redundant nature of the total strain, this processing route still gave rise to marked grain refinement. In the first extrusion cycle a deformation pattern will develop within each grain comprised of deformation bands and cell blocks. On rotating the shear plane, in the next extrusion cycle, new deformation bands and cell blocks will become established, each deforming individually on a different set of slip systems to those used in the previous orientation. In the third extrusion cycle, on rotating through a further 90°, the first shear will now be reversed, but this will change the stress state acting on the grain compared to the previous cycle and it will not be able to deform back with the same deformation pattern established during the immediately prior shear. It is therefore not possible for the deformation structure to be as readily "undone" by the reversed shear strains because they occur out of sequence, even though the total strain still becomes redundant every four extrusion cycles. Reversing the shears out of sequence thus makes it more difficult for cell blocks to rotate back to their initial orientation and causes the formation of submicron grains between the boundaries of deformation bands. In directional strain processes the cell blocks become refined with increasing strain as they become smeared out and subdivide as their aspect ratio increases. With +90° rotation, because the strain path is cyclic, a grain's geometry does not progressively change with strain and the overall deformation pattern within a grain will remain relatively stable. This will make it more difficult to refine coarse cell blocks formed early in the process. An inhomogeneous bimodal grain structure is therefore retained even at very high strains comprised of submicron grains and grains of the scale of the cell blocks.

5. Conclusions

The use of high resolution EBSD has enabled the boundary misorientations within the deformation structures of severely deformed alloys to be quantitatively analysed more rigorously than has previously been possible. This has revealed large differences in the evolution of the deformed state as a function of the strain path. Grain structures produced by severe deformation can be highly inhomogeneous and can not be simply described as submicron grained, without carefully considering the processing route used to produce them. It has been demonstrated that the most efficient route, for forming a submicron grain structure, is to maintain a constant strain path and the least effective is to reverse the shear each alternate cycle, by rotating the billet through 180°. Both forms of 90° rotation gave a greater extent of grain refinement than 180° rotation, but were less efficient, in terms of forming new high angle grain boundary area, than deformation with a constant strain path.

6. Acknowledgements

The authors acknowledge the support of the EPSRC (grants GR/L06997 and GR/L96779), Alcan Int, and British Steel. M.V. Markushev is grateful to the Royal Society/Nato Research Fellowship scheme for funding while carrying out this work.

7. References

1. Markushev, M.V., Bampton, C.C., Murashkin, M.Y., Hardwick, D.A., (1997) Structure and properties of ultra-fine grained aluminium alloys, *Mat. Sci & Eng.* **A234**, 927-931.
2. Yuan, S-Y., and Yeh, J-W., (1997) The improved microstructures and properties of 2024 alloy produced by reciprocating estrusion method, *THERMEC97* (ed. Chandre & Sakai) **1**, 1143-9.
3. Saito, Y., Tsui, N., Utsunomiya, H., Saki, T., Hong, R-G (1998) Production of ultra-fine grained bulk al alloys by accumulative roll-bonding (ARB) *6th Int. Conf. on Al-Alloys,* ICCA-6, (ed. Sato, Kumai, Kobayashi, Murakami) **3**, 1967-1972.
4. Segal, V.M., Reznikov, V.I., Drobyshevskiy, A.E., Kopylov, V.I., (1981) Plastic working of metals by simple shear, *Russian Metall.* **1**, 99-105.
5. Segal, V.M., (1996) Plastic deformation of crystalline materials, US Patent, 5513512, May 7.
6. Iwahashi, Y., Horita, Z., Nemoto, M., Langdon, T.G., (1998) The process of grain refinment in equal-*Acta Mater.* **46**, 3317-3331.
7. Vatne, H.E., Mousavi, M.G., Benum, S., Ronning, B., Nes, E., (1996) Texture and structure evolution in AlMn and Al deformed by monotonic and reversed tosion, *Mat. Sci. Forum* **217-222**, 553-558.
8. Davenport, S.B., Higginson, R.L., Sellars, C.M., (1998) The effect of strain path on material behaviour during hot rolling FCC metals, To appear in *Deformation Processing of Metals*, Trans of The Royal Society A, London.
9. Harris, C., Prangnell, P.B., and Duan, X., (1998) FE modelling of ECA extrusion of al and the study of its annealing behaivour, *6th Int. Conf. on Al-Alloys, ICAA-6,* (ed. Sato, Kumai, Kobayashi, Murakami), Toyohashi, Japan, **1**, p. 583-589 .
10. Humphreys, F.J. and Hatherly, M., (1995) *Recrystallisation and Related Annealing Phenomena*, Elsevier Science Ltd.
11. Hughes, D.A., Hansen, N., (1997) High angle boundaries formed by grain subdivision mechanisims *Acta Mater.* **45**, 3871-3886.

THERMOMECHANICAL CONDITIONS FOR SUBMICROCRYSTALLINE STRUCTURE FORMATION BY SEVERE PLASTIC DEFORMATION

F.Z. UTYASHEV, F.U. ENIKEEV, V.V. LATYSH[1], E.N. PETROV[2], V.A. VALITOV
Institute for Metals Superplasticity Problems, Khalturina 39, Ufa
[1]R and D Bureau "Iskra" Ufa State Aviation Technical University
K.Marx 12, Ufa, 450000, Russia
[2]The Russian Federal Nuclear Center - All-Russian Scientific Research
Institute of Technical Phisics, P.O. Box 245, 456770, Snezhinsk
Chelyabinsk Region, Russia

1. Introduction

This paper deals with thermomechanical conditions used for preparation of microcrystalline structures in metallic materials on the basis of the analysis of well known deformation methods. A new method for formation of microcrystalline structures and a device for realization of this method are proposed.

2. Main Thermomechanical and Technological Aspects of Formation of Micro-, Submicro-, and Nanocrystalline Structures

Submicrocrystalline and nanocrystalline structures may be obtained by torsion using the Bridgman technique (Figure 1) and the equal-channel angular (ECA) pressing method (Figure 2) [1–6]. For pure metals and single phase alloys it is necessary that the temperature not exceed $(0.25–0.3)$ T_m, to avoid the growth of the small grains that form as the true value of the deformation (according to the estimations of various investigators) approaches values as high as several tens. In order to obtain a nanocrystalline structure in metal billets several rotations of a sample during torsion under pressure and more than ten passes of a rod sample – during ECA pressing are necessary. Thermomechanical peculiarities of formation of microcrystalline structures that have been analyzed in more detail in [7, 8]. The present work considers a number of technological aspects which are important for using these deformation methods.

The above specified deformation methods are used with more or less success in the laboratory since they do not fully meet the requirements of commercial technologies. Thus, Bridgman's technique is apparently applicable for obtaining nanocrystalline structures in thin foil-like samples, while ECA pressing is a labor intensive process and cannot be practically implemented for preparation of microstructures in high-strength metals and large rod samples because of the insufficient strength of tooling. Moreover, with increasing lateral dimensions of rods the non-uniformity of their deformation increases.

T.C. Lowe and R.Z. Valiev (eds.), Investigations and Applications of Severe Plastic Deformation, 73–78.
© *2000 Kluwer Academic Publishers. Printed in the Netherlands.*

The problem of developing a technological process for obtaining samples with submicrocrystalline and nanocrystalline structures by deformation mode can be solved by the method based upon a combined deformation mode, which unites angular pressing and torsion.

3. Method and Device for Combined Deformation of Samples

The essence of this method is that a sample is subjected to deformation in a composite container, the inlet and outlet parts of which are joined along the peripheral sections of channels and accomplished with the possibility of rotation relative to each other in the joint plane. The channels from the joint plane completely or partially have an elliptic shape. It is also possible for these channels having different shapes to be joined in the joint plane along the peripheral sections, *e.g.* elliptic channel can be joined with a circular channel. Samples in such a device can be subjected to deformation by shear in the zone of channel intersection according to the following variants:

1. Torsion by rotating upper and lower parts of the die relative to each other, in this case a sample is displaced to the zone of channels junction when the channels are in a coaxial position. By successive application of torsion to each section and displacement of a sample by incremental steps, one can process a sample along its whole length. This variant is the most merciful for the tooling since deformation by torsion is achieved by rotation in the joint plane of strong parts of container due to elliptical shapes of channels and the sample. If a sufficient level of hydrostatic compression is imposed, this variant can be used for deformation of hard-to-deform materials.

2. Angular, including equal-channel, pressing of a sample. The required angle of channel intersection is set by rotating the container's parts, and pressing is performed. During multi-step pressing as sample hardening takes place the angle of channels intersection can be increased in order to decrease the loads on the punch.

3. Concurrent torsion and pressing. This is the most efficient variant, since it imposes large total strains in a sample in only one pass. Realization of this variant requires control over the rate or over the magnitude of force through the channels since the angle of their intersection changes during rotation of the container parts.

We stress here the possibility of practical realization of a combined deformation method which couples the capabilities of torsion and angular pressing. Figure 3 shows the scheme of the combined method as the sum of deformation schemes in Bridgman's anvil and ECA pressing. The approximate equality sign shows that the combined method not only unites the capabilities of these known methods but also expands them.

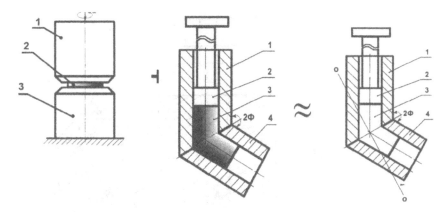

Figure 1. Torsion in Bridgman's anvil.　　*Figure 2.* Combined scheme.　　*Figure 3.* ECA pressing (torsion+pressing).

P- the axial Force
  M - the torque
  1- upper anvil
  2- sample
  3- lower anvil

1, 4 - parts of the composite
    container
  2 - punch
  3 - sample

1, 4 - parts of the composite
    container
  2 - punch
  3 – sample

Figure 4 shows a variant of the structural embodiment of the device. The peculiar features of this device are such that the plane of channel intersection is tilted with respect to its axes, but in such a way that in the cross-section plane the holes are limited by circles of equal diameters, the centers of which are lying practically at the same point. By O–O we designate the axis of rotation of the lower part of the container. The clearance between the conjugating planes is minimal, practically, these are surfaces lapped without any clearance but conjugate with the possibility of sliding.

Figure 5 shows various variants of channels junction with angles of their axes of intersection of 60°, 90°, 120° and 135°, respectively. The same figures show the shapes of cross-sections of channels and sections in the zones of channels junction. In addition, in Figure 5a a dashed line shows the position of channels after rotation of the lower part relative to the upper part of the container through an angle of 180°. From the above presented figures it follows that the device allows the intersection of channels at an angle less than 90° and correspondingly larger deformations of a sample are possible, as compared with ECA pressing at an angle of 90°.

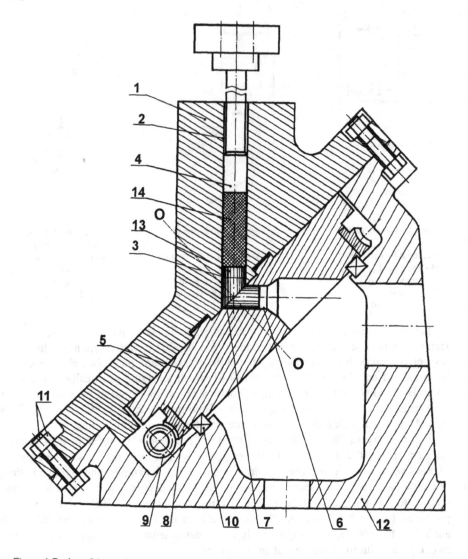

Figure 4. Design of the combined method device. List of parts: the stationary part of the container with a channel – 1, a section of the channel from the joint plane with an elliptical cross-section– 2, punch – 4, mobile part – 5 of the container with channel 6, a section of the channel – 7 from the joint plane with an elliptical cross-section, worm wheel – 8 rigidly connected with container 5, worm – 9, bearing – 10, bolted connection – 11, base – 12, sample – 13 and tablet – 14 used for ejecting a sample from channels and also as a lubricant.

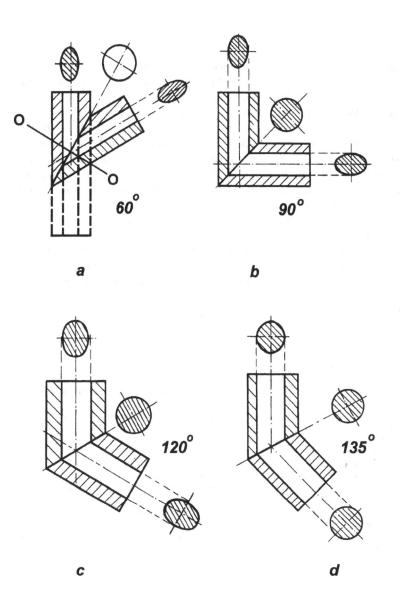

Figure 5. Variants of intersection of channel angles using the combined method.

4. Conclusion

We believe that the proposed method and device for combined deformation of samples is of interest because it could become a basis for development of commercial technology for fabrication of rather large billets with nanocrystalline structure. The method and device provide an intense accumulation of strain in billets and require less labor and energy consumption as compared with ECA pressing. At the same time the combined method of deformation has common features with methods of torsion and angular pressing which suggests that this method will induce deformations needed to form microcrystalline structures. Thus, this method, as well as the methods of torsion under pressure and ECA pressing, realizes shear and significantly non-monotonic deformation in each localized section of the billet. Complex loading of a billet and significant, *i.e.* non-monotonic change of material flow in the limited (on a macro level) center of deformation region, results (on a micro level) in an intense development of several mechanisms of deformation. This is the prerequisite for favorable accumulation of the required microstructural defects to form fine grains. The possibility of using the proposed device to explore different deformation methods extends our ability to obtain different types of microstructures in billets.

5. References

1. Bridgman, P.W. (1952) *Studies in Large Plastic Flow and Fracture with Special Emphasis on the Effects of Hydrostatic Pressure*, McGraw-Hill, New-York.
2. Valiev, R.Z., Korznikov, A.V., and Mulyukov, R.R. (1993) Structure and properties of ultrafine-grained metals produced by severe plastic deformation, *Mater. Sci. Eng.* **A168**, 141.
3. Valiev, R.Z., Krasilnikov, N.A., and Tsenev, N.K. (1991). Plastic deformation of submicron-grained alloys, *Mater. Sci. Eng.* **A137**, 35
4. Ahmadeev, N.H., Kopylov, V.I., Mulyukov, R.R., and Valiev, R.Z. (1992) Formation of submicrocrystallin structure in copper and nickel using severe plastic deformation, *Izv. Akad. Nauk RAN, Metally* **.5**, 96-101(in Russian).
5. Segal, V.M. (1995) Materials processing by simple shear, *Mater. Sci. Eng.* **A197**, 157-164.
6. Segal, V.M., Reznikov, V.I., Kopylov, V.I. (1994) *Processes of Plastic Structure Formation in Metals, Sciences and Engineering,* Publ. Haus, Minsk, Belorussia, (in Russian).
7. Utyashev, F.Z., Enikeev, F.U., Latysh, V.V. (1996) Comparision of deformation methods for ultrafine-graind structure formation, *Ann. Chim. Fr.* **21**, 379-389.
8. Utyashev, F.Z., Enikeev, F.U., Latysh, V.V (1998) Thermomechanical condition of formation of submicrocrystalline structure at large deformations, *Izv. Akad. Nauk RAN, Metally* **4,** 72-79 (In Russian).

II. MICROSTRUCTURAL CHARACTERIZATION AND MODELING OF SEVERE PLASTIC DEFORMATION MATERIALS

STRENGTHENING PROCESSES OF METALS BY SEVERE PLASTIC DEFORMATION

Analyses with Electron and Synchrotron Radiation

M.J. ZEHETBAUER
Institute of Materials Physics, University Vienna
Strudlhofgasse 4, A-1090 Wien, Austria

1. Definition and Features of Large Strain Deformation Stages

Since the early eigthies, large strain deformation has developed into a main field of research within the plasticity community. In the meantime the occurrence of stage IV and stage V has been revealed in all usual metals [1-5] (Fig. 1) and many alloys [6], for all deformation temperatures ([7, 8]), and for many different deformation modes including iterative ones [1]. Both stages are defined by a significant change in work hardening coefficient $\Theta = d\tau/d\gamma$ (τ, γ being the resolved shear stress / resolved shear strain) as compared with previous stage III where Θ strongly decreases: In stage IV a constant or even increasing hardening occurs whereas in stage V the work hardening coefficient Θ re-decreases as long as the macroscopic strength finally gets constant in a steady state deformation. Detailed thermal activation analyses (TAA) yielding the strain rate sensitivity and the density of thermally activatable obstacles as a function of deformation suggested that hardening in stage IV is governed by athermal storage of dislocations while hardening in stage V is characterized by thermally driven annihilation of dislocations [1]. TAA data in connection with measurements of total dislocation density also suggest to assume the storage & annihilation of edge dislocations in low temperature stage IV & V, and of screw dislocations in high temperature stage IV & V [3]. Applying suitable tools for microstructural analysis at large strains, *i.e.* residual electrical resistivity (RER, [9]) and/or X-ray Bragg Peak Profile analysis (XPA) with high intensity X-ray sources (rotating anode generator [10, 11]) it has been become possible to measure the (average) density and arrangement of dislocations as well as the related local long range internal stresses. Combining the dislocation density measurement with that of stress, the dislocation interaction parameter α can be derived, according to $\tau = \alpha\, G\, b \sqrt{\rho}$ [1] which is shown in Fig. 2 for the case of Al: It is apparent that the dislocation interaction (i.e., either the dislocation arrangement, or the type of interacting dislocations) changes dramatically when stage IV starts, but remains the same when stage V is entered. For cold work conditions $T \approx 0.2 - 0.3\, T_m$, $\alpha_{IV} = \alpha_V$ is of order ≈ 0.1 being independent on the metal; this point will be discussed at the end of this paper. The RER method also enables one to measure the evolution of deformation induced vacancies which grows to very high concentrations especially at the large strains: [12]. In detail, while there occurs a monotonic increase of vacancies in stages II – IV, this increase

T.C. Lowe and R.Z. Valiev (eds.), Investigations and Applications of Severe Plastic Deformation, 81–91.
© 2000 *Kluwer Academic Publishers. Printed in the Netherlands.*

82

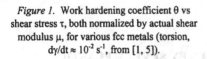

Figure 1. Work hardening coefficient θ vs shear stress τ, both normalized by actual shear modulus μ, for various fcc metals (torsion, dγ/dt ≈ 10^{-2} s^{-1}, from [1, 5]).

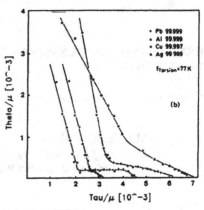

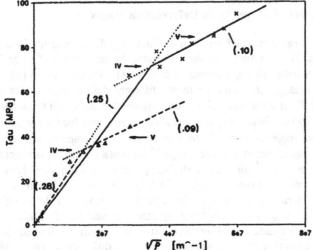

Figure 2. Res.shear stress vs.square root of total dislocation density ρ, in large strain cold worked Al. In parentheses: values of dislocation interaction parameter $α_i$. Upper curve: torsion T=77K; Lower curve: rolling, T=293K (from [1]).

deteriorates in stage V indicating the consumption of these vacancies by edge dislocations [1, 5] which start to climb just in this stage, for which clear evidence by TEM is given [1]. TEM investigations [13] also showed that in large strain cold working a marked fragmention of microstructure takes place, originally consisting of equally oriented lattice areas of high and low dislocation density ("cells") and gradually developing to misoriented areas shrinking in size with proceeding deformation. However, neither has been known on the dislocation structure of these walls, nor has been undertaken efforts in correlating them to macroscopic strengthening characteristics. New findings have been obtained by the application of new methods like quantitative contrast evaluation of X-ray Bragg Peak Analysis, the combination of XPA method with Synchrotron radiation (SXPA), as well as the Electron Back Scatter Patterning (EBSP) technique which are all described in the next section.

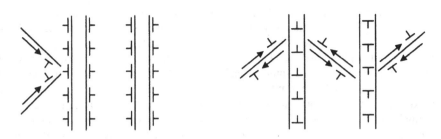

Figure 3. Sketch of a Polarized Dipole Wall ("PDW", left), and a Polarized Tilt Wall ("PTW", right); ref. [14]

2. New Experimental Methods (XPA, SXPA, EBSP) revealing New Microstructural Details of Severe Plastic Deformation

The X-ray Bragg Peak Profile Analysis (XPA) (see refs. [10, 11, 15, 16] and the paper of T. Ungar in this volume [17]) has proved to be the most efficient method in investigating details and/or evolution of microstructures occurring at large strains. From the terms of series expansion of the Fourier coefficients of a diffraction profile of a certain number and arrangement of dislocations, it is possible to determine the density (from broadening of the Bragg peak, ρ) but also the distribution of dislocations (from the tails of it, arrangement parameter M) and even the long range internal stresses connected to this distribution (from the asymmetry of the Bragg peak, $|\Delta(\Delta\sigma_i)|$). Moreover, the reflected intensity of a dislocation depends on the orientation of its Burgers vector **b** relative to the diffraction vector **g** chosen as has been exactly calculated by Wilkens [8]. Therefore, it is possible to evaluate the dominant Burgers vector **b** of dislocations

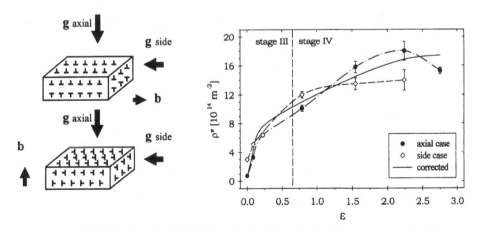

Figure 4. PDW- (left, above) and PTW-dislocations (left, below) with the Burgers vectors **b** relative to diffraction vector **g** for axial and side case. The rolling direction is parallel to **g** (side). At the right, it can be seen that according to contrast, the formal (as-measured dislocation density) is too high (low) as compared with the true one, depending on true strain ε applied and/or the direction of diffraction vector **g** (from [21]).

84

present either by applying two different reflections **g**, and/or by comparing the formal dislocation density with measurements from other methods, *i.e.* residual electrical resistivity [14, 19]. By this way it has been shown that while entering stage IV cold work hardening of Cu [10, 14] , Fe [11], Ni [19] and Al (this conference, [20]) there starts a transformation of dipole dominated dislocation cell walls (PDW's) to dislocation tilt walls (PTW's) which are similar to subboundaries (Fig. 3). The **b** - vectors of edge dislocations in PDW walls are *perpendicular to* **g** when the diffraction occurs on the normal section of rolled sample (Fig. 4 "axial case"), giving too *small* contrast and/or formal (= as measured) dislocation density as compared with average oriented dislocations and/or RER measurement (mean line in Fig. 4). When the PTW walls develop, the **b** – vector will rotate by 90° which means that it is now *parallel to* **g** giving a contrast being too *high* as compared with average oriented dislocations. For confirmation, also measurements with diffraction occuring across the transversal section of rolled sheet were performed ([21], Fig. 4, "side case"). Indeed, the formal dislocation density showed a behaviour contrary to axial case, *i.e.* too *high* formal densities in case of PDW walls in stage III, and too *low* ones for PTW structure in stage IV.

An additional confirmation of the PDW-PTW transformation to be characteristic for stage IV has been achieved by using the high intensity of Synchrotron radiation (being a factor 10^4 higher than even rotating anode intensity) in combination with XPA [19, 22]. Since this enables to reduce the spot size of investigation down to several ten microns, discovery of some substructural details as a function of large strain deformation can be expected since very local measurements of ρ, M and $|\Delta(\Delta\sigma_i)|$ within a single grain can be done *e.g.* in scanning mode. In Fig. 5 on the left, it is shown that these parameters start to fluctuate when stage IV of deformation is reached. The fluctuations are not of statistical nature but show certain correlations: while the dislocation density ρ takes a

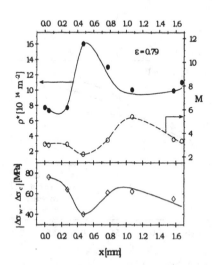
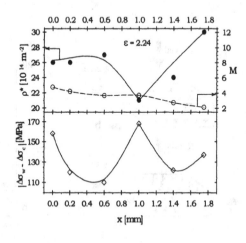

Figure 5. Scans made by SXPA in cold rolled polycrystalline Cu, showing the onset of PTW formation (left, stage IV) and their spreading (right, higher deformation) across the whole grain. PTW's are characterized by a high dislocation density ρ and a low internal stress $|\Delta(\Delta\sigma_i)|$ (from [22]).

maximum, the long range internal stress $|\Delta(\Delta\sigma_i)|$ is minimum at the same position inside the grain. Fig. 5 on the right demonstrates that these fluctuations successively spread over the whole grain when the deformation is continued, and only some positions with low ρ but high $|\Delta(\Delta\sigma_i)|$ are left. It is obvious that the fluctuation described first represents a PTW which allows for a locally higher dislocation density because of its particular low energy arrangement. It is also obvious that the fluctuations will change their sign the more PTW develop: then they represent the few PDW left in the microstructure exhibiting typically low values ρ and high ones of $|\Delta(\Delta\sigma_i)|$. Such a behaviour has not only be found in Cu as shown here but also in Ni [19], Al (this issue, [20]) and Fe [23], except in stage V of high stacking fault metals [19, 20] when static recovery processes will take place in between single passes of iterative deformation and prevent the formation of PTW's.

Another helpful method in analyzing large strain microstructures is available now through the Electron Back Scatter Diffraction technique which allows for an efficient and accurate determination of lattice (mis)orientation by a spatial resolution down to 0.2 μm (at least when either a well convergent electron beam, and/or a tilt facility of electron beam has been installed). Examples of the good efficiency of the method are given by the works of Schafler et al. on Al [20], and of M. Richert et al. [26] discussed below showing that also UFG structures can be studied under certain conditions.

3.   Modelling of Large Strain Hardening

Most of the models available which aim to describe the work hardening up to highest plastic deformation [28-33] make use of the well proved suggestion of Mughrabi to treat the heterogeneous ("cell") dislocation structure as a composite of soft and hard regions [27]. The most important models being designed by this principle are those (in chronological order) by *Zehetbauer & Les (ZL)* [28, 29], *Argon & Haasen (AH)* [30], (both being successfully applied even to high temperature stages IV & V [31, 30]), and recently, by *Nes & Marthinsen (NM)* [32] and *Estrin, Toth and co-workers (ET)* [33]. A detailed presentation and discussion have been carried out within a previous paper of the author [34] so that this paper will be restricted to a short introduction of the author's model and some comparison with the other models.

The theoretical concept of the ZL-model [28] has in common with the other models that, as a consequence of the composite ansatz, dislocations from soft and hard regions must be considered separately, and that there is no interaction in between them (with the ET-model as an exception [33]). The ZL-model, however, is the only one which distinguishes screw and edge dislocations and relates each of them to the soft and hard lattice areas. The quality of the model has been proven by very good fits to large strain strengthening characteristics of several fcc and bcc metals (for Cu see Fig. 6) and by the realistic output values for the physical parameters (see [34]). This is especially true for the concentration c and migration enthalpy δH of vacancies: c correctly follows the experimentally verified increase c(γ) of vacancies generated by deformation γ, and δH = 0.2 – 0.3 eV is typical of their migration enthalpy also provided by experiment [35]. It is thus *essential* for modelling to include the presence and kinetics of *deformation induced vacancies* which has not been done by any of the models quoted above.

86

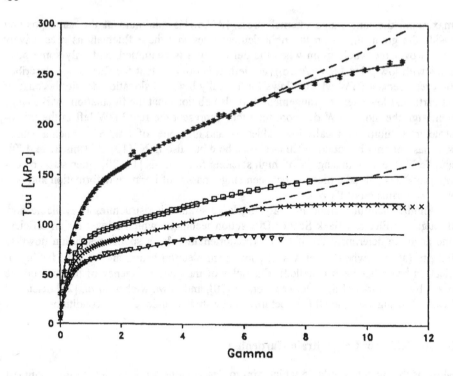

Figure 6. Res. shear stress τ, vs. res. shear strain γ, from torsion in Cu 99.95% for various deformation temperatures, performed at a res. shear strain rate dγ/dt = 10^{-2} s^{-1}. Signs: experimental data, full lines: calculation
by ZL-model (core diffusion, deform.ind.vac.), dashed lines: calculation by ZL-model (bulk diff., thermal
vac.), see also text. Temperatures: # 77K, □ 29 3 K, × 373 K, ∇ 473 K (from [1, 28]).

By example of large strain cold working of Cu, Fig. 6 shows the failure of the model at
even T = 77K if one uses the *bulk* migration enthalpy H^m = 1.2 eV for δH, and the respective *thermal* vacancy concentration for c. In this connection, for example, the question arises how the MN-model being designed very similarly to the ZL-model manages
its two climb mechanisms to operate at cold work conditions without taking into account deformation induced vacancies.

It has to be admitted that the ZL-model has not reflected so far details of the fragmentation process in microstructure as observed by TEM, XPA and EBSP and reported
above. Some merit has to be attested to the AH - model here: for the first time it used
the developing misorientation to describe the work hardening in stage IV, although the
continuous increase of internal stresses connected with that of misorientation could not
be confirmed by experiment [10, 11, 22, 34]. Recently, Zehetbauer & Les [34] gave an
idea how the microstructural fragmentation could combine with their model: before the
dislocations are incorporated as edges into a PTW wall, they have to pile up in front of
it, by a stress τ$_p$= α$_{pile}$ × n × μb/D ([36], α$_{pile}$ is the interaction parameter of piling-up
dislocations, n is their number, and D denotes the size of misoriented lattice area, *i.e.* of
a cell block). Such a relation may satisfy a well known empirical law which has been

frequently observed by combined strength/TEM investigations [37]. By taking into account that wall strength τ_w only contributes by $f_w\tau_w$ to the total strength, one arrives at the empirical description of stage IV strengthening mentioned before (for Cu: $\alpha_{pile}=0.5$, n=20, $f_w=0.3$). Since this stress must be equal to the local edge area stress in the ZL-model, one can derive $\alpha_2 = \alpha_{pile} \times n \times l_2/D$ (l_2 is the average dislocation distance in edge area). After inserting experimental values $l_2/D \approx 1/100$ and the other constants given above, $\alpha_2 = 0.1$ results which fairly coincides with the value obtained from fitting the strengthening data by the ZL-model (Fig. 2).

4. Relevancy of Stage IV and V for Evolution of Ultrafine Grained and/or Nanocrystalline Structure

In recent years, the group of R. Valiev has shown by numerous works that severe plastic deformation can even produce nanocrystalline structures which not only exhibit a strength being a factor 1.5 – 2 higher than usual polycrystalline material but also several other particular properties (see, *e.g.* [38, 39]). The precondition for that seems to be a high number of active slip systems accounting for subgrains sizes of a few 10 nm not only in one dimension as it is usually the case for a simple unidirectional deformation modes. Such conditions can be achieved either by a combination of two "conventional" deformation modes, *e.g.* torsion & compression, or by special deformation techniques like the Equal Channel Angular Extrusion (ECAE, [38]) or the Cyclic Extrusion Compression method (CEC, [40]) (Fig. 7). Already in 1987 J.Gil Sevillano and E.Aernoudt [41] evaluated combined torsion/compression tests by Sturges *et al.* [42] and reported that both stages IV and V will be extended for combined deformation as compared with unidirectional one. Recently, it has been shown with the example of Al-Mg5 that equiaxed ultrafine grained structures (down to 100 nm size) can be also achieved by the CEC method [26, 43]. M. & J. Richert [43] also performed intermittent compression

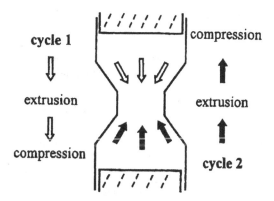

Figure 7. Sketch of the Cyclic-Extrusion-Compression method (CEC) as designed by J. & M. Richert [40].

tests during CEC deformation of Al-Mg5, results of which are shown as strengthening coefficient Θ versus resolved shear stress τ in Fig. 8. One can clearly recognize the typical sequence of deformation stages III, IV and V. In a recent cooperation with M.& J. Richert, A.G.H. Cracow, and H.P. Stüwe & R. Pippan, E.S.I. Leoben, Austria [43], both EBSP as well as XPA measurements have been performed on the same samples of Al-Mg5 which were subjected to the mechanical tests. In stage III, misoriented lattice areas were found which exhibited misorientation angles in the range 2-5°. Entering stage IV and higher deformations, however, about 1/3 of all misoriented areas with size 1 μm or smaller showed misorientation angles between 10° and 60° indicating the increasing development of PTW's [26]. From the XPA measurements the dislocation density ρ could be evaluated which is used to plot τ vs. $\sqrt{\rho}$ in order to determine the magnitude and/or changes of interaction parameters α in the case of CEC deformed Al-Mg5 (Fig. 9). In both stages IV & V, α is ≈ 0.1 which strongly suggests that the microstructural processes which govern stage IV (fragmentation) and stage V (climb of PTW-dislocations) in unidirectional deformation also apply to multidirectional deformation and, thus, rule the formation of ultrafine grained and/or nanocrystalline structure.

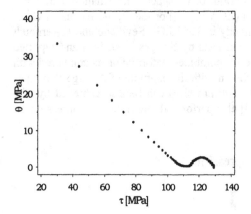

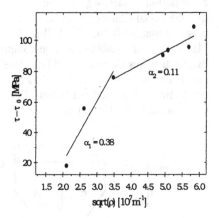

Figure 8. Strengthening plot of Al-Mg5 from CEC deformation ($\Theta \equiv d\tau/d\gamma$; τ denotes the res.shear stress, and γ the res. shear strain [43]).

Figure 9. Effective res.shear stress (τ - τ_o) vs. the res. square root of true dislocation density ρ, in Al-Mg5. From the slope, the dislocation interaction parameter α_i has been derived [43].

5. Conclusions

The main statements of the present paper may be summarized as follows:

(a) The formation of misoriented lattice areas called "fragmentation" is characteristic of plastic deformation at high strains. It consists of the successive transformation of dipolar dislocation walls ("polarized dipole walls – PDW") into dislocation tilt walls ("polarized tilt walls – PTW"). Thus, at first, compounds of cells with the same lattice orientation, *i.e.* "cell blocks" are formed which increasingly approach the size of a single cell as deformation proceeds. A direct investigation of the dynamics of cell block division will require *in situ* – deformation - tests by EBSP and/or microdiffraction-XPA by Synchrotron radiation.

(b) The occurrence of stage IV and V is universal in large strain plastic deformation irrespective of the deformation mode, including multidirectional ones.

(c) Strengthening of stage IV is caused by the successive restriction of mean free path of mobile dislocations due to decreasing cell block size. A misorientation angle up to about 10° allows the dislocations to pass from one cell/subgrain to another; higher misorientation will act as barriers and produce pile-ups of dislocations in the misoriented lattice area. The leading dislocation will experience sufficient local stress to be incorporated into still existing dislocation tilt walls.

(d) Multidirectional deformation is achieved either by new deformation methods like ECAE or CEC, or by combination of different conventional deformation modes. This leads to a marked extension of stages IV and V as concerns both stress and strain; and thus allows for the production of misoriented ultrafine grains down to some tens of nm. Accordingly, the strength of such materials is a factor 1.5 – 2 higher than that of a typical polycrystalline.

(e) In order to characterize microstructures from large strain deformation, the following experimental methods have been proved: (i) EBSP by a spatial resolution $\geq 0.2\ \mu m$, (ii) Synchrotron -XPA (currently $\geq 10\ \mu m$), and (iii) TEM for analyses of ultrafine- & nanocrystalline microstructures.

6. Acknowledgments

The author is indebted to Prof. H.P. Stüwe, ESI Leoben, Austria, for numerous fruitful discussions, and to Dr. Petr Les for performing many calculations necessary for this contribution. The Austrian Science Foundation is gratefully acknowledged for financial support under project 12945-PHY.

7. References

1. Zehetbauer, M., Seumer, V. (1993) Cold work hardening in stages IV and V of fcc metals. I. Experiments and interpretation, *Acta metall. mater.* **41**, 577-588
2. Zehetbauer, M., Les, P., Stüwe, H.P., and Fang, X.F. (1995) Model calculations of large strain strengthening characteristics of commercially pure bcc Iron, *phys.stat.sol. (a)* **151**, 305-311
3. Les, P., Stuewe, H.P., and Zehetbauer, M. (1997) Hardening and strain rate sensitivity in stage IV of deformation in fcc and bcc metals, *Mater.Sci.Eng.* **A234**, 453-455

4. Les, P., Zehetbauer, M., and Stüwe H.P.(1998) Stage IV and V work hardening of pure Ni: Experiments and modelling, *Kovove Materialy (Metallic Materials)* **36-3S**, 12-15
5. Schmidt, J.(1990) Untersuchung der Erholungs- u. Rekristallisationsvorgänge in tieftemperaturverformten Metallen mit Hilfe kalorischer Messungen, Ph.D.Thesis, Technical University Braunschweig, Germany
6. Gil Sevillano, J.(1991) Substructure and strengthening of heavily deformed single and two-phase metallic materials, *J.Phys.III* **1**, 967-988
7. Anongba, P.N.B., Bonneville, J., and Martin, J.L. (1993) Hardening stages of [112] Cu single crystals at intermediate and high temperatures I: Mechanical behaviour, II: Slip systems and microstructures, *Acta metall. mater.* **41**, 2897-2922
8. Les, P., Zehetbauer, M. and Stüwe, H.P. (1996) Characteristics of work hardening in late stages of high temperature deformation, *phys.stat.sol.(a)* **157**, 265-273
9. Kocer, M., Sachslehner, F., Müller, M., Schafler, E. and Zehetbauer, M. (1996) Measurement of dislocation density by residual electrical resistivity, *Mater. Sci. Forum* **210 – 213**, 133-140
10. Müller, M., Zehetbauer, M., Borbely, A., and Ungar, T. (1996) Stage IV work hardening in cell forming materials, pt.I: Features of the dislocation structure determined by X-ray line broadening, *Scripta mater.* **35**, 1461-1466
11. Schafler, E., Zehetbauer, M., Borbely, A. and Ungar, T. (1997) Dislocation densities and internal stresses in large strain cold worked pure Iron, *Mater.Sci.Eng.* A **234-236**, 445-448
12. Zehetbauer, M.(1994) Effects of non-equilibrium vacancies on strengthening, *Key Eng.Mater.* **97-98**, 287-306
13. Bay B., Hansen N., Hughes D.A. and Kuhlmann-Wilsdorf D.(1992) Evolution of fcc deformation structures in polyslip, *Acta metall.mater.* **40**, 205-219;
 Hughes, D.A. (1995) The evolution of deformation microstructures and local orientations, in: Proc.16th Int.Symp.Mater.Sci., ed. Hansen, N. et al., Riso National Laboratory, Roskilde (Denmark), p. 63-85.
14. Ungar, T. and Zehetbauer, M. (1996) Stage IV work hardening in cell forming materials, pt.II: A new mechanism, *Scripta mater.* **35**, 1467-1473
15. Ungar, T. (1994) Characteristically asymmetric X-ray line broadening, an indication of residual long-range internal stresses, *Mater.Sci.Forum* **166-169**, 23-43
16. Ungar, T., Ott, S., Sanders, P.G., Borbely, A. and Weertman, J.R. (1998) Dislocations, grain size and planar faults in nanostructured Cu determined by high resolution X-ray diffraction and a new procedure of peak profile analysis, *Acta mater.* **46**, 3693-3699
17. Ungar, T. (1999) Size distribution of grains or subgrains, dislocation density and dislocation character by using the dislocation model of strain anisotropy in X-ray line profile analysis, in: Proc. NATO Adv.Res.Workshop "Investigations & Applications of Severe Plastic Deformation", August 2-6 (Moscow, Russia), ed. Lowe T.C., and Valiev, R.Z., Kluwer Acad. Publ., The Netherlands, this volume.
18. Wilkens, M. (1987) X-ray line broadening and mean square strains of straight dislocations in elastically anisotropic crystals of cubic symmetry, *phys.stat.sol.(a)* **104**, K1-K6
19. Schafler, E., Zehetbauer, M., Kopacz, Ungar, T., Hanak, P., Amenitsch, H. and Bernstorff, S. (1999) Microstructural parameters in large strain deformed Ni-polycrystals as investigated by synchrotron radiation, *phys.stat.sol. (a)* **175**, No.2
20. Schafler, E., Zehetbauer, M., Hanak, P., Ungar, T., Hebesberger, T., Pippan, R., Mingler, B., Karnthaler, H.P., Amenitsch, H., and Bernstorff, S. (1999) Fragmentation in large strain cold rolled aluminium as observed by synchrotron X-ray Bragg peak profile analysis, EBSP and TEM, in: Proc. NATO Adv.Res.Workshop "Investigations & Applications of Severe Plastic Deformation", August 2-6 (Moscow, Russia), ed. Lowe, T.C. and Valiev, R.Z., Kluwer Acad. Publ., The Netherlands, this volume
21. Schafler, E. (1998) Untersuchung der mikrostrukturellen Entwicklung von hochverformten Metallen mittels Röntgen-Bragg-Profil-Analyse, Ph.D.Thesis, University Vienna, Austria
22. Zehetbauer, M., Ungar, T., Kral, R., Borbely, A., Schafler, E., Ortner, B., Amenitsch, H., and Bernstorff, S. (1999) Scanning X-ray diffraction peak profile analysis in deformed Cu-polycrystals by synchrotron radiation, *Acta mater.* **47**, 1053-1061
23. Zehetbauer, M., Ungar, T., Schafler, E., Bernstorff and H.Amenitsch, Microscale spatial distribution of dislocations and long range internal stresses in cold worked bcc Fe, *Key Eng.Mater.*, to be published
24. Juul Jensen, D. (1993) Automatic EBSP analysis for recrystallization studies, *Textures and Microstructures* **20**, 55-65
25. Hjelen, J., Oresund, R., Hoel, E., Runde, P., Furu, T. and Nes, E. (1993) EBSP, progress in technique and applications, *Textures and microstructures* **20**, 29-40

26. Richert, M., Stüwe, H.P., Richert, J., Pippan, R., and Motz, Ch. (1999) Characteristic features of micro-structure of AlMg5 deformed to large plastic strains, *Mater.Sci.Eng. A*, submitted for publication

27. Mughrabi, H.(1983) Dislocation wall land cell structures and long –range internal stresses in deformed metals crystals, *Acta metall.* **31**, 1367-1379

28. Zehetbauer, M. (1993) Cold work hardening in stages IV and V of fcc metals II: Model fits and physical results, *Acta metall.mater.* **41**, 589-599

29. Les, P. and Zehetbauer, M. (1994) Evolution of microstructural parameters in large strain deformation: Description by Zehetbauer's model, *Key Eng.Mater.* **97-98**, 335-340

30. Argon, A.S. and Haasen, P.(1993) A new mechanism of work hardening in the late stages of large strain plastic flow in fcc and diamond cubic crystals, *Acta metall. mater.* **41**, 3289-3306

31. Zehetbauer, M.and Les, P.(1996) Features & modelling of strengthening in late stages of high temperature deformation, in: Proc.35th Int.Conf.Met.Soc.CIM, ed.McQueen, H. *et al.*, Montreal (Canada), p.205-216

32. Marthinsen, K., and Nes, E.(1997) A general model for metal plasticity, *Mater. Sci. Eng.* **A 234-236**, 1095-1098 ;
Nes, E (1997) Modelling of work hardening and stress saturation in fcc metals,
Progr.Mater.Sci. **41**, 129-194

33. Estrin Y., Toth L.S.,Molinari A. and Brechet Y.(1998) A dislocation based model for all stages in large strain deformation, *Acta mater.* **46**, 5509-5522

34. Zehetbauer, M. and Les, P. (1998) Micromechanisms of plastic deformation in metals, *Kovove Materialy (Metallic Materials)* **36**, 153

35. Sassa, K., Petry, W. and Vogl, G. (1983) The nature of point defect in plastically deformed Aluminium *Phil.Mag.* **A 48**, 41-61

36. Hirth, J.P., and Lothe, J. (1992) Theory of Dislocations, Krieger Publ. Comp., Malaber (USA)

37. Argon, A.S. (1996) Mechanical properties of single-phase crystalline media: Deformation at low temperatures, ch.21 in: Physical Metallurgy, ed. Cahn, R.W. and Haasen, P., North Holland, Amsterdam The Netherlands, p. 1877-1955

38. Valiev R.Z. (1996) UFG Materials Produced by Severe Plastic Deformation,
A Thematical Issue, *Ann.Chim.Fr.* **21**, 369-556

39. Valiev R.Z. (1999) SPD processing and enhanced properties in metallic materials, in: Proc. NATO Adv.Res.Workshop "Investigations & Applications of Severe Plastic Deformation", August 2-6 (Moscow, Russia), ed. Lowe, T.C., and Valiev, R.Z., Kluwer Acad. Publ., The Netherlands, this issue; *Progr.Mater.Sci.*, to be published

40. Richert, J. and Richert, M. (1986) A new method fro unlimited deformation of metals and alloys, *Aluminium* **62**, 604-607

41. Gil Sevillano, J. and Aernoudt, E. (1987) Low energy dislocation structures in highly deformed materials, *Mater.Sci.Eng.* **86**, 35-51

42. Sturges, J.L., Parsons, B. and Cole, B.N. (1980) Dynamic stress-strain measurements under superimposed high hydrostatic pressure, in: Mechanical Properties at High Rates of Strain, ed. Harding, J., Conf.series **47**, The Institute of Physics, London (U.K.), p.35-48

43. Richert, M., Stüwe, H.P., Zehetbauer, M., Richert, J., Pippan, R., Motz, Ch., Schafler, E. (1999) *Acta mater.*, to be published

SIZE DISTRIBUTION OF GRAINS OR SUBGRAINS, DISLOCATION DENSITY AND DISLOCATION CHARACTER BY USING THE DISLOCATION MODEL OF STRAIN ANISOTROPY IN X-RAY LINE PROFILE ANALYSIS

T. UNGÁR

Department of General Physics, Eötvös University Budapest, H-1088 Múzeum krt. 6-8, Budapest, Hungary

Abstract

Anisotropic strain broadening in X-ray line profile analysis means that the breadth or the Fourier coefficients of diffraction profiles are *not* a monotonous function of the diffraction angle. The lack of a physically sound model makes the interpretation of line broadening difficult or even impossible. Dislocations are anisotropic lattice imperfections with anisotropic contrast effects in diffraction. It has been suggested recently that anisotropic X-ray line broadening is caused by dislocations. The classical procedures of Williamson and Hall, and Warren and Averbach are suggested to be replaced by the *modified* Williamson-Hall plot and the *modified* Warren-Averbach method in which the modulus of the diffraction vector or its square, g or g^2, are replaced by $g\bar{C}^{1/2}$ or $g^2\bar{C}$, respectively, where $\bar{C}$ are the average dislocation contrast factors. A straightforward procedure has been elaborated to separate size and strain broadening which enables to determine particle size and size distribution and the structure of dislocations in terms of dislocation densities and the character of dislocations in polycrystalline or submicron grain size materials.

1. Introduction

Strain anisotropy in X-ray line-profile analysis or in Rietveld structure refinement is a well known phenomenon [1, 2]. This means that neither the FWHM nor the integral breadth in the Williamson-Hall plots [3] nor the Fourier coefficients in the Warren-Averbach method [4] are monotonous functions of the modulus of the diffraction vector or its square, g or g^2, respectively. In the field of structure refinement different solutions have been suggested to handle this problem [1, 5-9]. The Caglioti formula [5] relates the FWHM to the diffraction angle and the Miller indices

$$\text{FWHM}^2 = U(hkl)\,\text{tg}^2\theta + V(hkl)\,\text{tg}\theta + W(hkl) , \qquad (1)$$

where $U(hkl)$, $V(hkl)$ and $W(hkl)$ are three experimental fitting parameters depending individually on the Miller indices. The Rietveld code well known as FULPROF

93

T.C. Lowe and R.Z. Valiev (eds.), Investigations and Applications of Severe Plastic Deformation, 93–102.
© 2000 *Kluwer Academic Publishers. Printed in the Netherlands.*

determines the physical part of anisotropic strain broadening as an *hkl* dependent part of the parameter *U(hkl)*. It is the task of the experimetator to attribute physical meaning to these experimentally obtained parameters [6]. This means that a physically sound model of strain anisotropy, which relates microstructural quantities to the parameters *U*, *V* and *W* is still subject of research [7-11]. Several attempts have been done to relate strain anisotropy to the elastic anisotropy of crystals [7,11]. In these, however, it is assumed that lattice distortions are caused by random displacements of atoms. Recently is has been shown that the dislocation model of the mean square strain accounts automatically for strain anisotropy [12]. It has been suggested that the classical procedures of Williamson and Hall [3], and Warren and Averbach [4] are replaced by the *modified* Williamson-Hall plot and the *modified* Warren-Averbach methods in which the modulus of the diffraction vector or its square, g or g^2, are replaced by $g\bar{C}^{1/2}$ or $g^2\bar{C}$, respectively, where $\bar{C}$ are the average dislocation contrast factors [12]. The average dislocation contrast factors have been shown to be simple functions of the *hkl* indices if the specimen are more or less texture free polycrystals or all possible slip systems in the crystal are more or less equally populated by dislocations [13]. A comprehensive scheme has been worked out for the average dislocation contrast factors enabling the evaluation of strain anisotropy i) in terms of dislocation densities and ii) the evaluation of dislocation character in terms of edge and screw components in the cubic crystal system [14]. Since this whole procedure enables the separation of strain effect from line broadening, the size of crystals or coherent domains can be determined in a reliable and straightforward manner even in the presence of strain [12-14]. Three stable size parameters, *D*, *d* and L_0 can be determined by the *modified* Williamson-Hall plots of the FWHM and the integral breadths and by the *modified* Warren-Averbach procedure, respectively. The width and the mean size, σ and *m*, of a log-normal size distribution, *f(x)*, can be obtained from these size parameters in a simple least squares procedure [15] (see also the contribution by Ungár, Alexandrov and Hanák in the present proceedings). In the present work a brief summary of the procedure will be presented with an example of its application to severely deformed submicron grain size copper.

2. The average contrast factors of dislocations in an untextured ideally polycrystalline cubic material

The Fourier coefficients of the line profile of a Bragg reflection are the product of the "size" and "distortion" coefficients, A_L^S and A_L^D [4]:

$$A_L = A_L^S A_L^D, \tag{2}$$

where *L* is the Fourier length defined as $L=na_3$ and n are integers. a_3 is the unit of the Fourier length in the direction of the diffraction vector **g**

$$a_3 = \lambda/[2(\sin\theta_2 - \sin\theta_1)] \quad , \tag{3}$$

where the line profile is measured from θ_1 to θ_2, λ is the wavelength of the X-rays. The basic equation of the Warren-Averbach [4] analysis is obtained by taking the logarithm of equation (2):

$$ln\, A_L(g) \cong ln\, A_L{}^S - 2\pi^2 L^2 g^2 <\varepsilon_g{}^2> . \qquad (4)$$

In the case of dislocations the mean square strain has the following form [16-18]:

$$<\varepsilon_g{}^2> \cong \frac{\rho C b^2}{4\pi} ln(R_e/L), \qquad (5)$$

where ρ is the density, R_e the effective outer cut-off radius, b the Burgers vector and C the contrast factor of dislocations, respectively. Equation (5) shows that for dislocations $<\varepsilon_g{}^2>$ is L dependent. This L dependence is usually observed in the experiments of line profile analysis [4]. The contrast factor depends on the relative positions of the diffraction vector and the line and Burgers vectors of the dislocations, in a similar way as in transmission electron microscopy [19]. For example, a dislocation for which $bg=0$ has no considerable contribution to line broadening, or in other words, it is invisible. The values of C can be obtained by numerical methods taking into account the elastic constants of the material [14, 17, 20, 21]. Recently it has been shown that the average dislocation contrast factors can be given as [13]:

$$\overline{C} = A - B\, (h^2 k^2 + h^2 l^2 + k^2 l^2)/(h^2 + k^2 + l^2)^2 , \qquad (6)$$

where A and B are constants depending on the elastic constants of the crystal. The value of A is the average contrast factor corresponding to the $h00$ reflection: $\overline{C}_{h00} = A$. Denoting the fourth order ratio in the above equation by H^2 eq. (6) can be written as:

$$\overline{C} = \overline{C}_{h00}\,(1 - qH^2), \qquad (7)$$

where $q = B/A$. This equation shows that in the case of untextured polycrystals or randomly populated Burgers vectors the average contrast factors can be obtained if $\overline{C}_{h00}$ and q are known.

3. The average dislocation contrast factors in cubic crystals [14]

The dislocation contrast factors corresponding to the most common type of edge and screw dislocations in fcc and bcc crystals were evaluated as a function of the elastic constants of crystals in a wide range of the elastic anisotropy, A_i, (where $A_i = 2c_{44}/(c_{11} - c_{12})$ and c_{ij} are the elastic constants) ranging from 0.5 to 8 [14]. Typical values of $\overline{C}_{h00}$ and q *versus* A_i in a few representative cases are shown in Figs. 1 and 2. The solid lines through the calculated datapoints are empirical functions of the following form:

$$\overline{C}_{h00} \text{ or } q = a[1 - exp(-A_i/b)] + cA_i + d, \tag{8}$$

where the parameters a, b, c and d were fitted to the calculated datapoints and are given in tables in [14].

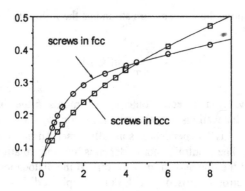

Figure 1. The average dislocation contrast factors of the *h00* type reflections as a function of the elastic anisotropy for the basic screw dislocations in fcc and bcc crystals [14]. Note that the elastic anisotropy in copper is 3.21. Further values of $\overline{C}_{h00}$ for edge dislocations in cubic crystals can be found in [14].

4. The *modified* Williamson-Hall and Warren-Averbach procedures [12]

The Williamson-Hall plot [13] can be solved for dislocated crystals and has been called the *modified* Williamson-Hall plot in [12]:

$$\Delta K \cong 0.9/D + (\pi M^2 b^2/2)^{1/2} \rho^{1/2} K \overline{C}^{1/2} + O(K^2 \overline{C}), \tag{9}$$

where $K=2\sin\theta/\lambda$, $\Delta K=2\cos\theta(\Delta\theta)/\lambda$, θ and λ are the diffraction angle and the wavelength of X-rays and $g=K$ at the exact Bragg position. D, ρ and b are the average particle size, the average dislocation density and the modulus of the

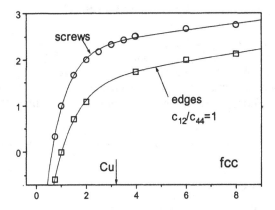

Figure 2. Calculated values of the q parameters for screw and edge dislocations versus the elastic anisotropy in fcc crystals [14]. (In the case of edge dislocations q depends also on c_{12}/c_{44} [14].) Further q values for bcc crystals can be found in [14].

Burgers vector of dislocations. M is a constant depending on the effective outer cut-off radius of dislocations. Since M can only be obtained from the tails of the profiles, taking into account the shape of their decay [17], it is more appropriate to introduce the notation:

$$\Delta K \cong \alpha + \beta K \, \overline{C} + \gamma K^2 \overline{C} \,, \tag{10}$$

where α and β can be related to the parameters in equation (9) and γ is an additional constant not interpreted physically. Inserting $\overline{C}$ from equation (7) into (10) the following is obtained:

$$\Delta K \cong \alpha + \beta K [\, \overline{C}_{h00}(1-qH^2)]^{1/2} + \gamma K^2 \overline{C}_{h00}(1-qH^2) \,. \tag{11}$$

Equation (11) can be solved numerically for α, β, γ and q using the experimental values of ΔK. The value of $\overline{C}_{h00}$ can be obtained from numerically calculated plots like Fig. 1 or tables given in [14]. On the other hand, since the values of $\overline{C}_{h00}$ are only a multiplicative factor in equation (11), they have no effect on the value of α which is the size parameter in the *modified* Williamson-Hall plot. The value of q obtained by the numerical solution of equation (11) is denoted as the experimental value: q_{exp}. It can also be obtained by a simple graphical method by using the quadratic form of the *modified* Williamson-Hall plot [13] and taking equation (6) for $\overline{C}$:

$$(\Delta K)^2 \cong \alpha + \beta A K^2 + \beta B K^2 H^2 \,, \tag{12}$$

This equation can be written as:

98

$$[(\Delta K)^2 - \alpha]/K^2 \cong \beta A + \beta B H^2 . \tag{13}$$

Plotting the left hand side of equation (13) versus H^2 and fitting the value of α to obtain a linear relation, the ratio $B/A=-q$ can be obtained experimentally. In a paper by Ungár, Alexandrov and Hanák in the present proceedings the size distribution and dislocation structure of copper deformed by equal channel angular pressing (ECA) has been studied [22]. The FWHM, taken by courtesy from [22] for specimen No. 3

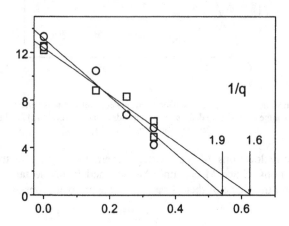

Figure 3. The left hand side of equation (13) plotted versus H^2 for specimens No. 3 (open circles) and 6 (open squares) deformed by 2 ECA passes and 12 ECA passes plus 20% compression, respectively. The best values of α in order to obtain a linear relation were about 2×10^{-5} and 6×10^{-5} for specimens No. 3 and 6, respectively. For details of specimen see the paper of Ungár, Alexandrov, Hanák in this proceedings [22].

and 6 deformed by 2 ECA passes and 12 ECA passes plus 20% compression, respectively, are plotted according to equation (13) in Fig. 3. The best values of α in order to obtain a linear relation were about 2×10^{-5} and 6×10^{-5} for the two specimens, respectively. The experimental q values are indicated in the figure at the zero intersections of the abscissa by the fitted linear regressions. The values of q for screw or edge dislocations in copper are 2.37 or 1.68, respectively, as it can be seen in Fig. 2 and as it can be obtained more precisely from data presented in [14]. The random mixed character of dislocations would mean that $q=2.02$. An experimental value, q_{exp}, lower or higher than this value indicates more edge or more screw character, respectively. As it can be seen from Fig. 3 specimen No. 3 contains dislocations of random mixed character, whereas in specimen No. 6 the dislocations have more edge character. (For more details about the q_{exp} values of the submicron copper specimen deformed by ECA see [22].) Specimen No. 6 has been deformed by 12 ECA passes plus 20% compression which is considerably larger than the deformation in specimen No. 3. Our results indicate that after extremely large deformations the average dislocation character shifts toward more edge in accordance with the easier annihilation of screw dislocations. The *modified* Warren-Averbach method is discussed in detail by Ungár, Alexandrov and Hanák in the present proceedings [22].

5. Size distribution

According to equation (9) in the present work and equation (5) in [22] three different particle sizes can be obtained: i) D from the FWHM, ii) d from the integral breadths and iii) L_0 from the Fourier coefficients. These three gauges sample the largest, the medium and the smallest lengths in the specimen since they are related to the central part, the integral and the outermost tails of the profiles, respectively, as it has been shown earlier [15, 23].

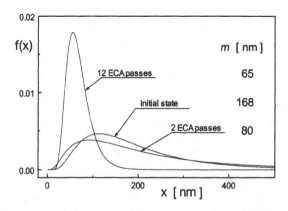

Figure 4. The size distribution functions of i) subgrains in specimen No. 1 and 3, the extruded initial state and deformed by 2 ECA passes, respectively, and of grains in specimen No. 5 deformed by 12 ECA passes. For details of specimen see the paper of Ungár, Alexandrov, Hanák in this proceedings [22].

At the same time, these three gauges are rather stable and only restrictedly sensitive to experimental errors or fluctuations [15]. A simple numerical procedure has been worked out recently to relate the experimentally determined D, d and L_0 values to the parameters of a size distribution function, $f(x)$, especially for a log-normal size distribution [15]:

$$f(x) = \frac{1}{\sqrt{2\pi}\sigma} \frac{1}{x} \exp\left\{ -\frac{[\ln(x/m)]^2}{2\sigma^2} \right\}, \tag{14}$$

where x is the average size of a particle in the direction of the diffraction vector, σ the width of size distribution and m the momentum of the probability density function. (In the literature of line profile analysis the average particle size in the direction of the diffraction vector is called average "column length" [2].) Denote the size function of a particle (in the direction of the diffraction vector) by $V(x)$: $V(x)=1$ inside and $V(x)=0$

outside the particle. If the particle is distortion free the scattered intensity is the square of the Fourier transform of $V(x)$ [13]:

$$I(K, x_0) = [(\sin Kx_0)/Kx_0]^2 , \qquad (15)$$

where x_0 is the average size of the particle (or the average column length in the g direction in the particle). Assuming that the particles in the specimen have the size distribution $f(x_0)$ the scattered intensity is:

$$I^s(K) = \int_0^\infty f(x_0) I(K, x_0) dx_0 , \qquad (16)$$

where the superscript s refers to "size". Denote the FWHM, the integral breadths and the particle size determined by the Warren-Averbach method of $I^s(K)$ by $A_{\sigma,m}$, $B_{\sigma,m}$ and $C_{\sigma,m}$, respectively. The following least squares equation provides the values of σ and m corresponding to the measured values of D, d and L_0:

$$(A_{\sigma,m} - D)^2 + (B_{\sigma,m} - d)^2 + (C_{\sigma,m} - L_0)^2 = minimum \; for \; \sigma \; and \; m . \qquad (17)$$

The σ and m values determined by the above procedure provide the size distribution function $f(x)$ corresponding to the measured values of D, d and L_0. The $f(x)$ functions for three typical severely deformed specimen are shown in Fig. 4. A more detailed account of the dislocation structure and the size-distribution of subgrains or grains in severely deformed copper specimens is given in the paper of Ungár, Alexandrov, Hanák in this proceedings [22]. In the present work only three typical size distribution functions are presented for the specimen: 1) an extruded copper sample which is considered as the "initial state" for further deformation by ECA, 2) a sample deformed by 2 ECA passes (starting from the "initial state"), 3) a sample deformed by 12 ECA passes (starting from the "initial state"). After 2 ECA passes the subgrain size distribution is shifted to smaller values. TEM investigations have indicated that in the "initial state" and after 2 ECA passes the smallest coherent units of the material are subgrains. After 12 ECA passes the size distribution shifts to considerably smaller values and becomes narrow. TEM investigations have indicated that in this state the smallest coherent units of the material are fragmented grains of submicron size. The size distributions determined by the present X-ray procedure were found to be in good correlation with TEM investigations. Details of the TEM studies will be published elsewhere.

6. Conclusions

1.   It has been shown that strain anisotropy can be well interpreted by the dislocation model of the root mean square of lattice distortions [12].
2.   The root mean square is proportional to the product of the dislocation density and the contrast factors of dislocations [16, 17].

3.   In more or less texture free polycrystals or if all possible slip systems are populated randomly by dislocations the average dislocation contrast factors are a linear function of the fourth order invariant of the *hkl* indices [13].

4.   Based on the results of conclusion 3. the average dislocation contrast factors have been evaluated numerically as a function of the elastic constants in cubic crystals for the case of edge and screw dislocations. The results have been summarized in terms of simple analytic functions [14].

5.   The numerically evaluated average dislocation contrast factors enable the experimental determination of dislocation types in terms of screw or edge character [14].

6.   The *modified* Williamson-Hall plot and Warren-Averbach procedures have been introduced which enable to determine correctly the strain and size contribution to X-ray line broadening [12].

7.   The density and the arrangement parameter of dislocations, ρ and M can be obtained from the Fourier coefficients using the *modified* Warren-Averbach procedure on all available diffraction profiles [12,23].

8.   Three stable size parameters are suggested to be determined from the FWHM, the integral breadths and the Fourier coefficients by the *modified* Williamson-Hall and Warren-Averbach procedures. A straightforward method provides the size-distribution function of subcells, subgrains or grains directly related to the three stable size parameters [15].

9.   The size distributions of the coherently scattering domains are in good correlation with the microstructures observed in TEM (to be published elsewhere).

7.   Acknowledgements

The author is indebted to Professors I. Alexandrov and R. Valiev for providing the ECA pressed specimen for the X-ray investigations. Thanks are due to the Hungarian National Science Foundation, OTKA T022968, T022976 and the Hungarian Government Fund FKFP 0116/1997 for financial support of this research.

8.   References

1.   Suortti, P., (1993) Bragg Reflection Profile Shape in X-ray powder diffraction patterns, in *The Rietveld Method*, edited by R. A. Young, IUCr Monographs on Crystallography, Vol. 5, Oxford University Press, p. 167-185.

2.   Delhez, R., deKeijser, Th. H., Langford, J. I., Louër, D., Mittemeijer, E. J. & Sonneveld, J., (1993) Crystal Imperfection Broadening and Peak Shape in the Rietveld Method in *The Rietveld Method*, edited by R. A. Young, IUCr Monographs on Crystallography, Vol. 5, Oxford University Press, p. 132-166.

3.   Williamson, G. K. and Hall, W. H., (1953) X-Ray Line Broadening from filed aluminium and wolfram, *Acta Metall.*, **1**, 22-31.

4.   Warren, B. E. (1959) X-ray studies of deformed metals, *Progr. Metal Phys.* **8**, 147-202.

5.   Caglioti, G., Paoletti, A. & Ricci, F. P., (1958) Choice of collimators for a crystal spectrometer for neutron diffraction, *Nucl. Instrum.* **3**, 223-228.

6.   Latroche, M., Rodrigues-Carvajal, J., Pecheron-Guegan, A. and Bouree-Vigneron, F., (1995) Structural studies of $LaNi_4CoD_{6.11}$ and $LaNi_{3.55}Mn_{0.4}Al_{0.3}Co_{0.75}D_{5.57}$ by means of neutron powder diffraction, *J. Alloys and Compounds*, **218**, 64-72.

7.   Stephens, P. W., (1998) Phenomenological model of anisotropic peak broadening in powder diffraction, *J. Appl. Cryst.*, **32**, 281-289.

102

8. Le Bail, A. & Jouanneaux, A., (1997) A Qualitative account for anisotropic broadening in whole-powder-diffraction-pattern fitting by second-rank tensors, *J. Appl. Cryst.* 30, 265-271.
9. P. Scardi, (1999) A new whole-powder pattern-fitting approach, in *Analysis of Real Structure of Matter*, eds. H.-J. Bunge, J. Fiala and R. L. Snyder, IUCr series, Oxford Univ. Press, in press.
10. Fischer, J. E., Bendele, G., Dinnebier, R., Stephens, P. W., Lin, C. L., Bykovetz, N. and Zhu, Q., (1995) Structural analysis of fullerene and fulleride solids from synchrotron x-ray powder diffraction, *J. Phys. Chem. Solids,* **56**, 1445-1457.
11. Reimann, K. and Wuerschum, R. (1997) Distribution of internal strains in nanocrystalline Pd studied by x-ray diffraction, *J. Appl. Phys.,* **81**, 7186-7192.
12. Ungár, T. and Borbély, A. (1996) The effect of dislocation contrast on X-ray line broadening: a new approach to line profile analysis, *Appl. Phys. Lett.,* **69**, 3173-3175.
13. Ungár, T. and Tichy, Gy., (1999) The effect of dislocation contrasts on X-ray line profiles in untextured polycrystals, *Phys. Stat. Sol.,* **171**, 425-434.
14. Ungár, T., Dragomir, I., Révész, Á. and Borbély, A., (1999) The contrast factors of dislocations in cubic crystals: the dislocation model of strain anisotropy in practice *J. Appl. Cryst.,* in press.
15. Ungár, T., Borbély, A., Goren-Muginstein, G. R., Berger, S. and Rosen, A. R. (1999) Particle-size, size distribution and dislocations in nanocrystalline tungsten-carbide, *Nanostructured Mater.,* **11**, 103-113.
16. M. A. Krivoglaz, (1969) in *Theory of X-ray and Thermal Neutron Scattering by real Crystals*, Plenum Press, N. Y.; and in *X-ray and Neutron Diffraction in Nonideal Crystals*, Springer-Verlag, Berlin, Heidelberg, New York, 1996.
17. Wilkens, M., (1970) The determination of density and distribution of dislocations in deformed single crystals from broadened x-ray diffraction profiles, *Phys. Stat. Sol.* (a) **2**, 359-370.
18. Groma, I., Ungár, T. and Wilkens, M. (1988) Asymmetric X-Ray Line Broadening of Plastically Deformed Crystals. Part I: Theory, *J. Appl. Cryst.,* **21**, 47-53.
19. Hirsh, P. B., Howie, A., Nicholson, R. B., Pashley, D. W. and Whealen, M. J., *Electron Microscopy of Thin Crystals*, Butterworths, London.
20. Klimanek, P. and Kuzel Jr., R. (1988-89) X-ray diffraction line broadening due to dislocations in non-cubic materials. I-III., *J. Appl. Cryst.,* **21**, 59-66, *ibid.* **21**, 363-368, **22**, 299-307.
21. Wilkens, M., (1987) X-ray line broadening and mean square strains of straight dislocations in elastically anisotropic crystals of cubic symmetry, *Phys. Stat. Sol.* (a) **104**, K1-K6.
22. Ungár, T., Alexandrov, I. and Hanák, P. (1999) Grain and Subgrain size-distribution and dislocation densities in severely deformed copper determined by a new procedure of X-ray line profile analysis, *in the present Proceedings.*
23. Ungár, T., Ott, S., Sanders, P. G., Borbély, A. and Weertman, J. R., (1998) Dislocations, grain size and planar faults in nanostructured copper determined by high resolution x-ray diffraction and a new procedure of peak profile analysis, *Acta Mater.,* **10**, 3693-3699.

X-RAY STUDIES AND COMPUTER SIMULATION OF NANOSTRUCTURED SPD METALS

I.V. ALEXANDROV
Institute of Physics of Advanced Materials
Ufa State Aviation Technical University, K. Marksa 12, 450000 Ufa
Russia

Abstract

The results of investigations of structure peculiarities in metals subjected to severe plastic deformation (SPD) are presented. Special attention is paid to X-ray investigations of microstructure evolution in bulk nanostructured samples of pure Cu obtained by SPD processing by torsion under high applied pressure and equal channel angular (ECA) pressing. Computer simulation is used to analyze the X-ray results.

1. Introduction

SPD methods, *i.e.* large deformation under high imposed pressure, are among the most innovative techniques for creating nanostructured states in materials. These states allow the development of new physical and mechanical properties in different materials [1]. SPD methods have advantages in comparison with other methods of processing nanostructures. An important one is the ability to create bulk samples without porosity, thereby allowing thorough structural characterization and investigation of different physical and mechanical properties. SPD methods are also promising for producing commercially viable products. To understand observed properties and the design of optimal SPD processing methods it is necessary to use detailed structural characterization to define the underlying microstructural mechanisms and develop models of microstructure evolution during SPD.

There exists voluminous data concerning the evolution of microstructure during large deformations obtained by rolling, extrusion, and torsion to true strain $\varepsilon = 2$–5, [2–8]. Such large strains typically cause fragmentation of structure, enhancement of misorientation angles between fragments, and decrease of the dislocation density within fragments with increasing strain rate [2, 3]. Corresponding theories and models have been developed to explain these results[4–8].

At the same time, very intense deformation has been less studied. In particular, the role of applied high pressure in altering the evolution of microstructure and enabling the attainment of large strain (true strain $\varepsilon \geq 10$) before fracture is most important. We already clearly understand that SPD forms ultrafine grained (UFG) nanostructures [9–18]. This process includes the increase of misorientation angles between grains with increasing strain rate. We also understand that the applied high pressures and low

T.C. Lowe and R.Z. Valiev (eds.), Investigations and Applications of Severe Plastic Deformation, 103–108.
© 2000 *Kluwer Academic Publishers. Printed in the Netherlands.*

deformation temperatures prevent recovery and recrystallization during SPD [10]. Thus it is necessary carefully examine the microstructures to reveal which mechanisms allow the formation of nanostructures. Toward this end, X-ray diffraction (XRD) analysis can be successfully applied along with transmission electron microscopy (TEM).

It is shown that XRD analysis is among the most informative methods of investigation of UFG materials structure [13]. This method provides statistically reliable information about size of coherently scattering domains (CSD) which can correspond to grain size in nanomaterials, elastic microdistortions, lattice parameters, static and dynamic atomic displacements, preferred grains crystal lattice orientation and so on [14–18].

The purpose of the present investigation is to provide X-ray data for the determination and development of concepts for modeling the evolution of UFG nanostructures in pure Cu (99.98%) during SPD.

2. Experimental Procedures

Samples were subjected to SPD by High Pressure Torsion (HPT) and ECA pressing with different strain rates [1]. For HPT processing, disc-shaped bulk samples were placed in-between the heads of a 400 ton press and subject to applied pressure P of several GPa while simultaneously rotating the heads to apply torsion, up to 6 complete 360 degree rotations. The resulting samples were of disc shape with diameters of 10–20 mm and thicknesses of 0.3–1 mm.

During ECA pressing the initial ingots were pressed repeatedly in a special die-set through two equal cross section channels with an intersecting angle of 90°. The samples were 80–100 mm long and 20 mm in diameter. The maximum number of passes was 10.

The X-ray investigations were conducted at temperatures between 85–295 K using RIGAKU D/MAX 2400 and DRON–4–07 diffractometers with Cu K_α radiation.

3. Experimental Results

3.1. TORSION STRAINING

X-ray patterns evolve substantially from the initial pattern for the coarse grained state during HPT up to 2 turns, but remain almost unchanged by additional torsional deformation. Comparing X-ray patterns of coarse grained (CG) and SPD Cu one can see significant redistribution of intensity of different X-ray peaks, increasing the (111) X-ray peak, considerable broadening, shift of centroids, long tails of X-ray peaks, and small enhancement of the diffusion background (Figure 1) [14, 16].

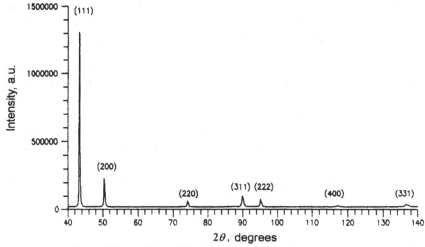

Figure 1. Typical X-ray diffraction pattern of Cu subject to torsion straining ($\gamma \approx 10$).

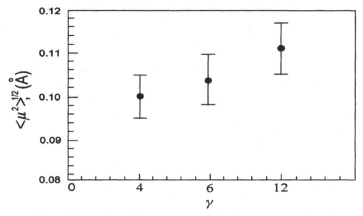

Figure 2. Dependence of the average atomic displacements on a deformation degree γ.

Calculations show that an averaged increase of the Lorentz component in the Voigt function, which describes the shape of X-ray peaks as a linear combination of standard Lorentz (long tails) and Gauss (short tails) functions, increases gradually from 46% in the CG state up to 100% with increasing torsional strain [16].

The broadening of X-ray peaks is due to the decrease of the mean size of CSD down to less than 50 nm and the increase of averaged elastic microdistortions up to 0.12%. In this case the size itself and the character of its evolution depend on grains belonging to different texture components. Elastic microdistortions depend on belonging to one or another texture component as well. The levels of microdistortions in the direction <200> are more than twice those in the direction <111> [14].

The averaged dislocation density increases with increasing number of HPT rotations and reaches $(2.0 \pm 0.1) \times 10^{14}\,\text{m}^{-2}$ after 6 rotations [16].

The lattice parameter and volume of the unit cell of crystal lattice in Cu samples, subjected to HPT, are less than in CG Cu. The deviation value of lattice parameter also increases with increasing torsional rotation and reaches $(0.019 \pm 0.004)\%$ after 6 rotations [14].

The Debye-Waller parameter and the value of atomic displacements of the crystal lattice develops with increasing rotation (Figure 2). After 6 rotations the atomic displacements reach $(4.3 \pm 0.2)\%$ of the shortest distance between atoms compared to $(3.7 \pm 0.2)\%$ in CG Cu [16].

The integral background intensity on the X-ray pattern increases by $(6 \pm 3)\%$ after 6 rotations. This indicates the increase of density of crystal lattice defects in nanostructured Cu processed by SPD [14].

A weak axial texture forms in Cu during HPT. The inverse pole figure of the deformation axis is characterized by enhanced intensity for orientations in-between pole (111) and pole (110), with maxima close to pole (111) [18].

3.2. ECA PRESSING

The process of nanostructure formation during ECA pressing has been less studied by XRD analysis than has been nanostructure evolution during torsional straining. This investigation shows that the relative intensity of X-ray peaks from Cu X-ray patterns for samples prepared from ECA pressings with different number of passes (from 1 to10) significantly differ from each other.

After the first ECA pass the mean size of CSD is less than 60 nm and averaged microdistortions are as high as 0.19%. Transformation of CSD size and of elastic microdistortions occurs with increasing number of ECA passes. Obviously, this is due to the complex character of stress and ingots orientation changes between passes through channels during ECA pressing.

In the incomplete Cu pole figure, three intense nonsymmetric texture maxima are discovered after 10 passes. The inverse pole figure indicates a texture component of (331) plane parallel to the deformation axis. This fact is consistent with the observation of a strong peak (331) on a corresponding X-ray pattern [18].

4. Discussion

The experimental XRD results indicate complex structural transformations during SPD of pure Cu. The general trends for increasing deformation are: decreasing of the size of the CSD, increasing of microdistortions, decreasing of lattice parameters, enhancement of Debye-Waller parameter and of atomic displacements from cites of an equilibrium perfect crystal lattice, increasing of integral intensity of X-ray diffusive scattering background, and development of crystallographic texture. In this case the CSD size and elastic microdistortions in grains belonging to different texture components are different.

The XRD results concerning the reduction of CSD sizes and enhancement of elastic microdistortions are in conformity with TEM results [9]. On the other hand, the XRD results complement measurements obtained by other methods, providing additional quantity and quality of data. In particular, X-ray methods reveal the decrease of lattice parameter, the enhancement of atomic displacements, the development of crystallographic texture and others useful information.

Data from detailed X-ray investigations enables the development of structural models of the SPD nanostructured state. One such model supported by this investigation was proposed by Valiev [19] in which UFG nanostructures are characterized as having extremely small grains and high elastic long-range stress fields that are created by extrinsic grain boundary dislocations of high density.

The application of XRD to determine values of atomic displacements caused by high elastic long range stress fields and to evaluate an averaged size of elastic stresses has not been done before. These atomic displacements are very significant and they exceed the same parameters in CG Cu. In this case not only static atomic displacements increase, indicating the development of a very distorted defective structure, but also dynamic ones increase, indicating changes in the thermal properties and phonon spectrum caused by SPD. The analysis of XRD experimental data by computer simulation of nanostructured materials with regard to different assemblies of extrinsic grain boundary dislocations allows estimation of their density, which can be as large as 10^9m^{-1} [20, 21].

The results of this investigation do not contradict, but support, the scheme of microstructure evolution during SPD developed in [10]. It is shown by XRD analysis that the saturation stage comes quickly during SPD by HPT. Near this stage structural parameters change slowly and smoothly or remain unchanged. Preferred grain orientation also does not change. Changes of CSD sizes and elastic microdistortions take place during ECA pressing. Significant transformations of crystallographic texture happen during each pass, as indicated by the extreme changes in X-ray peaks. The observed differences coincide with changes of sample orientation during subsequent passes. This orientation change causes activation of alternative slip systems [22] and leads to observed changes in crystallographic texture during ECA pressing.

In conclusion, note a special significance of investigations of crystallographic texture during SPD of metals. Such investigations deepen our understanding of the role of grain boundaries during microstructure evolution, reveal SPD mechanisms, help us predict the grain boundary misorientation spectrum, and allow us to make suppositions about the anisotropy of structural properties.

5. Summary

These X-ray investigations provide new information about the character of microstructure evolution in pure Cu during SPD by different methods.

108

6. References

1. Valiev, R.Z (1996) Ultrafine-grained materials produced by severe plastic deformation, Valiev, R.Z. ed., *Annales de Chimie. Science des matériaux* **21**, 369–520.
2. Bykov, V.M., Lihachev, V.A., Nikonov, Yu.A., Serbina, L.L., Shibalova, L.I. (1978) Fragmentation and dynamic recrystallization in copper at large and very large plastic deformations, *Physics of Metals and Metal Science* **45**, 163–169.
3. Gindin, I.A., Starodubov, Ya.D., Aksenov, V.K. (1980) Structure and strength properties of metals with extremely distorted crystal lattice, *Metalphysics* **2**, 49–67.
4. Rybin, V.V. (1986) *Large Plastic Deformations and Fracture of Materials*, Metallurgy, Moscow.
5. Zehetbauer, M., Seumer, V. (1993) Cold work hardening in stages IV and V of F.C.C. metals – I. Experiments and interpretation, *Acta metall. mater.* **41**, 577–588.
6. Zehetbauer, M. (1993) Cold work hardening in stages IV and V of F.C.C. metals – II. Models fits and physical results, *Acta metall. mater.* **41**, 589–599.
7. Ungár, T., Zehetbauer, M. (1996) Stage IV work hardening in cell forming materials. Part II: A new mechanism, *Acta metall. mater.* **35**, 1467–1473.
8. Estrin, Y., Tóth, L.S., Molinari, A., Bréchet, Y. (1998) A dislocation-based model for all hardening stages in large strain deformation, *Acta mater.* **46**, 5509–5522.
9. Smirnova, N.A., Levit, V.I., Pilyugin, V.I., Kuznetsov, R.I., Davydova, L.S., Sazonova, V.A. (1986) Structural evolution of F.C.C. single crystals at large plastic deformations, *Physics of Meatals and Metal Science* **61**, 1170–1177.
10. Valiev, R.Z., Ivanisenko, Yu.V., Rauch, E.F., Baudelet, B. (1996) Structure and deformation behaviour of ARMCO iron subjected to severe plastic deformation, *Acta mater.* **44**, 4705–4712.
11. Akhmadeev, N.A., Valiev, R.Z., Kopylov, V.I., Mulyukov, R.R. (1992) Development of subgranular structure in copper and nickel at intense shear deformation, *Metals* **5**, 96–101.
12. Iwahashi, Y., Horita, Z., Nemoto, M., Langdon, T.G. (1998) The process of grain refinement in equal-channel angular pressing, *Acta metall. Mater.* **46**, 3317–3331.
13. Alexandrov, I.V., Valiev, R.Z. (1994) Investigation of nanocrystalline materials by X-ray analysis, *Physics of Metals and Metal Science* **77**, 77–87.
14. Zhang, K., Alexandrov, I.V., Valiev, R.Z., and Lu, K. (1996) The structural characterization of a nanocrystalline Cu by means of the X-ray diffraction, *J. appl. phys.* **21**, 407–416.
15. Zhang, K., Alexandrov, I.V., and Lu, K. (1997) The X-ray diffraction study of a nanostructured Cu processed by equal-channel angular pressing, *NanoStructured Materials* **9**, 347–350.
16. Zhang, K., Alexandrov, I.V., Kilmametov, A.R., Valiev, R.Z., Lu, K. (1997) The crystallite-size dependence of structural parameters in pure ultrafine-grained copper, *J. Phys. D: Appl. Phys.* **30**, 3008–3015.
17. Zhang, K., Alexandrov, I.V., Valiev, R.Z., Lu, K. (1998) The thermal behavior of atoms in ultrafine-grained Ni processed by severe plastic deformation, *J. Appl. Phys.* **84**, 1924–1927.
18. Alexandrov, I.V., Wang, Y.D., Zhang, K., Lu, K., Valiev, R.Z. (1996) X-ray analysis of the textured nanocrystalline materials. *Proceed. of the Eleventh Intern. Conf. on Textures in Materials (ICOTOM-11), Xi'an, China*, 929–940.
19. Valiev, R.Z. (1995) Approach to nanostructured solids through the studies of submicron grained polycrystals, *NanoStructured Materials* **6**, 73–82.
20. Alexandrov, I.V., Valiev, R.Z. (1996) Computer simulation of X-ray diffraction patterns of nanocrystalline materials, *Philos. Magaz.* **B 73**, 861–872.
21. Alexandrov, I.V., Enikeev, N.A., Valiev, R.Z. (1999) Investigation of Assemblies of Grain Boundary Dislocations in Nanostructured Copper by Computer Simulation, *Mater. Sci. Forum* **294–296**, 207–210.
22. Gibbs, M.A., Hartwig, K.T., Cornell, L.R., Goforth, R.E., Payzant, E.A. (1998) Texture formation in bulk iron processed by simple shear, *Scr. Mater.* **39**, 1699–1704.

AN ANALYSIS OF HETEROPHASE STRUCTURES OF TI₃AL, TIAL, NI₃AL INTERMETALLICS SYNTHESIZED BY THE METHOD OF THE SPHERICAL SHOCK WAVE ACTION

B.A. GREENBERG, E.P. ROMANOV, S.V. SUDAREVA, O.V.
ANTONOVA, N.D. BAKHTEEVA, T.S. BOYARSHINOVA, A.V.
NEMCHENKO, E.V. SHOROKHOV*, V.I. BUZANOV*
Institute of Metal Physics, Ural Division, Russian Academy of Sciences
18 S. Kovalevskoi str., Ekaterinburg, 620219, Russia
*Russian Federal Nuclear Centre-VNIITF
Chelyabinsk region, Snezhinsk, Russia

Abstract

Synthesis of the intermetallics from pure metal powders was performed in a spherical capsule of conservation, with explosions of charges symmetrically distributed over its surface. X-ray diffraction, metallography, and scanning and transmission electron microscopy, were used to study typical structures. These were found to be dendrite, martensite and swirl-like antiphase domain structures. Phases were found to be enriched with iron. The microstructures thus revealed can be used as indications of the phenomena that occur in the capsule of conservation under explosive loading.

1. Introduction

Just as the method of shock wave loading can be used to compact powders, so it can be applied to obtain new materials. For example, such new materials can be rapidly quenched metal alloys, either amorphous or microcrystalline [1]. We were trying to use this method to produce intermetallics from metal powders [2–7]. This method is characterized by ultra-rapid rates of heating and subsequent cooling and with a high deformation rate. The aim of our study is to reconstruct the time sequence of the processes taking place during synthesis of the intermetallics and to try to understand "the diverse complexity of phenomena" [8] that can be observed under shock wave compression.

2. Experimental Procedure

A quasi-spherical system of loading was used (Figure 1). Pure metal powders were placed into a steel spherical capsule 18 mm in diameter with the wall thickness of 0.2 mm. The capsule was placed in the center of a massive sphere of 60 mm in diameter. The spherical layer adjoining the capsule was made of brass to prevent the sphere from

T.C. Lowe and R.Z. Valiev (eds.), Investigations and Applications of Severe Plastic Deformation, 109–114.
© 2000 Kluwer Academic Publishers. Printed in the Netherlands.

destruction. A spherical charge, 10 mm thick, was initiated at 12 points uniformly distributed over the sphere surface; the initiation time in different points differed by no more than by 10^{-7} s. The peak pressure on the capsule wall was about 50 ± 5GPa. The samples after loading had the shape of a ball of about 15 mm in diameter. In the central part of the samples, a cave of irregular shape (of 5 - 7 mm in size) was formed as a result of accumulation of energy. Powders of titanium and aluminum (supplied by Starck and the Eckart) with particles of irregular form less than 63 µm were used with composition ratios: 50 at.% Ti to 50 at.% for the first sample, and 75 at.% Ti to 25 at.% for the second sample. In addition, the powders of nickel and aluminum were used in ratio 75 at.% Ni to 25 at.% Al.

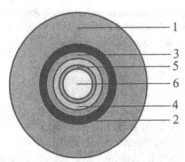

Figure 1. Scheme of the device for loading powders: 1 - steel casing; 2 - explosive; 3 - steel; 4 - brass; 5 - steel cover; 6 – powder.

3. Results

Our main goal was to ascertain the mechanisms of phase formation: either (1) the melting on the periphery of the initial particles resulted in their compacting, or (2) the melting of the bulk of powder with new phases forming in solid solution. If the latter mechanism took place, then the phase transformations need to be further clarified, that is, we would need to know what high temperature phases were formed and what subsequent transformations they were undergoing. The task seemed to be similar to that of an equilibrium phase diagram of Ti-Al [9], yet the transformation paths proved to be different.

3.1. Ti-50 AT. % Al COMPOSITION

Figure 2 shows a dendritic structure. Such dendritic structures and the X-ray diffraction lines of the pure metals (Ti and Al) give evidence to the fact that the powder has melted. Only the intermediate layer of some 500 µm in thickness that is immediately adjacent to the steel casing, reveals individual particles of nearly pure titanium powder. As seen from the optical microphotographs (Figure 2a), the dendrite branches are non-orthogonal to its trunk. Morphology of a dendrite is known to to indicate the phases that have formed (see [10], for example). The SEM microphotograph (Figure 2b) contrast the dendrite and interdendritic layers, indicating that they are different phases. The TEM micrographs shows the swirl-like structure of the domain boundaries inside the dendrite.

On the basis of the data obtained, one can decide on the dendrite-forming phase. It would be unlikely to have a bcc β phase, since none of the investigation techniques used in the study could identify this phase. Besides, had the dendrite contained a cubic phase, its branches would have been orthogonal to its trunk, for the dendrite axes would be parallel to the crystallographic directions <001>. So, neither could it be a fcc phase. We suggest that the initial high temperature phase growing from the melt in the form of dendrite was a disordered α phase with a hcp lattice.

According to the microanalysis results there is a phase enriched with iron in the interdendritic space. It is this phase that is shown in Figure 2c as white regions stretched out. According to the microanalysis in Fe radiation, the content of Fe in the interdendritic layers reaches 5–10 at.%, while inside the dendrite it amounts to only 1–3 at.%. A total absence of iron was found in the intermediate layer. As the central cave is approached, the content of iron increases. These data give evidence to the fact that under explosive loading the iron particles are "flying" from the steel casing into the depth of the sample synthesized.

The evolution of the microstructure could be described by a schematic reaction as follows: $L \to L+\alpha \to \alpha+X_{Fe}$, where α phase forms the dendrite, and X_{Fe} phase fills into the interdendritic space. It is this structure that would have been solidified as a result of the super-fast quenching, but for the stage of slow cooling that follows (so-called residual temperatures). At this stage the ordering occurs that results in the transformation $\alpha \to \gamma + \alpha_2$. Both phases have a great number of domains with distorted boundaries. This bears witness to the fact that these phases are slightly ordered. Instead of the usual lamellar-type structure, an unstable heterophase structure was observed that contained individual domains of γ and α_2 phases. Therefore, under the given conditions of explosive loading, the disordered phases are formed from the melt, and it is only at the residual temperatures that these phases undergo further transformations.

3.2. Ti-25 AT.% Al COMPOSITION

Microanalysis has demonstrated a small (1–3 at.%) addition of iron. There was no trace of a dendrite structure revealed. This structure contains martensite lamellas (Figure 3). It is these lamellas that proved to bear witness to the $\beta \to \alpha$ transformation, when the initial high-temperature β phase was transformed into the α phase. The superstructural lines show evidence of very small domains that are, in fact, nuclei of α_2 phase in the disordered α phase. As in the previous case, ordering seems to take place only at the residual temperatures. That there was no superstructural lines on the diffraction patterns indicates that the integral degree of the long-range order is very small.

Evolution of the microstructure can be described as follows. $L \to L+\beta \to \alpha+\beta$. With further gradual cooling, a transformation $\alpha+\beta \to \alpha_2 + B2$ is quite possible, but the phases thus formed are slightly ordered.

3.3. Ni-25 AT.% Al COMPOSITION

A dendritic structure was revealed. The optical microphotograph (Figure 4a) shows clearly that the branches of dendrite are orthogonal to the trunk. This indicates a cubic phase has formed the dendrite. Inner structure of the dendrite is seen from a TEM image (Figure 4b). Outside the dendrite, there are islands formed by the martensite lamellas. TEM image of the martensite lamellas is shown in Figure 4c. X-ray microanalysis was used to demonstrate that the dendrite has an average composition about 79 at.% Ni and 21 at.%. Aluminum content in the interdendritic layers proved to be considerably greater (35–50 at.%) than in the dendrite itself. Like in the previous cases, iron was found in the sample. As the center of sample is approached, the iron concentration grows to reach 3.2–3.5 at.%.

Evolution of the microstructure can be described as follows. A dendrite, that has been formed from a Al-poor cubic phase, grows from the melt. We suggest that it was a disordered γ phase, so that the schematic reaction takes the form $L \rightarrow L+\gamma$. The interdendritic space enriched with Al would fill with the β phase that can undergo martensite transformation into γ phase, so we have $L \rightarrow L+\beta \rightarrow \gamma+\beta$. At the residual temperatures, cuboid γ' phase is formed inside the dendrite as a result of the $\gamma \rightarrow \gamma'$ transformation. However, γ martensite can not undergo the $\gamma \rightarrow \gamma'$ transition and remains disordered, as is the β phase inside the martensitic islands (Figure 4c).

4. Discussion

The experimental results presented in the study were obtained at a pressure of 50 GPa and they differ considerably from those [1] obtained at a pressure less than 25 GPa. In this latter case the changes that took place there were due to the melting of the surface layer of the particles. Therefore, as the pressure grows, a transition takes place from compacting to alloying (see also [11]).

Strictly speaking, what the explosive loading causes can not be called an intermetallic reaction exactly, if by such an intermetallic reaction one means the formation of an ordered phase from the melt with no intermediate disordered phase. Due to the high rate of cooling, a metastable state is realized. A disordered state has come to be such a metastable state. Circulation of shock waves causes mixing of the melt that, in turn, provides better uniformity of the solid solution. The whole system under study falls in a sort of trap, not reaching its equilibrium state. This phenomenon is known as "disorder trapping" (see, for example [12, 13]).

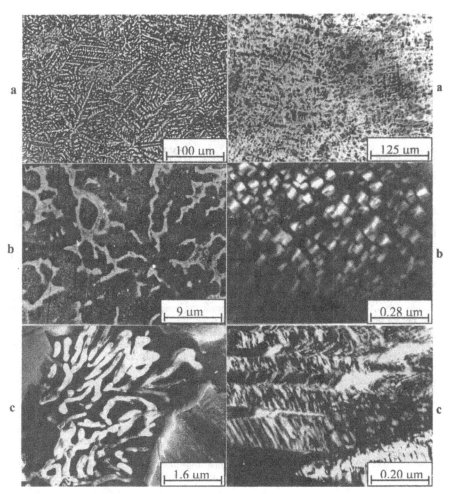

Figure 2. Dendritic structure of the synthesized Ti-50 at.% Al sample: a - an optical micrograph; b - SEM secondary electron micrograph;.c - dark-field image of a region between dendrites.

Figure 4. Dendritic structure of the Ni-25 at.% Al sample: a - an optical micrograph; b - γ'-phase cuboids inside a dendrite (dark-field image); c - martensite plates in the interdendritic space (dark-field image).

Figure 3. Martensite structure of the Ti-25 at.% Al sample (bright-field image).

114

5. Acknowledgement

The authors are grateful to the Russian Foundation of Fundamental Research (the grants 95-02-05656, 97-02-26671, 98-02- 17278) for financial support.

6. Reference

1. Nesterenko, V.F. (1992) *Impulse Loading of Heterogeneous Materials*, Novosibirsk: Nauka, Sibir. Otdelenie.
2. Shorokhov, E.V., Greenberg, B.A., Boyarshinova, T.S., Sudareva, S.V., Buzanov, V.I., Panova, E.V., Romanov, E.P. (1997) Analysis of heterophase structures of intermetallic compounds Ti_3Al and TiAl prepared by the method of the spherical shock wave action: I. X-ray diffraction and scanning microscopy.- *Phys. Met. Metallogr.* **83**, No. 4, 428-434.
3. Sudareva S.V., Antonova O.V., Boyarshinova T.S., Greenberg B.A., Romanov E.P., Shorokhov E.V. (1997) Analysis of heterophase structures of titanium aluminides prepared by the method of the spherical shock wave action. III. electron-microscopic study of the TiAl structure., *Phys. Met. Metallogr.* **83**, No. 6, 633-639
4. Boyarshinova T.S., Sudareva S.V., Greenberg B.A., Antonova O.V., Panova E.V., Shorokhov E.V. (1998) analysis of heterophase structures of titanium aluminides prepared by the method of the spherical shock wave action. III. electron-microscopic study of the Ti_3Al structure., *Phys. Met. Metallogr.* **85**, No. 4, 463-468.
5. Shorokhov E.V., Greenberg B.A., Boyarshinova T.S., Sudareva S.V., Buzanov V.I., Antonova O.V., Panova E.V.(1997) Shock-wave synthesis of intermetallic compounds Ti_3Al, TiAl and analysis of heterophase structure formation, *J. Phys IV France* **7**, C3-7 - C3-12.
6. Shorokhov E.V., Greenberg B.A., Sudareva S.V., Antonova O.V., Boyarshinova T.S., Romanov E.P. (1997) An Analysis of Heterophase Structures of Ti_3Al and TiAl Intermetallics Synthesized by the Method of the Spherical Shock Wave Action, *Mat. Sci. Eng. A* **239-240**, 688-693.
7. Bakhteeva, N.D., Sudareva, S.V., Greenberg, B.A., Nemchenko, A.V., Boyarshinova, T.S., Shorokhov, E.V., Buzanov, V.I. (1999, in press) Analysis of heterophase structure of Ni_3Al intermetallics synthesized by the method of the spherical shock wave action, *Phys. Met. Metallogr.*
8. Dremin, A.N., Breusov, O.N. Processes (1968) Occurring in solids under the action of strong shock waves, *Rus. J. Chem. Success* **37**, No. 3, 898–916.
9. Yamabe, Y., Takeyama, M., Kikuchi, M. (1995) Microstructure evolution through solid-solid phase transformations in gamma titanium aluminides, *Gamma Titanium Aluminids, ed. Kim Y.-W. et.al, TMS, Warrendale, PA, USA*, 111–129.
10. Jung, J.Y., Park, J.K., Chun, C.H. (1995) Solidification structures of Ti – (35–58) at.%Al intermetallic compounds, *Gamma Titanium Aluminids, ed. Kim Y.-W. et.al, TMS, Warrendale, PA, USA*, 459–466.
11. Kochsiek, D., Preummer, R., Brunold, A. (1995) Synthesis of intermetallic aluminides by explosive reaction pressing, *Metallwissenschaft und Technik* **49**, No. 3, 168–172.
12. Shao, G., Tsakiropoulos, P., Miodownik, A.P. (1995) Study of phase transformations in Ti-47Al-3V, *Gamma Titanium Aluminids, ed. Kim Y.-W. et.al, TMS, Warrendale, PA, USA*, 173–180.
13. Assari, H., Barth, M., Greer, A.L., Herlach, D.M. (1998) Kinetics of Solidification of intermetallic compounds in the Ni-Al system, *Acta Mater.* **46**, No. 2, 491–500.

STRUCTURAL CHANGES INDUCED BY SEVERE PLASTIC DEFORMATION OF FE- AND CO-BASED AMORPHOUS ALLOYS

N. NOSKOVA, L. KORSHUNOV, A. POTAPOV, N. TCHERNENKO
Institute of Metal Physics, UD RAS, Ekaterinburg, 620219, Russia

Abstract

The amorphous alloys $Fe_{81}Si_7B_{12}$, $Fe_{81}Si_4B_{13}C_2$, $Fe_{64}Co_{21}B_{15}$, $Fe_{73.5}Cu_1Nb_3Si_{13.5}B_9$, and $Fe_5Co_{70}Si_{15}B_{10}$ prepared as strips by fast melt quenching on the rotating Cu disc were examined. The strips were 6–12 mm wide and 25–40 μm thick. The structure, coercivity and friction coefficient of these amorphous alloys after plastic deformation by tension, rolling, shear pressure, and dry sliding friction were analyzed. The samples underwent mechanical loading at 293 K in air at rates excluding heating of the material. Coercivity was shown to increase by a factor of ten or more at all types of deformation. The electron microscopic study of the structure of deformed strips revealed the presence of crystalline precipitates 2 to 50 nm in size having a relatively equiaxed shape in the amorphous matrix of the Fe-based alloys and a nearly plate shape in the Co-based alloys. The phases were identified from electron microdiffraction patterns of the alloys.

1. Introduction

Amorphous metal alloys based on transition metals are of great interest thanks to the combination of some unique properties. One of the principal issues in the study of such alloys is to examine specific features of their transformation from the amorphous state to the crystalline one, particularly the nanocrystalline state. For example, it has been found [1] for the $Fe_{73.5}Cu_1Nb_3Si_{13.5}B_9$ amorphous alloy that crystallization at 823 K for 0.5 h caused formation of the nanocrystalline structure with the average grain size of 10 nm. However, it was shown [2] that the average grain size of the $Fe_{73.5}Cu_1Nb_3Si_{13.5}B_9$ alloy could be decreased after nanocrystallization if the amorphous alloy was heated rapidly to 933 K and was held at this temperature just for 5 to 10 s. In this case the average size of grains did not exceed 8 nm. The $Fe_5Co_{70}Si_{15}B_{10}$ amorphous alloy was treated under similar conditions (923 K, 10 s) [3, 4] providing grains 15 to 50 nm in size, as distinct from the alloy crystallized at 873 K for 1 h when the grain size ranged between 50 and 200 nm. Moreover, it was shown that grains could be refined still further (6 nm) by rapid crystallization [2–4] if the amorphous alloy was deformed beforehand or underwent low-temperature annealing [5].

Severe plastic deformation of alloys in the amorphous state was shown [6–8] to lead to the appearance of crystallization centers in the amorphous matrix and, hence, development of crystallization during deformation. The $Pd_{81}Cu_7Si_{12}$ amorphous alloy was nanocrystallized [9, 10] under creep conditions at 773 K. The average grain size did

115

not exceed 10 nm. The same alloy, which was crystallized at 823 K for 10 s, had grains with the average size of 50 nm.

In this connection, the goal of this study was to analyze the effect of severe plastic deformation on the structure and properties of amorphous alloys.

2. Experimental

Alloys of the formulas $Fe_{64}Co_{21}B_{15}$, $Fe_{81}Si_4B_{13}Cr_2$, $Fe_{81}Si_7B_{12}$ $Fe_5Co_{70}Si_{15}B_{10}$, and $Fe_{73.5}Cu_1Nb_3Si_{13.5}B_9$ were examined in the amorphous and nanocrystalline states. The amorphous metal alloys in the form of strips 10 to 20 mm wide and 30 μm thick were prepared by rapid melt quenching. The samples underwent severe plastic deformation by external sliding friction (Figure 1a) and high-pressure (8 GPa) shear (Figure 1b). Different degrees of pressure-shear deformation were obtained in Bridgeman anvils by turning to various angles the upper anvil relative the fixed lower anvil. The sample was clamped between the anvils. The degree of deformation ε was calculated from the formula $\varepsilon = 1 \, n\varphi\xi/l$, where φ is the turning angle in radians, ξ is the radius from the rotation axis, and l is the sample's thickness. Depending on the number of turns, the true deformation varied between 5.3 and 7.1. The tribological loading of the amorphous alloys was realized under conditions of dry sliding using the cylinder-plane scheme. Fragments of the amorphous strip were fixed on the surface of a cylindrical indenter and a plate made of steel (Figure 1a). The alloy-on-alloy friction was realized during reciprocal movement of an indenter at the average speed of 0.075 m/s under the load of 98 N. These conditions of rubbing almost eliminated friction heating of the strips. The mean volume temperature in the friction zone did not exceed 30° C. The friction coefficient of the $Fe_{64}Co_{21}B_{15}$ amorphous alloy was K = 0.15.

The strip samples of $Fe_{73.5}Cu_1Nb_3Si_{13.5}B_9$ and $Fe_5Co_{70}Si_{15}B_{10}$ underwent tension in vacuum (10^{-5} torr) at the rate of $7 \times 10^{-5} \, s^{-1}$ and the temperature of 300 K. The alloys were examined in two different states: the amorphous state (rapid quenching from melt) and after annealing of the amorphous strip at 813 K (Figure 1c).

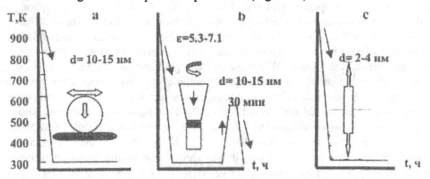

Figure 1. Schemes of plastic deformation of amorphous and nanocrystalline alloys: a – $Fe_{64}Co_{21}B_{15}$ and $Fe_{81}Si_4B_{13}Cr_2$; b – $Fe_{81}Si_7B_{12}$ [8]; c – $Fe_{73.5}Cu_1Nb_3Si_{13.5}B_9$, $Fe_{73}Ni_{0.5}Cu_1Nb_3Si_{13.5}B_9$, and $Fe_5Co_{70}Si_{15}B_{10}$ [4–7].

Structural changes in the alloys during annealing and tension were studied by the method of transmission electron microscopy. Note that subsequent to the external friction loading the strip samples were thinned using chemical and electrochemical methods on the side opposite to the surface exposed to friction. As a result, structural changes in the friction surface of the strips were preserved.

3. Results and Discussion

The effect of severe plastic deformation on the structure of amorphous alloys was analyzed using the following alloys: $Fe_{64}Co_{21}B_{15}$, $Fe_{81}Si_4B_{13}Cr_2$, and $Fe_{81}Si_7B_{12}$. As was mentioned in the foregoing, strips of $Fe_{64}Co_{21}B_{15}$ and $Fe_{81}Si_4B_{13}Cr_2$ were exposed to external sliding friction, while the sample of $Fe_{81}Si_7B_{12}$, which was shaped as a disk 6 mm in diameter and 30 μm thick, underwent high-pressure shear deformation. The structure of the $Fe_{64}Co_{21}B_{15}$ alloy before and after the tests is given in Figure 2 and Table 1. From the electron microdiffraction patterns it is seen that the alloy was in the amorphous state before the test (Figure 2a). A nanocrystalline structure with nanophases 15 nm in size is observed near the friction surface after the test (Figure 2b). One may think that friction causes development of the first stage of crystallization in the surface of this amorphous alloy [11]. This is followed by appearance of nanophases in the amorphous matrix.

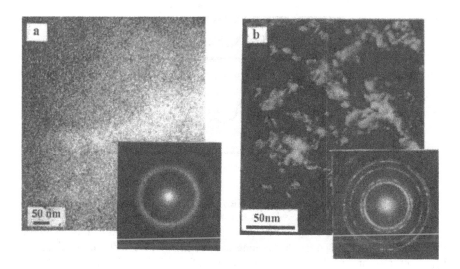

Figure 2. Electron microscopic dark-field photographs of the structure and microdiffraction patterns of the $Fe_{64}Co_{21}B_{15}$ alloy before (a) and after the external friction test (b).

The coercive force H_c of the alloys was measured: H_c was 6.4 A/m and 64.0 A/m for the $Fe_{64}Co_{21}B_{15}$ alloy, and H_c was 8.8 A/m and 38.4 A/m for the $Fe_{81}Si_4B_{13}Cr_2$ alloy before and after the surface friction tests respectively. The coercive force of the

$Fe_{81}Si_7B_{12}$ alloy before and after the pressure-shear test was equal to 13.5 A/m and 1800.3 A/m respectively. The alloy structure was altered. The electron microdiffraction pattern and the X-ray diffraction pattern (Figure 3) exhibit the second amorphous halo at small reflection angles in the deformed alloy. The structure becomes inhomogeneous and the amorphous matrix is preserved during subsequent low-temperature (723 K) annealing of the alloy. Nanocrystalline regions with nanophases 10 to 15 nm in size appear in the amorphous matrix. The alloy acquires the nanocrystalline state [7].

TABLE 1. Phases in deformed amorphous alloys

Composition	Test	Phases
$Fe_{64}Co_{21}B_{15}$	Friction	α-Fe, α-Co, β-Co
$Fe_5Co_{70}Si_{15}B_{10}$	Tension	α-Co, β-Co, Fe_3S
$Fe_{81}Si_7B_{12}$	High-pressure shear	α-(Fe-B), α-(Fe-S), Fe_3S
$Fe_{73.5}Cu_1Nb_3Si_{13.5}B_9$	Tension	α-(Fe-Si), Fe_3S

The structure of the deformed $Fe_{73.5}Cu_1Nb_3Si_{13.5}B_9$ amorphous alloy is shown in Figure 4. As is seen from Figure 4, very fine nanophases (2 to 4 nm) are present and an additional diffusion halo appears in the electron microdiffraction pattern at small angles.

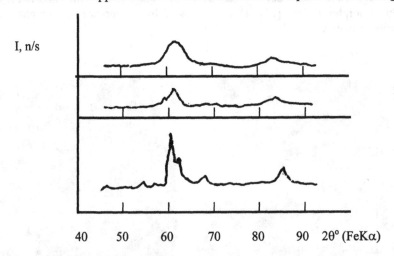

I, n/s

40   50   60   70   80   90 $2\theta°$ (FeKα)

Figure 3. X-ray diffraction patterns of the $Fe_{81}Si_7B_{12}$ alloy: a – amorphous state; b – deformed state, ε = 6.5; c – subsequent annealing at 723 K for 1 h

The structure of the deformed $Fe_{73.5}Cu_1Nb_3Si_{13.5}B_9$ nanocrystalline alloy is presented in Figure 4b. One can see banded luminescent points (shown with an arrow in Figure 4b), which also are nanophases having the size of 2 to 4 nm. This distinguishes them from nanograins of the initial structure, which are 8 to 10 nm in size. It may be thought that in this case localized deformation bands, which pass at boundaries of the initial grains, are formed. If the nanophase boundaries are assumed to be amorphous, one may expect the appearance of finer nanophases in the boundaries during deformation.

4. Conclusion

Plastic deformation (external friction, high-pressure shear, and tension) may cause nanocrystallization of amorphous alloys. Crystallization of the $Fe_{64}Co_{21}B_{15}$ amorphous alloy, which results from severe plastic deformation by dry friction, is due to the fact that α-Fe, α-Co, and β–Co nanophases start crystallizing in some areas of the amorphous sample. Crystallization of the $Fe_{81}Si_7B_{12}$ amorphous alloy, which is caused by severe pressure-shear deformation, is associated with formation of α–Fe–B and α–Fe–Si nanophases of solid solutions in the amorphous matrix. When this alloy is annealed at low temperature, it crystallizes and solid-solution α–Fe–Si and Fe_2B nanophases are formed.

Deformation of the $Fe_{73.5}Cu_1Nb_3Si_{13.5}B_9$ and $Fe5Co_{70}Si_{15}B_{10}$ amorphous alloys by tension leads to appearance of nanophases in the amorphous structure.

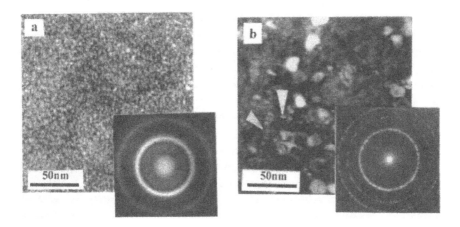

Figure 4. Electron microscopic dark-field photographs of the structure and microdiffraction patterns of the $Fe_{73.5}Cu_1Nb_3Si_{13.5}B_9$ alloy after tension: a – amorphous state; b – annealing at 813 K for 30 min.

5. References

1. Yoshizawa, Y., Yamauchi, K. (1990) Fe-based solf magnetic alloys composed of ultrafine grain structure. *Mater.Trans. JIM* **31**, 307–314.
2. Glazer, A., Lukshina, V., Potapov,., Noskova, N. (1992) Nanocrystalline $Fe_{73.5}Cu_1Nb_3Si_{13.5}B_9$ alloy rapidly crystallized from on amorphous state at elevated temperatures, *The Physics of Metals and Metallography* **74**, 163–166.
3. Glazer, A., Noskova, N., Lukshina, V., Potapov, A., Tagirov, R., Ponomareva,. A. (1993) Effect of rapid crystallization of $Fe_5Co_{70}Si_{15}B_{10}$ glass on its magnetic properties, *The Physics of Metals and Metallography* **76**, No. 2, 222–224.
4. Noskova, N.I., Ponomareva, E.G., Lukshina, V.A., and Potapov, A.P. (1995) Effect of rapid crystallization of $Fe_5Co_{70}Si_{15}B_{10}$ glass on its properties, *NanoStructured Materials* **6**, 969–972.

5. Noskova, N., Ponomareva, Glazer, A., Lukshina, V., Potapov, A. (1993) The Effect of preliminary deformation and low-temperatusre annealing on grain size in nanocrystalline $Fe_{73.5}Cu_1Nb_3Si_{13.5}B_9$ alloy produced by crystallization of amorphous ribbons, *The Physics of Metals and Metallography* **76**, . 5, 535–537.

6. Noskova, N.I.,Vildanova, N.F., Filipov, Yu.I., and Potapov, A.P. (1985) Strength, plasticity, and fracture of ribbons of $Fe_5Co_{70}Si_{15}B_{10}$ amorphous alloy, *Phys. Stat. Sol. (a)* **87**, 549–557.

7. Noskova, N., Vildanova, N., Kuzneshov R., Tagirov, R.., Glazer,.A., Effect of pressure-shear deformation of $Fe_{81}Si_7B_{12}$ amorphous alloy on its structure and properties, *The Physics of Metals and Metallography* **65**, No 3, 594–601.

8. Trudeau, M. (1997) *Nanophase Materials Using High Mechanical Energy,* Programme and Abstracts, NATO ASI, Science and Technology, St-Petersburg, Russia, 20.

9. Noskova, N., Ponomareva, E., Kuznetshov, V., Glazer, A., Lukshina, V., Potapov, A. (1994) Crystallization of the amorphous Pd-Cu-Si alloy under creep conditions, *The Physics of Metals and Metallography* **77**, No. 5, 509–512.

10. Kuznetsov, V.N., Ponomareva, E.G., Noskova, N.I. (1996) Nanocrystallization of the amorphous Pd-Cu-Si alloy under creep conditions, *J. of Non-Crystalline Solids* **205-207**, 829–832.

11. Boldyrev, V., Veksler, V., Noskova, N.,Gavrilyk, V., Vildanova, N. (1999) Study of the crystallization kinetics of the $Fe_{64}Co_{21}B_{15}$ amorphous alloy, *The Physics of Metals and Metallography* **87**, 439–442.

STRUCTURE OF GRAINS AND INTERNAL STRESS FIELDS IN ULTRAFINE GRAINED NI PRODUCED BY SEVERE PLASTIC DEFORMATION

N.A. KONEVA, N.A. POPOVA, L.N. IGNATENKO,
E.E. PEKARSKAYA, YU. R. KOLOBOV, E.V. KOZLOV
Tomsk State University of Architecture and Building
Solyanaya Sq. 2, Tomsk 634003, Russia

1. Introduction

The prospect of obtaining large volumes of materials with submicron grain size has become increasingly real. The equal channel angular (ECA) pressing appears to be among the most successful methods for producing such materials [1]. The structure of the materials obtained by this method has been the subject of intensive investigation [2–5]. In most cases the description of the structure of such materials has been qualitative. Recently detailed quantitative investigations of materials produced by ECA pressing, were reported [6, 7]. At the same time investigations of the structure of materials deformed by tension, compression, rolling and torsion up to great degrees of deformation have also become more quantitative as well (see, for example, [8–10]). For practical application of ultrafine-grained materials data on their structural stability is rather important. The later is closely connected, firstly, with the internal stress fields and, secondly, with the size distribution function for the substructure parameters [11]. Consequently, in such investigations the measurement of such parameters should be carried out statistically.

The present work is devoted to detailed quantitative statistical investigation of grain and subgrain structure, the dislocation structure, and the stress fields in ultrafine-grained Ni produced by ECA pressing.

2. Material and experimental procedure

Material was prepared in the laboratory headed by Prof. R.Z. Valiev (Ufa) and kindly presented to us for investigation. The material was studied in the as-received initial state and after tempering (T = 398 K, t = 3 h). Transmission diffraction electron microscopy was the main method of investigation. Two investigation procedures were used: the method of thin foils and the method of replicas. Thin plates were cut by the electrospark method and then electrochemically thinned. The foils were studied in the electron microscope EM–125 at the accelerating voltage of 125 kV with the systematic use of the goniometer. A suitable region of the electrochemically polished sample surface, was chosen for revealing the figures of etching. Carbon replicas, shaded by the platinum, were taken off the sample surface. Measurements were made of the sizes the grains, subgrains and cells, and the boundary densities (average and local) of different types of dislocations. Grains were separated in accord with its internal dislocation structure during the process of measurement of the grain sizes. The internal stress fields were deter-

T.C. Lowe and R.Z. Valiev (eds.), Investigations and Applications of Severe Plastic Deformation, 121–126.
© 2000 *Kluwer Academic Publishers. Printed in the Netherlands.*

mined by three methods [12]: 1) by the dislocation lines curvature; 2) by the curvature-torsion of the crystal lattice; 3) by broadening of the X-ray Bragg reflections. In the latter case the X-ray diffractometer DRON–1.5, monochromatic CuK_α-radiation, was used. The first method allows determination of the tangent stresses, the second – the moment stresses and the third – the normal components of the stresses.

3. Investigation results and discussions

3.1. STRUCTURE OF GRAINS AND SUBGRAINS

Grain and subgrain structures produced by severe plastic deformation appear to be rather complex [2–10]. They are the result of the dynamical evolution of ensembles of dislocations, subboundaries, and the high angle boundaries, mostly frozen at the moment deformation is stopped. In different regions of the material the microprocesses of deformation and recovery occur with different intensity [13]. This is why different states of the defect substructure can be seen throughout the material volume. In relatively pure Ni the following formations in the defect substructure should be expected: 1) grains, limited to the high-angle boundaries; 2) subgrains, limited to low- and high-angle boundaries; 3) cells, limited to the dislocation walls; 4) separate dislocations and their groups. Electron microscopic images of the ultrafine-grained nickel are shown in Figures 1 and 2.

Figure 1. Typical electron microscope images of the ultrafine grained Ni substructure observed in replicas (a) and foils (b).

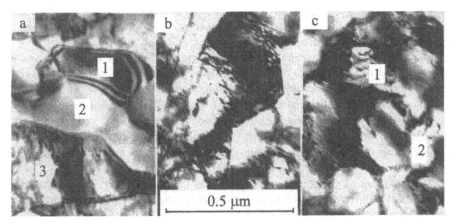

Figure 2. TEM images of subgrains of the three types with the different dislocation structure and of the three types of subboundaries: (a) the groups (1, 2) of the subgrains without dislocations limited low – angle defec tive boundaries with striped contrast and the micrograin (3) divided by the subgrains with dense dislocation boundaries; (b) groups of subgrains containing dislocations and limited by subboundaries with dislocations and with striped contrast; (c) the group of subgrains (1, 2) containing dislocations and divided by dense dislocations walls.

The substructures in local regions were very anisotropic. They consisted of high-angle and geometrically necessary boundaries, being approximately parallel to each other. With a separation distance of 0.2 μm. Similar boundaries were observed by others [8–10]. The space between them, approximately 1 μm (1.00–1.25 μm), is divided by the high-angle boundaries into anisotropic ultrafine grains. Most of the ultrafine grains were divided by low-angle boundaries into subgrains. The subgrains were also anisotropic.

The available subgrains can be classified into several types according to the character of the dislocation structure formed in them. In accord with the order of decreasing scalar dislocation density one can distinguish: 1) subgrains with chaotic and network dislocation structure; 2) subgrains with cells; and 3) almost dislocation cell-free subgrains. The sizes of the defect formations decrease in the following succession: grains → subgrains with dislocations → subgrains without dislocation cells. The anisotropy coefficient (k) and the dispersion of the sizes decrease in the same succession ($k = l/h$, l is length, h is width of subgrains). Figure 3 shows this distinctly. A typical distribution of subgrain sizes is given in Figure 4. The analysis of Figure 1–4 shows a complex hierarchy of substructures. The misorientations across subgrain boundaries are 6 ± 2^0. If there are several subgrains in a micrograin, the misorientations can be as large as 14 ± 4^0. One can suppose that on the basis of only average sizes of cells, subgrains, and grains the total picture of transformations in the defect sub-

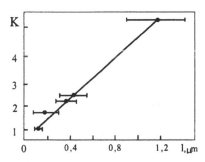

Figure 3. The dependencies of anisotropy coefficient (k) on the length of grains, subgrains, and cells (initial state).

124

system cannot be given at all. For this purpose the analysis of the size distribution of subgrains as well as the description of the processes taking place in the center and on the edges of the distribution are necessary.

Subgrains refine via division by the new boundaries and they grow via boundary movement. At different stages of the plastic deformation either of these processes can be predominant. In this investigation, subgrain refinement was the leading process.

The average scalar dislocation density (ρ) was 1.3×10^{14} m^{-2}. The lowest value of the dislocation density was observed in so-called "subgrain without dislocations" ($\rho = 0.3 \times 10^{14}$ m^{-2}). An intermediate value existed in subgrains with dislocation cells and the highest ρ was found in subgrains with dislocation networks and chaotic dislocation structure.

The evolution of microstructure probably was likely oscillatory. Locally the dislocation density is increased by active dislocation glide and then it is decreased in the process of forming new boundaries.

3.2. THE STRUCTURE OF GRAIN AND SUBGRAIN BOUNDARIES

Visually, all the boundaries can be divided into three types: 1) the boundaries with the striped contrast, small defects; 2) the boundaries with the striped contrast containing the individual dislocations and the dislocation networks; and 3) dense dislocation boundaries. Typical patterns of these boundaries are given in Figure 1 and 2. Similar boundary types have been already described for greatly deformed materials [8–10]. Hence for the separation of grains and subgrains, the same part of a foil was observed using different electron beam orientations in the method of replicas and by direct observation. Initially, 40-45% of the boundaries were of the first type, about 30% of the second type, and 25% of the third type. Annealing increases the proportion of the first boundary type, while

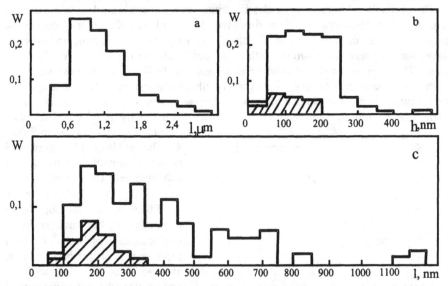

Figure 4. The grain (a) and subgrain (b, c) size distribution in the initial state (a, c – the length, b – the width, a – replicas, b, c – foils). Dislocation-free subgrains are shaded, subgrains with net and cell dislocation substructure are nonshaded.

decreasing the proportions of the second and third types, especially at higher temperature.

3.3. INTERNAL STRESSES

Average values of internal stresses between 190–250MPa were measured by the X-ray method. Boundaries are main source of these internal stress fields. Measurements of dislocation line curvature showed the spatial inhomogeneity of internal stress. The role of non-equilibrium boundaries and their intersections as sources of stress fields has been discussed in literature repeatedly for materials with submicron grain size (see, for example, [13]). As a rule, the stress amplitude appears to be greater near the boundary and decreases at a distance from it (Figure 5). The cell boundaries appeared to be the most highly stressed, 300–400 MPa compared to 240–260 MPa near grain boundaries and subgrains. However, boundaries of different types had nearly the same value of the average stresses. The stress fields decreased by 100–150 MPa in the the subgrain center. The internal stress fields appear to play an important role in substructure evolution during deformation because of the prevailing tendency to locally minimize internal energy [14].

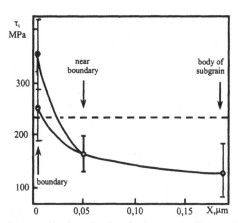

Figure 5. The dependencies of anisotropy coefficient (k) from the length, subgrains and cells (initial state)

Analyzing the average values of the internal stresses, the internal local stress field can exceed the average values by 2–3 times. It has been reported that grain boundary intersections turned to be one more source of the maxima [12, 15]. This is also where the joint disclinations exist [2, 16]. These sources of internal stress have been repeatedly discussed in theoretical investigations. The quantity of these sources is not large in the material studied here, but the amplitude of the field created by them is high, $(1–5) \times 10^3$ MPa.

4. Acknowledgements

The authors thank professor R.Z. Valiev who brought our attention to the problem of materials with the ultrafine grain size and kindly presented us the Ni samples for investigation and discussed with the authors the results.

5. References

1. Valiev, R.Z. (1997) Structure and mechanical properties of ultrafine-grained metals, *Mater. Sci. and Eng.* **A243–236**, 59–66.

2. Valiev, R.Z. (1996) Ultrafine-grained materials produced by severe plastic deformation: an introduction, *Annales de Chimie. Science des materiaux* **6–7**, 369–378.
3. Iwahashi, Y., Horita, M., Landgon, T.G. (1998) The process of grain refinement is equal-channel angular pressing, *Acta mater* **46**, 3317–3331.
4. Liu, Z.Y., Liong, G.X., Wang, E.D., Wang, Z.R. (1998) The effect of cumulative large plastic strain on the structure and properties of a Cu-Zn alloy, *Mater. Sci. And Eng.* **A242**, 137–140.
5. Chokkingal, U., Suriadi, A.B., Thomson, P.F. (1998) Microstructure development during equal channel angular drawing of Al at room temperature, *Scr. Mater.* **39**, 667–684.
6. Agnew, S.R., Weertman, S.R. (1998) Cyclic softening of ultrafine grain copper, *Mater. Sci. and Eng.* **A244**, 145–153.
7. Furukawa, M., Horita, Z., Nemoto, M., Valiev, R.Z., Langdon, T.G. (1998) Factors influencing the flow and hardness of materials with ultrafine grain sizes, *Phil. Mag.* **A78**, 203–215.
8. Rybin, V.V. (1986) *Large plastic deformatrion and fracture of metals,* Metallurgia, Moscow.
9. Winther, G., Jensen, D.J., Hansen, N. (1997) Dense dislocation walls and microbands aligned with slip planes – theoretical considerations, *Acta mater.* **45**, 5059–5068.
10. Goderey, A., Jensen, D.J., Hansen, N. (1998) Slip pattern, microstructure and local crystallography in an aluminium single crystall of copper orientation {112} <111>, *Acta mater.* **46**, 835–848.
11. Martin, J.W., Dohetry, R.D., Cantor, B. (1997) *Stability of microstructure metallic systems,* Cambridge University Press, Cambridge.
12. Kozlov, E.V., Popova, N.A., Ivanov, Yu.F., Ignatenko, L.N., Koneva, N.A., Pekarskaya, E.E. (1996) Structure and sources of long-range stress fields in ultrafine-grained copper, *Annales de Chimie. Science des materiaux* **21**, 427–442
13. Valiev, R.Z. (1995) Approach to nanostructured solids through the studies of submicron grained polycrystals, *Nanostructured Materials* **6**, 73–82.
14. Koneva, N.A., Kozlov, E.V., Trishkina, L.I., Pekarskaya, E.E. (1997) Thermodynamics of substructure transformations under plastic deformation of metals and alloys, *Mater. Sci. And Eng.* **A243–236**, 614–616.
15. Nazarov, A.A., Romanov, A.E., Valiev, R.Z. (1994) On the nature of high internal stresses in ultrafine grained materials, *Nanostructured materials* **4**, 93–101.
16. Gryaznov, V.G., Trusov, L.I. (1993) Size effects in micromechanics of nanocrystals, *Progress in Mater. Sci.* **37**, 289–401.

CRYSTAL LATTICE DISTORSIONS IN ULTRAFINE-GRAINED METALS PRODUCED BY SEVERE PLASTIC DEFORMATION

A.N. TYUMENTSEV, M.V. TRETIAK[1], A.D. KOROTAEV[1], YU.P. PINZHIN[1], R.Z. VALIEV[2], R.K. ISLAMGALIEV[2], A.V. KORZNIKOV[2]
Institute of Strength Physics and Materials Technology, RAS, 2/1 Akademichesky av., Tomsk 634050, Russia
[1]*Siberian Physicotechnical Institute, 1 Novosobornaja Sq., 634050 Tomsk, Russia*
[2]*Institute of Physics of Advanced Materials Ufa State Aviation Technical University, 12, K. Marks str., Ufa 450000, Russia*

1. Introduction

Much evidence [1–3] suggests that the unusual properties of ultrafine-grained (UFG) metallic materials are determined not only by the small grain size, but also by the high defective concentration in the grain boundaries. The grain-boundary defects in this case may be not only dislocations, but partial disclinations as well [4]. Direct structural analysis [5] has shown a high density of these defects in the grain boundaries of UFG nickel produced by equal-channel angular (ECA) pressing. In the grain bulk, high-energy structural states with high continuous misorientations (high curvature of the lattice) have been detected as well. This paper deals with revealing via comparative electron microscopy the above structural states in distinct UFG materials (copper, nickel, and the complexly doped intermetallic compound Ni–18at.% Al–8at.% Cr–1at.% Zr–0.15at.% B) produced by various methods of severe plastic deformation (SPD).

2. Investigation techniques

SPD was accomplished by ECA pressing (Ni) and by torsion at a high quasi-hydrostatic pressure (Cu, Ni$_3$Al). ECA pressing was performed at room temperature. A test specimen was passed many times at a high pressure through two channels of diameter 20 mm intersecting at an angle of 90^0. Subjected to deformation by torsion (at a pressure of 40 tons) were disk specimens of thickness 0.2 mm and diameter 10 mm. The features of the defect substructure in UFG copper and in Ni$_3$Al were investigated at a number of disk revolutions N = 5 and 1/2, 1, 3, and 5, respectively.

T.C. Lowe and R.Z. Valiev (eds.), Investigations and Applications of Severe Plastic Deformation, 127–132.
© 2000 *Kluwer Academic Publishers. Printed in the Netherlands.*

3. Results

3.1. SUBSTRUCTURES WITH CONTINUOUS MISORIENTATIONS

Structural states with continuous misorientations (Figure 1) have been revealed in all studied materials and deformation conditions.

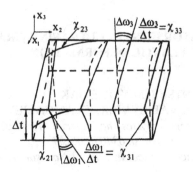

Figure 1. Sketch of misorientations in a structural state with a high curvature of the lattice.

It has been shown [5] that the curvature parameters of the lattice for such states can conveniently be analyzed using the bending-torsion tensor χ_{ij} [6]. In doing so, however, quantitative data about the horizontal components of the lattice curvature can be obtained only for the planes normal to the foil plane (χ_{31} in Figure 1) and are insensitive to the processes inherent in thin foils subjected to uncontrolled bending. We have obtained such data by measuring the horizontal components of the angular dimensions of effective reflections ($\Delta\gamma_h$) whose vectors were normal to the projection of the inclination axis of the goniometer. For reflections like these the values of $\Delta\gamma_h$ correspond to the ranges of inclination angles of a specimen in the goniometer for which the microregion under observation shows only slight changes in the intensity of the dark-field electron microscopy contrast. For this case we have $\chi_{31} = (\Delta\gamma_h - \Delta\gamma_0)/\Delta t$, where $\Delta\gamma_0 \approx (0.5 - 1)^0$ is the angular dimension of the diffraction maximum in a defectless crystal and Δt is the thickness of the foil. The measurements performed have shown that the highest values of the lattice curvature are observed for the Ni_3Al-based alloy. For this alloy, even after torsion with $N = 1/2$, probable values of $\Delta\gamma_h$ range up to $(4 - 8)^0$. This corresponds to $\chi_{31} \approx (20 - 40)$ deg/μm for $\Delta t \approx 0.15$ μm. With that, the density of dislocations of like sign geometrically necessary for such a curvature to appear is $\rho_\pm = \rho_+ - \rho_- \approx \chi_{31}/|b| \approx (1.5 - 3) \times 10^{11}$ cm^{-2}. For nickel and copper, the lattice curvature parameters are substantially lower, being, respectively, $\chi_{ij} \le 20$ deg/μ ($\rho_\pm \le 1.5 \times 10^{11}$ cm^{-2}) and $\chi_{ij} \le 10$ deg/$\mu\mu$ ($\rho_\pm \le 7 \times 10^{10}$ cm^{-2}).

As we have shown earlier [5, 7, 8], this difference results from the fact that these materials, in which individual dislocations have different mobilities, are dissimilar in the capacity for dislocation relaxation of intense local stresses inherent in substructures with high continuous misorientations. These stresses can be estimated by the formula $\sigma_{loc.} \approx (1 + v) \times \Delta h \times |\chi_{ij}| \times G$, where Δh is the characteristic dimension of the dislocation charge zone, v is the Poisson coefficient, and G is the shear modulus. For $\Delta h \approx 0.1$ μm and χ_{ij} varying from 10 (for copper) to 40 deg/μm (for Ni_3Al) $\sigma_{loc.}$ increases from $(1 + v)$ G/60 to $(1 + v)$ G/15, approaching the theoretical strength of crystal for the high-strength intermetallic compound.

3.2. PILES UP OF PARTIAL DISCLINATIONS IN GRAIN BOUNDARIES

A high curvature of the lattice of a material is responsible for a critically important feature of the defect structure of the grain boundaries. The matter is that for the lattice orientation continuously varying along the submicrograin boundaries these boundaries (except in the particular case that the components of the bending-torsion tensor on both sides of a boundary are equal in magnitude) are boundaries with continuously varying misorientation vectors. This is exemplified by a fragment of UFG nickel containing a boundary of this type, shown in Figure 2. A structural state with a high curvature of the lattice ($\chi_{21} \approx 20$ deg/μm) has been detected here only in grain I. This state is illustrated in Figure 2 by the extinction contour $g = [111]$ (shown by arrows) continuously shifting along a grain boundary.

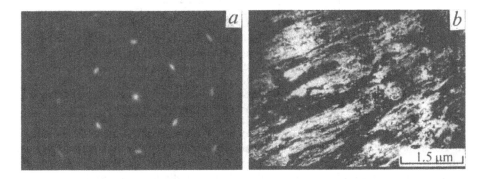

Figure 2. Electron microscopic contrast changed with the inclination angle of a UFG nickel specimen in a goniometer. Bright-field image. $\varphi = 9^0$ (a) and 6^0 (b).

As can be inferred from Figure 3, since no substructure with continuous misorientations has been detected in grain II, the change of the misorientation vector in direction 2 is entirely determined by the component χ_{21} of the bending-torsion tensor for grain I, and so we have $\partial\theta/\partial x_2 = \chi_{21} \approx 20$ deg/μm. In accordance with the theory of defects [6], to describe grain boundaries with a variable misorientation vector, it is advantageous to invoke the notion of partial disclinations. In doing this, the above fragment of a submicrograin boundary can be considered as a combination of a conventional (θ = Const) grain boundary and a pile up of partial disclination of like sign enclosed in this boundary. It seems that the parameter most convenient for characterization of the density of such grain-boundary disclinations is the gradient of the misorientation vector at the boundary, $\partial\theta/\partial r$.

All studied materials show a high density of boundaries with variable θ values, the effective density of grain-boundary disclinations in these materials being substantially

130

different. The latter circumstance is due to the fact that this quantity is directly related to the curvature parameters of the lattice. The highest values of $\partial\theta/\partial x_i$ ($\approx$ 30 deg/μm) have been found for the Ni$_3$Al alloy showing the highest values of χ_{ij}, while for nickel and copper we had $\partial\theta/\partial x_i \leq$ 20 and 10 deg/μm, respectively. Obviously, like for the lattice curvature (see above), the increase of the effective density of grain-boundary disclinations in the sequence Cu – Ni – Ni$_3$Al is due to the decrease of the efficiency of the dislocation relaxation of high-defect dislocation-disclination substructures in materials showing a lower mobility of dislocations.

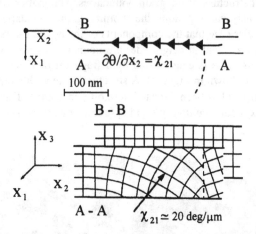

Figure 3. Sketch of the misorientations in the neighborhood of the grain boundary with a high density of continuously distributed partial disclinations, presented in Figure 2.

Thus, the structural state of UFG materials produced by SPD methods contain an unusually high defective densities. With that, a substructure with a high curvature of the lattice is formed in the grain bulk, and a high density of partial disclinations is observed at the grain boundaries. The simulation of partial disclinations by ensembles of dislocations is, as a rule, unproductive and often is impossible in principle. The latter primarily refers to the boundaries where the effective disclination density is high. Thus, for $\partial\theta/\partial x_i \approx$ 30 deg/μm the value of θ within the boundaries of submicrograins of size 0.3 μm may change by $\Delta\theta \sim 10^0$, and thus we shall have large-angle boundaries which cannot be described in terms of a dislocation model.

Moreover, partial disclinations are sources of high local internal stresses whose fields differ qualitatively from the fields generated by individual dislocations and by ensembles of dislocations in substructures with a highly curved lattice. Dislocation models are of little use for analysis of these stresses, while the theory of disclinations makes such an analysis substantially easier. Estimates [5] have shown, in particular, that the local internal stresses resulting from grain-boundary piles up of partial disclinations with the effective disclination density $\partial\theta/\partial x_i \approx$ 20 deg/μm may range from G/50 to G/15 away from the disclination nuclei and in their neighborhood, respectively. It is of importance that in the grain boundaries both wedge and screw components of partial disclinations show a high density. The former are known to be sources of high hydrostatic components of the stress tensor (and stress gradients) in the boundary plane, while the latter are responsible for high shear components of this tensor [9].

Thus, pile ups of partial disclinations give rise to very high local internal stresses in the near-boundary zone. A great dilation of the lattice in this zone may result in experimentally observed variations in atomic density, and fundamental physical properties of the crystals [1]. Diffusion over the nuclei of partial disclinations may be responsible for the substantial increase of the coefficients of grain-boundary diffusion in UFG materials [3]. The latter circumstance is known to be a factor activating the phenomena of "low-temperature" and "high-rate" superplasticity.

3.3. ANISOTROPY OF THE MICROSTRUCTURE OF UFG MATERIALS PRODUCED BY TORSIONAL DEFORMATION

One of the most important results of the dark-field analysis of misorientations is that in foils subjected to torsional deformation, if the foil planes are normal to the torsion axis, the horizontal (lying in the foil plane and normal to the torsion axis) misorientation components are much greater than the azimuthal ones. This refers to both the discrete misorientations and the horizontal components (χ_{31} and χ_{21} in Figure 1) of the lattice curvature. Thus, for the Ni_3Al alloy subjected to torsional deformation with N = 1/2 and 1, the (mostly continuous) azimuthal misorientations ($\Delta\alpha$) within the boundaries of the selector diaphragm are in most cases not over (8 - 10) deg. For a selector diaphragm of diameter ~ 1.5 μm this corresponds to the curvature azimuthal component χ_{23} (see Figure 1) or $\chi_{13} \approx (5 - 7)$ deg/μm. As shown above, the horizontal components of the lattice curvature for the given alloy reach about (20 - 40) deg/μm. For copper specimens the azimuthal spreads of the diffraction maxima ($\Delta\alpha$) in the majority of submicrograins are within the limits of their experimental detection at maximum values of $\Delta\alpha \approx 3$ deg, which corresponds to $\chi_{i3(i\neq3)} \leq 2$ deg/μm. The horizontal components (χ_{31} and χ_{21}) of the bending curvature of the lattice reach ~ 10 deg/μm for a considerable number of submicrograins.

The total (discrete plus continuous) horizontal misorientations are even greater: for the Ni_3Al-base alloy they reach $(20 - 30)^0$ on typical scales of the order of the foil thickness (0.2 μm). Thus, the above total values of horizontal misorientations are over the respective azimuthal components by no less than an order of magnitude.

The revealed high anisotropy of the field of rotations is related to the anisotropy of the field of displacements and displacement gradients in torsion. This can be realized qualitatively if we represent the field of rotations as the asymmetric part of the distortion tensor, namely, by vectors ω with components $\omega_1 = 1/2 \ (\partial u_3/\partial x_2 - \partial u_2/\partial x_3)$, $\omega_2 = 1/2 \ (\partial u_1/\partial x_3 - \partial u_3/\partial x_1)$ and $\omega_3 = 1/2 \ (\partial u_2/\partial x_1 - \partial u_1/\partial x_2)$. For the coordinate system of Figure 1 we have ω_1 and ω_2 are the horizontal components of the vectors ω on the x_1 and x_2 axes, respectively, lying in the specimen (foil) plane, and ω_3 are their azimuthal component. For the torsion about the x_3 axis with the displacements along this axis $u_3 = 0$, the modules of the vectors ω_1 and ω_2 are equal in magnitude to the shear components of the stress tensor $\gamma_{23} = 1/2 \ (\partial u_3/\partial x_2 + \partial u_2/\partial x_3)$ and $\gamma_{13} = 1/2 \ (\partial u_1/\partial x_3 + \partial u_3/\partial x_1)$, respectively. In view of the radial symmetry of the shear on torsion , we have $\partial u_2/\partial x_3 = \partial u_1/\partial x_3$ and $\gamma_{23} = \gamma_{13} = 2\pi NR/h$ (with N being the number of disk revolutions, h the disk thickness, and R the distance from the disk center), and the magnitudes of the horizontal components of the vector of rotations, ω_h, are independent of their directions in the torsion plane. We then have $\omega_h = \omega_1 = \omega_2$, which is substantially (R/h times) greater than the maximum value of $\omega_3 = 2\pi N$.

Another important feature of the grain structure of copper and of the Ni$_3$Al alloy subjected to deformation by torsion is that the reduction in grain size is more pronounced along the axis of torsion. Electron microscopy of cross sections normal to the plane of torsion has shown that this feature is related to the microshear bands of width ≤ 0.1 μm developing predominantly in the maximum shear stress directions lying in the plane of torsion.

4. Summary

A characteristic feature of the UFG materials produced by SPD methods is the presence of structural states with a high (tens of deg/μm) curvature of the lattice and of grain boundaries with variable misorientation vectors. The defect structure of such boundaries can be described in terms of pile ups of continuously distributed partial disclinations being sources of high local internal stresses in the neighborhood of the grain boundaries.

The high anisotropy of the fields of displacements and displacement gradients, characteristic of deformation by torsion, is responsible for considerable structural and crystallographic anisotropy of the resulting UFG states. This is apparent in the fact that the grain size reduction is more pronounced along the axis of torsion and in the high anisotropy of the field of rotations which are maximum in the directions normal to this axis.

5. References

1. Valiev, R.Z., Korznikov, A.V., and Mulyukov, R.R. (1992) The structure and properties of submicrocrystalline-structure metallic materials, *Fizika Metallov i Metallovedenie* 73, 373–384.
2. Valiev, R.Z, Mulyukov, R.R., Ovchinnikov, V.V., *et al.* (1990) The physical width of intercrystallite boundaries, *Metallofizika* 12, No. 5, 124–126.
3. Kolobov, Yu.R., Grabovetskaya, G.P., Ratochka, I.V., *et al.* (1996) Effect of grain-boundary diffusion fluxes of cooper on the acceleration of creep in submicrocrystalline nickel, *Ann. Chim. Fr.* 21, Nos. 6–7, 483–493.
4. Nazarov, A.A., Romanov, A.E., Valiev, R.Z. (1996) Random disclination ensembles in ultrafine-grained materials produced by severe plastic deformation, *Scripta Materialia*, 34, No. 5, 729–734.
5. Tyumentsev, A.N., Pinzhin, Yu.P., Korotaev, A.D. *et al.* (1998) Electron-microscopic investigation of grain boundaries in ultrafine-grained nickel produced by severe plastic deformation, *Physics of Metals and Metallography* 86, 110–120.
6. Likhachev, V.A., Volkov, A.E. and Shudegov, V.E. (1986) *The Continual Theory of Defects*, Leningrad University Publishers, Leningrad.
7. Korotaev, A.D., Tyumentsev, A.N. and Sukhovarov, V.F. (1989) *Dispersion Hardening of Refractory Metals*, Nauka, Novosibirsk.
8. Korotaev, A.D., Tyumentsev, A.N. and Pinzhin, Yu.P. (1998) Activation of the mesolevel plastic flow in high-strength materials and characteristic types of defective substructures, *Physical Mesomechanics*, 1, No. 1, 21–32.
9. Likhachev, V.A. and Khairov, R.Yu. (1975) *Introduction into the Theory of Disclinations*, Leningrad University Publishers, Leningrad.

GRAIN AND SUBGRAIN SIZE-DISTRIBUTION AND DISLOCATION DENSITIES IN SEVERELY DEFORMED COPPER DETERMINED BY A NEW PROCEDURE OF X-RAY LINE PROFILE ANALYSIS

T. UNGÁR, I. ALEXANDROV*, P. HANÁK

*Department of General Physics, Eötvös University Budapest
Pázmány Péter sétány 1/A, H-1518, P.O.B 32, Budapest, Hungary
*Inst.of Physics of Advanced Materials, Ufa State Aviation Technical
University, 12, K. Marks str., Ufa, 450000, Russia*

Abstract

The size-distribution of grain size and subgrain size and the dislocation density were determined in submicron grain size copper by a new procedure of X-ray line profile analysis based on the dislocation model of strain anisotropy. Taking into account the anisotropic dislocation contrast factors in the *modified* Williamson-Hall plots of the FWHM and the integral breadths and in the *modified* Warren-Averbach procedure enables to determine three stable size parameters, D, d and L_0. The width and the mean size, σ and m, of a log-normal size distribution is obtained from these size parameters in a simple least squares procedure. The grain or subgrain size of the copper specimen was reduced well below 100 nm by equal channel angular pressing (ECA) in the bulk. It has been found that with increasing the number of ECA passes the width of the size distribution of grains or subgrains becomes narrower, however, the average dislocation density decreases. The data indicate as if 40 to 50 nm would be a lower size barrier which cannot be overcome by applying ECA at room temperature alone.

1. Introduction

During the evolution of plastic deformation through subsequent stages the macroscopic grains fragment. First into dislocation cells with low or no misorientation followed by the formation of cell-blocks where the blocks are separated by larger misorientation angles. Further deformation reduces the cell-block size which, at the end of stage IV reduces to the size of dislocation cells producing submicron grain-size material. The size distribution of subgrains or submicron grains can either be determined by transmission electron microscopy (TEM) or by X-ray line profile analysis. The success of the X-ray method depends on the separation of size and strain effecting line profiles simultaneously. Strain anisotropy, which means that neither the FWHM nor the integral breadth nor the Fourier coefficients in the Warren-Averbach plot are monotonous functions of the diffraction vector or its square, g or g^2 [1-4], respectively, makes the separation of size and strain difficult. Recently is has been shown that the anisotropic contrast effect of dislocations can well account for strain anisotropy and the classical

133

T.C. Lowe and R.Z. Valiev (eds.), Investigations and Applications of Severe Plastic Deformation, 133–138.
© *2000 Kluwer Academic Publishers. Printed in the Netherlands.*

procedures have been *modified* in order to separate size and strain effects in a straightforward manner [5]. The dislocation contrast factors have been shown to be simple functions of the *hkl* indices if the specimen are more or less texture free polycrystals or all possible slip systems in the crystal are more or less equally populated [6]. A comprehensive scheme has been worked out for the average dislocation contrast factors enabling the evaluation of strain anisotropy in terms of dislocation densities, dislocation character, *i.e.* edge and screw components in the cubic crystal system [7]. Three stable size parameters, D, d and L_0 can be determined form the *modified* Williamson-Hall plots of the FWHM and the integral breadths and from the *modified* Warren-Averbach procedure, respectively. The width and the mean size, σ and m, of a log-normal size distribution, $f(x)$, can be obtained from these size parameters in a simple least squares procedure [8]. In the present work the above procedures have been applied to bulk copper deformed by ECA at room temperature.

2. Experimental

2.1. SPECIMEN

Polycrystalline rods of 80 mm length and 20 mm diameter were prepared by extrusion from 99.98% copper of about a few micrometer initial grain size. Deformation was carried out by the method of ECA, *cf.* [9], producing submicron average grain size. Altogether six specimen were investigated which are numbered from 1 to 6. No. 1 is the initial extruded state, No. 2 has been compressed by 20%. No. 3 and 4 were deformed by 2 ECA passes and No. 4 was additionally compressed by 20%. No. 5 and 6 were deformed by 12 ECA passes and No. 6 was additionally compressed by 20%. Additional compression after the ECA passes was carried out in order to introduce well defined directional deformation in order to determine long-range-internal stresses related either to dislocation cell walls or to submicron grain boundaries. In order to avoid machining effects an approximately 100 μm surface layer was removed form the specimen surface by chemical etching before the X-ray experiments.

2.2. X-RAY DIFFRACTION TECHNIQUE

The diffraction profiles were measured by a special double crystal diffractometer with negligible instrumental broadening [5, 10]. A fine focus rotating cobalt anode, Nonius FR 591, was operated as a line focus at 35 kV and 50 mA. The symmetrical 220 reflection of a Ge monochromator was used in order to have wavelength compensation in the lower angle region. The $K\alpha_2$ component of the Co radiation was eliminated by an 0.16 mm slit between the source and the Ge crystal. By curving the Ge crystal sagittally in the plane perpendicular to the plane-of-incidence the brilliance of the diffractometer was increased by a factor of 5. The profiles were registered by a linear position sensitive gas flow detector, OED 50 Braun, Munich. The instrumental line broadening was kept below 5% of the physical broadening by setting the distance between specimen and detector to 0.5 m. For avoiding air scattering and absorption this distance was overbridged by an evacuated tube closed by mylar windows.

2.3. EVALUATION OF DIFFRACTION PROFILES

The FWHM and the integral breadths of the diffraction profiles were evaluated by the *modified* Williamson-Hall plot [1]:

$$\Delta K \cong 0.9/D + (\pi M^2 b^2/2)^{1/2} \rho^{1/2} K \overline{C}^{1/2} + O(K^2 \overline{C}), \qquad (1)$$

where $K=2\sin\theta/\lambda$, $\Delta K=2\cos\theta(\Delta\theta)/\lambda$, θ and λ are the diffraction angle and the wavelength of X-rays. D, ρ and b are the average particle size, the average dislocation density and the modulus of the Burgers vector of dislocations. M is a constant depending on the effective outer cut-off radius of dislocations. M can only be obtained from the tails of the profiles [11] therefore it will be determined form the Fourier coefficients as shown below. (Note that in the classical Williamson-Hall plot [1] ΔK is plotted *vs.* K as shown in Fig. 1.) In a recent work it was shown that the average dislocation contrast factors can be written as [6]:

$$\overline{C} = \overline{C}_{h00} (1-qH^2), \qquad (2)$$

where $\overline{C}_{h00}$ is the average dislocation contrast factor for the *h00* reflections (in cubic crystals), q is a parameter depending on the elastic constants of the crystal and the character of dislocations in the crystal [7] and

$$H^2 = (h^2k^2+h^2l^2+k^2l^2)/(h^2+k^2+l^2)^2 . \qquad (3)$$

Inserting eqs. (2) and (3) into (1) yields:

$$\Delta K \cong \alpha + \beta K^2 \overline{C}_{h00}(1-qH^2), \qquad (4)$$

where $\alpha=(0.9/D)^2$ and $\beta=\pi M^2 b^2\rho/2$. Equation (4) can be solved by the method of least squares for q and α. The values of $\overline{C}_{h00}$ can be obtained from the tables in [7], in the case of copper $\overline{C}_{h00}$=0.3065. On the basis of Fig. 2 in [7] the experimental values of q provide information about the screw, edge or mixed character of dislocations. The dislocation density can be obtained by the *modified* Warren-Averbach method [5]:

$$ln A(L) \cong ln A^S(L) - \rho B L^2 ln(R_e/L) (K^2\overline{C}) + O(K^4\overline{C}^2), \qquad (5)$$

where $A(L)$ is the real part of the Fourier coefficients, A^S is the size Fourier coefficient as defined by Warren [2], $B=\pi b^2/2$, R_e is the effective outer cut-off radius of dislocations and O stands for higher order terms in $K^2\overline{C}$. L is the Fourier length defined as [2]: $L=na_3$, where $a_3=\lambda/2(\sin\theta_2-\sin\theta_1)$, n are integers starting from zero and $(\theta_2-\theta_1)$ is the angular range of the measured diffraction profile. The average particle size corresponding to the Fourier coefficients is denoted by L_0. It is obtained from the size Fourier coefficients, A^S, by taking the intercept of the initial slope at A^S=0 [2].

136

2.4. SIZE DISTRIBUTION

According to equations (4) and (5) three different particle sizes can be obtained: i) D from the FWHM, ii) d from the integral breadths and iii) L_0 from the Fourier coefficients. These three gauges sample the largest, the medium and the smallest lengths in the specimen since they are related to the central part, the integral and the outermost tails of the profiles, respectively, as it was shown earlier [8,12]. In a recent work it has been shown that assuming a log-normal size distribution, $f(x)$, the two parameters, the width and the mean value of the size, σ and m, respectively, can be obtained from the experimental values of D, d and L_0 by a simple least squares procedure [8]. A brief summary of the procedure is given in the paper by Ungár in the present proceedings.

3. Results and discussion

A typical classical Williamson-Hall plot of the FWHM (in 1/nm units) is shown in Fig. 1 for specimen No. 3 deformed by 2 ECA passes. A strong strain anisotropy can be well observed. The values of q were determined by the method of least squares and are listed for all investigated specimen in Table 1. The average contrast factors were determined according to eq. (2) by using the experimental q values and $\overline{C}_{h00}=0.3065$. In reference [7] it has been shown that the value of $\overline{C}_{h00}$ is almost the same for edge and screw dislocations and the average is 0.3065. With the $\overline{C}$ values determined in this way the *modified* Williamson-Hall plots according to eqs. (1) and (4) have been constructed. A typical *modified* Williamson-Hall plot of the FWHM and the integral breadths is shown in Fig. 2 for the specimen deformed by 2 ECA passes. The classical Warren-Averbach plot reveals the same kind of strain anisotropy as the classical Williamson-Hall plots and the *modified* Warren-Averbach plot becomes well-behaved in a similar manner as the *modified* Williamson-Hall plots in Fig. 2 The particle sizes D, d and L_0 were determined for all investigated specimen and the two parameters, σ and m of a log-normal size distribution $f(x)$ were derived as described in paragraph 2.4 and are listed in Table 1.

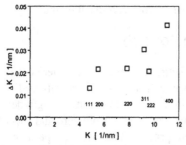
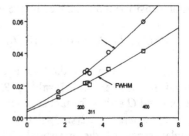

Figure 1. Classical Williamson-Hall plot of the FWHM in 1/nm units of specimen No. 3 deformed by 2 ECA passes. The Miller indices of the reflections are also indicated.

Figure 2. Modified Williamson-Hall plot of the FWHM and the integral breadths in 1/nm units of specimen No. 3 deformed by 2 ECA passes. The Miller indices of the reflections are also indicated.

TABLE 1. The average dislocation densities, the experimental values of q and the two parameters, σ and m of the log-normal size distribution function $f(x)$.

No.	Specimen	$\rho\ [10^{14}\,m^{-2}]$	q	σ	m [nm]
1	Initial state (I.S.)	4.4	2.1	0.62	169
2	I.S. + 20% compression	7.2	1.7	0.43	62
3	2ECA passes	11.1	2.1	0.81	79
4	2ECA p. + 20% comp.	9.4	1.95	0.74	92
5	12ECA passes	4.3	1.75	0.37	65
6	12ECA p. + 20% comp.	12.8	1.7	0.61	110

The values of q for screw or edge dislocations in copper are 2.37 or 1.68, respectively [7]. The random mixed character means that $q=2.02$. An experimental value lower or higher than this value indicates more edge or more screw character, respectively. As it can be seen from Table 1. the specimens extruded plus compressed, deformed by 12 ECA passes and compressed after 12 ECA passes, i.e. No. 2, 5 and 6, respectively, contain dislocations with more edge than screw character. The other three specimen contain random mixed character dislocations. The dislocation densities and the mean values of grain or subgrain size have been related to each other in Figs. 3 and 4. The dislocation density is shown *vs. m* in Fig. 3. The present data indicate that with decreasing grain or subgrain size the dislocation density decreases to extremely low values when the extent of deformation becomes large indicating dynamical recovery. At the same time, there seems to be a lower barrier of fragmentation by plastic deformation under the present experimental conditions. The dimensionless quantity $m\rho^{1/2}$ is plotted *vs. m* in Fig. 4. It seems to decrease to zero when the grain or subgrain size decreases to about 50 nm, indicating the exhaustion of dislocations as m decreases.

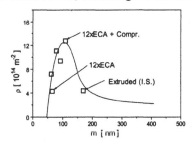

Figure 3. The dislocation density *vs. m*. In three particular cases the datapoints are labelled, the other data can be identified from Table 1.

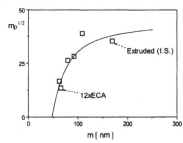

Figure 4. The dimensionless quantity $m\rho^{1/2}$ *vs. m*. In a few cases the datapoints are labelled, the other data can be identified from Table 1.

138

4. Acknowledgements

T. U. and P. H. are grateful for the support of the Hungarian National Science Foundation, OTKA, Grant No. T 022968 and T 022976. Thanks are due to F. Székely for his assistance in carrying out the X-ray diffraction experiments.

5. References

1. Williamson, G. K. and Hall, W. H., (1953) X-Ray Line broadening from filed aluminium and wolfram, *Acta metall.*, 1, 22-31.
2. Warren, B. E. (1959) X-ray studies of deformed metals, *Progr. Metal Phys.* 8, 147-202.
3. Stephens, P. W., (1998) Phenomenological Model of Anisotropic Peak Broadening in Powder Diffraction, *J. Appl. Cryst.* 32, 281-289.
4. Reimann, K. and Wuerschum, R. (1997) Distribution of internal strains in nanocrystalline Pd studied by X-ray diffraction, *J. Appl. Phys.* 81, 7186-7192.
5. Ungár, T. and Borbély, A. (1996) The effect of dislocation contrast on X-ray line broadening: a new approach to line profile analysis, *Appl. Phys. Lett.* 69, 3173-3175.
6. Ungár, T. and Tichy, Gy., (1999) The effect of dislocation contrasts on X-ray line profiles in untextured polycrystals, *Phys. Stat. Sol.* 171, 425-434.
7. Ungár, T., Dragomir, I., Révész, Á. and Borbély, A., (1999) The Contrast Factors of Dislocations in Cubic Crystals: the Dislocation Model of Strain Anisotropy in Practice *J. Appl. Cryst.*, in press.
8. Ungár, T., Borbély, A., Goren-Muginstein, G. R., Berger, S. and Rosen, A. R. (1999) Particle-size, size distribution and dislocations in nanocrystalline Tungsten-Carbide, *Nanostructured Mater.*, 11, 103-113.
9. Valiev, R. Z., Kozlov, E. V., Ivanov, Yu. F., Lian, J., Nazarov, A. A. and Baudelet, B., (1994) Deformation Behaviour of Ultra-Fine-Grained Copper, *Acta Metall. Mater.* 42, 2467-2474.
10. Wilkens, M. and Eckert, H., (1964) Roentgenographische Untersuchungen ueber die Versetzungsanordnung in Plastisch Verformten Kupfereinkristallen, *Z. Naturforschung*, 19a, 459-470.
11. Wilkens, M., (1970) The determination of density and distribution of dislocations in deformed single crystals from broadened X-ray diffraction profiles, *Phys. Stat. Sol.* (a) 2, 359-370.
12. Ungár, T., Ott, S., Sanders, P. G., Borbély, A. and Weertman, J. R., (1998) Dislocations, grain size and planar faults in nanostructured copper determined by high resolution x-ray diffraction and a new procedure of peak profile analysis, *Acta Mater.* 10, 3693-3699.
13. Guinier, A. (1963) *X-ray Diffraction*, Freeman, San Francisco, CA.

CALCULATION OF ENERGY INTENSITY AND TEMPERATURE OF MECHANOACTIVATION PROCESS IN PLANETARY BALL MILL BY COMPUTER SIMULATION

E.V. SHELEKHOV, V.V. TCHERDYNTSEV, L.YU. PUSTOV, S.D. KALOSHKIN, I.A. TOMILIN
Moscow State Steel and Alloys Institute, Leninsky prosp., 4, Moscow 117936, Russia.

Abstract

Mechanical alloying (MA) is one of the novel processes of alloy formation by severe plastic deformation of metallic components in high-energy ball mills. Analysis of the energy intensity and average temperature of the milling process in planetary ball mill was carried out by computer simulation. The dependences of energy dissipation and average temperature in a vial on the fill fraction of the vial by balls, on the elasticity of ball's collision and their friction coefficient are determined. The results of computer simulation were compared with ones calculated using the analytical formula. The obtained results allow one to choose optimal modes of ball milling with regard to the specific character of concrete tasks.

1. Introduction

MA process techniques of production of new materials with unique properties. The interest to MA process was caused by an opportunity to obtain nonequilibrium phases and strongly defective structures represents both scientific and practical importance. Severe plastic deformation of metallic components in high-energy ball mills can result the formation of amorphous structures, supersaturated solid solutions, new metastable intermetallic compounds. At the same time very often the MA process is an optimal way to produce precursor materials for subsequent consolidation.

Results of MA processes depend on mill design: geometric and dynamic parameters, character of motion of milling bodies *etc.* [1–4]. Energy dissipation was offered as a criterion for a comparison of results obtained under various conditions of MA [5]. Milling temperature is an important parameter influencing the MA process too [6–8]. It is necessary to know these two parameters to characterize the MA process quantitatively.

There are few studies on the experimental determination of these parameters [9–11]. Such procedures are applicable only to few types of mills. For example, the estimation of temperature of MA in the planetary mill can be carried out only indirectly [10]. Therefore computation of these parameters has gained wide acceptance [12–16].

T.C. Lowe and R.Z. Valiev (eds.), Investigations and Applications of Severe Plastic Deformation, 139–145.
© 2000 *Kluwer Academic Publishers. Printed in the Netherlands.*

The aim of the present study is to analyze dependencies of energy dissipation and average temperature of planetary ball mill on various parameters of MA process, such as the fill fraction of a vial ξ, elasticity of ball collision η and their friction coefficient f.

2. Simulation Procedure

The theoretical basis of the model used here was described previously [15, 16]. Only its main concepts are presented here.

The balls' motion was considered within the limit of a single layer of a thickness equal to the ball's diameter. Such an approximation is valid in the case when all balls used are of the same diameter [16]. The layerwise character of the balls motion in a planetary mill was recorded photographically [17]. To determine a total dissipation of energy in the vial, it is necessary to multiply the measured energy dissipation by the number of layers in which the balls move.

The center of vial was accepted as the origin of coordinates. The trajectories of ball motion in the field of three forces – two centrifugal forces and Coriolis force – were computed. The gravity force was neglected due to the fact that the acceleration of milling bodies in the mills of this type is 25–60 times greater than free fall acceleration [18]. Other factors were taken into account, including the elastic coefficient of centric impact, determined as a portion of kinetic energy, which is conserved under the head-on impact and the coefficient of sliding friction, which is equal to a ratio between tangential and normal components of force of response during a short time of collision. The determination of energy dissipation was carried out by summation of energy losses of each ball per unit time at collisions with the vial wall or other balls [19].

To calculate the temperature in a vial, it was supposed that the heat exchange occurs only through the contact area of collision of the balls. Determination of contact area and time was carried out within the framework of the theory of elasticity. Heat exchange through the gas environment in the vial was neglected due to the fact that heat conductivity of gas is 3–4 orders less than that for the materials of the balls or vial.

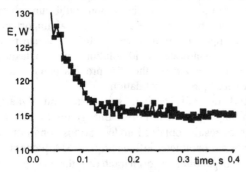

Figure 1. Process of the steady-state conditions reaching (ball's number is 20, ball size $D_b = 8.8$ mm, fill fraction $\xi = 45\%$, friction coefficient f = 0.9, elastic coefficient of the centric impact $\eta = 0.25$.

To determine an average temperature in the vial, the superposition of the solutions of uniform and nonuniform equations of thermal conductivity was employed [16].

On the basis of this model, a computer program for calculations was formulated. Calculation of the energy parameters was performed with a time step of 10^{-4} seconds.

The starting velocities of balls were taken as zero. On the initial stage of computation, the high values of energy dissipation, which are caused by irregular interactions of balls, were observed; but a short time later, when the motions of balls became ordered, the energy dissipation decreased and, than, became constant. Figure 1 illustrates the process of the steady-state conditions reaching. The calculated values of energy dissipation and temperature are accepted, if the successively computed values have a relative difference of no more than 0.1%.

For calculations, the following parameters of mill were accepted: radius of the planet carrier R = 50 mm; interior radius of the vial R_{vi} = 31 mm; exterior radius of vial R_{ve} = 37 mm; rotation speed of the planet carrier v_p = 20 s^{-1}; temperature of the cooling liquid (water) T_w = 6° C. All of these parameters correspond to a AGO–2U planetary ball mill.

3. Results and Discussion

3.1. DEPENDENCES OF MA ENERGY PARAMETERS ON THE FILL FRACTION

As it is known (see, for example, [14]), the character of ball motion depends on the degree of filling of a vial by balls, which is characterized by a fill fraction ξ. In the two-dimensional model ξ was established as a percent of the area, which is occupied by balls, from general area of vial. It was shown that, depending on filling factor ξ, processing materials can be subjected to actions of various types character of action: from weak action – abrasive wearing of powder, to a stronger action - strain by a collision [19]. In this case, an examination of dependence of energy dissipation E on the fill fraction is of great interest. A variation of ξ was carried out by increasing the number of balls at their

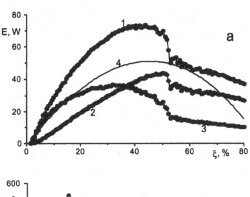

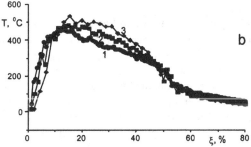

Figure 2. a - A plot of E vs. ξ varied by raising number of balls from 1 to 121 (ball diameter D_b = 4.8 mm, η = 0.25, f = 0.9): (1) - total value of E; (2) - E_n; (3) - E_t, (4) - E calculated by formula [12], b – A plot of T vs ξ varied by raising number balls from 1 to 121 (D_b: (1) 4.8, (2) 6.6, and (3) 8.8 mm. η = 0.25, f = 0.9

constant size.

Figure 2a (curve 1) shows a plot of E vs. ξ. It represents a curve with a maximum at a fill fraction of 45–48%. At $\xi \approx 52\%$, an abrupt decrease in the E was observed. This decrease corresponds to the rise of the «spread» effect This effect consist in the formation of close-packed layer of practically fixed balls on the vial wall [19]. Dependence of E on ξ by the analytical equation, which was proposed in [12], is presented in Figure 2a, curve 4. These calculated results agree satisfactorily with our data. However, the effect of «spread» was not considered in [12]; that is why the curve is continuous.

It is possible to assume a difference between the contributions into the energy dissipation, which are made of frontal shocks (normal component of energy dissipation E_n and tangential shocks (tangential component of energy dissipation E_τ). The obtained dependencies (Figure 2a, curves 2 and 3) testify that at small fill fractions, the dissipation of energy occurs predominantly at the tangential shocks, and at large fill fractions, it occurs primarily at frontal shocks. The rise of the «spread» effect at $\xi \approx$ 52% results in abrupt decreases in both E_n and E_τ. A comparison of the contributions of the «ball-ball» and «ball-wall» interactions to the energy dissipation has shown the predominance of the first one above the second at all ξ.

In general, the obtained data correlate well with the calculated results [14]. The differences may be caused by the fact that summation of energy dissipation in [14] was performed after one revolution of a vial. The model used in the present study gives a more precise result, because it allows one to calculate the energy after the steady state conditions are reached.

From these results we can conclude that it is possible to change a character of mechanical load influence on the processed material considerably by changing ξ. At small ξ, material will undergo abrading loadings, while at higher ξ, the normal deformation prevails.

The dependence of temperature on ξ (Figure 2b) is a curve with a maximum at $\xi = 15\%$. Temperature in a vial depends only slightly on the ball's diameter, and it increases with increasing ball's diameter. In detail, the influence of the ball's size on the energy parameters was considered in [19].

Comparing the curves on Figures 2a and 2b, it is seen, that the maximums of T and E do not coincide. The maximum of temperature is located at $\xi = 15\%$, *i.e.* in the region where rather high values of energy dissipations are in optimal combination with rather low heat removal.

Thus, to reach the most non-equilibrium state of processed material in the minimum time, it is preferable to treat it at $\xi \sim 45\%$. This mode provides the maximum energy dissipation and rather low temperatures, *i.e.* this mode can prevent thermal decomposition of the obtained non-equilibrium structure.

3.2. DEPENDENCE OF MA ENERGY PARAMETERS ON THE FRICTION COEFFICIENT

A ball motions character and, correspondingly, energy parameters of MA process, depends strongly on the friction of ball against other balls and vial wall. The processes of the ball rolling over another ball at small ξ and ball's tearing off from a wall of a vial at medium values of ξ, which were observed in [19], may be associated only with a friction. So, the friction coefficient f must have a significant effect on energy parameters of milling. The magnitude of f depends both on the balls and vial materials and on the powder layer on the ball's surface. That is why the investigation of the friction coefficient influence on the energy parameters of MA is of significant interest.

Figure 3a shows a dependence of energy dissipation on the friction coefficient. At the absence of the friction (f = 0), the energy dissipation in the system is insignificant. At the f values up to 0.2, normal component of energy dissipation E_n practically absent, and increase of E is determined by an increase of the tangential component E_τ. At small

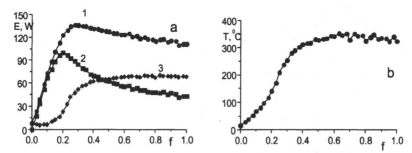

Figure 3. a - Plots of E vs. f: (1) - total value of E; (2) - E_n; (3) - E_τ. b - A plot of T vs. f Ball diameter D_b = 8,8 mm, η = 0.25, ξ = 0.45, number of ball 20.

f ball's tearing off from a vial wall are absent. At significant magnitudes of f ball can be "engaged" with the nearest balls, which facilitate it's tearing off from the vial wall. At f = 0.45, magnitudes of E_τ and E_n become equal. At the posterior increasing of f, the E_n vs. f curve is flattened out.

Dependence of medial temperature in a vial on a friction coefficient is shown on the Figure 3b. Initial increase of the temperature is due to the fact that the increase of f leads to an increase of energy dissipation and, correspondingly, to significant heating. This curve flattens out at f ≈ 0.6. Similarity of the T(f) and E_n(f) curves suggests that, at least at small and medial f, the temperature in the vial determined mainly by the normal collisions of balls.

3.3. DEPENDENCE OF MA ENERGY PARAMETERS ON THE ELASTIC COEFFICIENT OF THE CENTRIC IMPACT

Elastic coefficient of centric impact η is determined as a ratio between a height of free-falling ball kickback from any surface to the total height of its fall. This coefficient changes from 0 (absolutely inelastic collision) up to 1 (absolutely elastic collision). This coefficient depends on the elastic properties of balls material as well as on the powder layer on the ball's or vial's surface [20].

Figure 4 shows a dependence of E on η. This curve have a wide maximum at η = 0.2 – 0.5. At η close to zero, kinetic energy of ball dissipates completely at normal collision. That is why the contribution of E_n into general magnitude of energy dissipation is about 80% at the small η. Balls are moving in the compact aggregation. A value of E_n decreases with increasing of η.

At η > 0.5, the ball's motions become disordered, they are chaotically distributed throughout the vial. Energy dissipation for normal collisions at large η is insignificant. That is why a prevalence of E_r. at η > 0.5 was observed. At η close to 1, the mobility of balls is very high, and the number of tangential collisions decreases.

The observed dependence of the character of ball motion on η is in a good agreement with the supposition that it is a transition from the cooperative to chaotic character of the ball motion with increase of η [21].

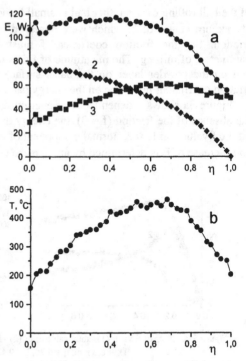

Figure 4. a - Plots of E vs. η: (1) - total value of E; (2) - E_n; (3) - E_r. b - a plot of T vs. f ball diameter D_b = 8,8 mm, f = 0.9, ξ = 0.45, number of ball 20s

Figure 4b shows the dependence of T on η, which is a curve with a wide maximum at η = 0.5 – 0.7. At small η, the heat exchange between the balls and the vial wall affects the temperature strongly. Aggregation of the balls on a vial wall leads to its cooling.

4. Conclusions

The analysis performed allows one to conclude that by varying the filling factor, it is possible to change the character of action on the material from abrasion to shock strain. The "spread" effect is accompanied by an abrupt decay of energy dissipation and temperature. The coefficient of friction and elastic coefficient of centric impact affect the energy parameters strongly.

The obtained results allow one to choose optimal modes of MA processing.

5. References

1. Kurt, C. and Schultz, L. (1993) Phase formation and martensitic transformation in mechanically alloyed nanocrystalline Fe-Ni, *J. Appl. Phys.* **73**, 1975–1980.
2. Hong, L.B. and Fultz, B. (1996) Two-phase coexistence in Fe-Ni alloys synthesized by ball milling, *J. Appl. Phys.* **79**, 3946–3954.
3. Börner, I. and Eckert, J. (1997) Nanostructure formation and steady-state grain size of ball-milled iron powres, *Mater. Sci. Eng.* **226-228**, 541–545.
4. Liu, L. and Magini, M. (1997) Correlation between energy transfers and solid state reactions induced by mechanical alloying on the $Mo_{33}Si_{66}$ system, *J. Mater. Res.* **12**, 2281–2287.
5. Butyagin, P.Yu. (1987) Energy aspects of mechanochemistery, *Izv. SO AN SSSR, ser. khim. nauk* **5**, 48–59.
6. Koch, C.C., Pathak, D., and Yamada, K. (1993) Effect of milling temperature on structure of intermetalllic compounds after mechanical milling, *Pr. 2 Int. Conf. Str. Appl. Mech. All.,* Vancouver, pp. 205–212.
7. Xu, J., He, J.H., and Ma, E. (1997) Effect of milling temperature on mechanical alloying of the immiscible Cu-Ta system, *Metall. Mater. Trans.* **28**, 1569–1580.
8. Klassen, T., Herr, U., and Averback, R.S. (1997) Ball milling of systens with positive heat of mixing: effect of temperature in Ag-Cu, *Acta Mater.* **45**, 2921–2930.
9. Streletskii, A.N. (1993) Measurements and calculation of main parameters of powder mechanical treatment in different mills, *Pr. 2 Int. Conf. Str. Appl. Mech. All.,* Vancouver, pp. 51–58.
10. Gerasimov, K.B., Gusev, A.A., Kolpakov, V.V., and Ivanov, E.Yu. (1991) Temperature control during mechanical alloying in planetary ball mill, *Sib. khim. J.* **3**, 140–145.
11. Calka, A., Wexler, D., and Li, Z.L. (1997) Estimation of milling temperature from phase transformation under low energy milling conditions, *Proc. Int. Conf. RQ-9 (suppl.),* Elsevier, Bratislawa, pp. 191–194.
12. Burgio, N., Iasonna, A., and Magini, M. (1991) Mechanical Alloying of Fe-Zn system: the correlation between imput energy and end product, *Il Nuovo Cimento* **13D**, 459–476.
13. 13.Abdellaoui, M. and Gaffet, E. (1994) Mechanical alloying in a planetary ball mill: kinematic description, *J. de Phis. IV* **4**, 291–296.
14. Dallimore, M.P. and McCormick, P.G. (1997) Distinct element modelling of mechanical alloying in a planetary ball mill, *Mater. Sci. Forum* **235–238**, 5–14.
15. Shelekhov, E.V. and. Salimon, A.I. (1997) The computer simulation of planetary and vibratory mills, *Aerosol* **2**, 61–69.
16. Shelekhov, E.V. and Sviridiova, T.A. (in press) Simulation of ball motion and heating in planetary mill. Effect of milling conditions on the products of Ni and Nb powder mixture mechanoactivation, *Materialoved.*
17. Watanabe, R., Hashimoto, H., Lee, G.G. (1995) Computer simulation of milling ball motion in mechanical alloying, *Mater. Trans. JIM* **36**, 102–109.
18. Boldyrev, V.V., Pavlov, S.V., Poluboyarov, V.A., and Dushkin, A.V. (1995) Evaluation of the efficiency of various machines in mechanical activation, *Inorg. Mater.* **31**, 1035–1042.
19. Shelekhov, E.V., Tcherdyntsev, V.V., Pustov, L.Yu., Kaloshkin, S.D., and. Tomilin, I.A. (1999) Computer simulation of mechanoactivation process in the planetary ball mill: determination of the energy parameters of milling, *Proceedings of ISMANAM99*, to be published.
20. Huang, H., Dailimore, M.P., Pan, J., and McCormic, P.G. (1998) An investigation of the effect of powder on the impact characteristics between a ball and a plate using free falling experiments, *Mater. Sci. Eng.* **241**, 38–47.
21. Magini, M. (1992) The role of energy transfer in mechanical alloying powder processing, *Mater. Sci. Forum* **88–90**, 121–128.

III. MICROSTRUCTURE EVOLUTION DURING SEVERE PLASTIC DEFORMATION PROCESSING

MICROSTRUCTURAL EVOLUTION DURING PROCESSING BY SEVERE PLASTIC DEFORMATION

TERENCE G. LANGDON[1], MINORU FURUKAWA[2], ZENJI HORITA[3], MINORU NEMOTO[3]
[1]Departments of Materials Science and Mechanical Engineering
University of Southern California, Los Angeles, CA 90089-1453, USA
[2]Department of Technology, Fukuoka University of Education, Munakata
Fukuoka 811-4192, Japan
[3]Department of Materials Science and Engineering, Kyushu University
Fukuoka 812-8581, Japan

Abstract

The procedure of equal-channel angular (ECA) pressing may be used to subject a material to severe plastic deformation by pressing repetitively through a special die without any concomitant change in the cross-sectional dimensions of the sample. This processing method gives grain refinement in polycrystalline materials with as-pressed grain sizes typically in the submicrometer range. This paper describes the factors influencing the development of a homogeneous microstructure through ECA pressing including the influence of the pressing speed and the angle contained within the special die, the effect of rotating the sample between consecutive pressings and the significance of pressing to a high total strain. An experimental example is presented to illustrate the potential for using ECA pressing to achieve a superplastic forming capability at very rapid strain rates.

1. Introduction

Equal-channel angular (ECA) pressing is a procedure in which a material is subjected to a very high strain by simple shear but without any change in the cross-sectional area of the sample [1]. This processing technique has the capability of producing large bulk pore-free samples having ultrafine grain sizes which lie typically within the submicrometer range [2]. There is a potential for using ECA pressing both to strengthen a material at low temperatures through the Hall-Petch relationship and to achieve exceptional superplastic ductilities at high temperatures where diffusion-controlled processes become important. Although ECA pressing provides opportunities for improving the mechanical characteristics of materials, relatively little attention has been given to the various factors influencing the development of homogeneous microstructures. These factors are examined in this report.

149

T.C. Lowe and R.Z. Valiev (eds.), Investigations and Applications of Severe Plastic Deformation, 149–154.
© 2000 *Kluwer Academic Publishers. Printed in the Netherlands.*

2. The principles of ECA pressing

Figure 1 illustrates the principle of ECA pressing. The die contains two channels, equal in cross-section, which intersect at an angle near the center of the die. The test sample is machined to fit within the channels and it is pressed through the die with a plunger. Three planes, labelled x, y and z, may be defined within the pressed sample. Repetitive pressings of the same sample can be undertaken and the strain accrued on

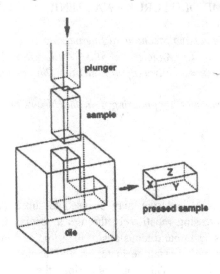

Figure 1. Principle of ECA pressing.

each passage through the die is dependent upon the angle Φ between the two channels and the angle Ψ at the outer arc of curvature where the two channels intersect [3]. It can be shown that if $\Phi = 90°$, so that the two channels intersect at right angles, the strain is close to ~1 for any value of Ψ.

When samples are pressed repetitively, it is possible to define four distinct processing routes, as illustrated in Fig. 2. In route A the sample is not rotated between consecutive pressings, in route B it is rotated by 90° between each pressing, where route B_A denotes rotations in alternate directions between each pressing and route B_C denotes rotations in the same direction between each pressing, and route C denotes a rotation of 180° between each pressing. The significance of these different routes and the shearing characteristics are discussed elsewhere [4].

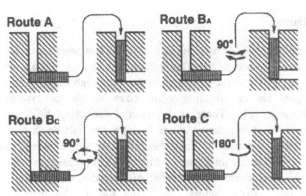

Figure 2. The four processing routes for ECA pressing.

3. Factors influencing microstructural evolution

High strains are introduced in ECA pressing by repetitively pressing the same sample through the die. The four different processing routes illustrated in Fig. 2 give rise to different shearing characteristics within the sample and this leads to different microstructures. Figure 3 shows the microstructures and associated selected area electron diffraction (SAED) patterns for samples of pure (99.99%) Al subjected to four pressings to a total strain of ~4, using a die with $\Phi = 90°$ and with the samples pressed using the four different processing routes: the SAED patterns were taken with a diameter of 12.3 μm, the initial grain size of the material was ~1.0 mm and the photomicrographs were taken on the x plane as defined in Fig. 1. Inspection shows the diffracted beams form net patterns after pressing via routes A, B_A and C so that the microstructures consist of subgrains separated by boundaries having low angles of misorientation, whereas the diffracted beams lie randomly around rings when using route B_C so that the grains in this microstructure are separated by boundaries having high angles of misorientation. In addition, it is apparent that the grains are essentially equiaxed and homogeneous in route B_C, with an average size of ~1.3 μm, whereas elongated grains are visible for the other three routes. Thus, route B_C is preferable for producing an equiaxed grain structure with high angle grain boundaries [5, 6].

The angle Φ subtended by the two channels within the die is critical in determining the total strain accrued in the sample on each separate passage through the die. The

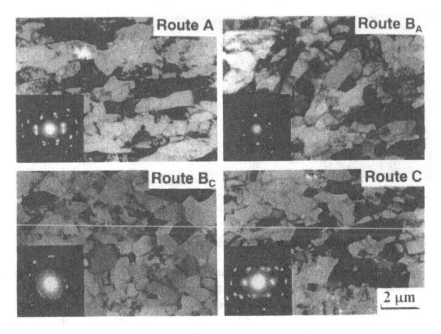

Figure 3. Microstructures and SAED patterns in pure Al after 4 pressings to a strain of ~4 using processing routes A, B_A, B_C and C.

results in Fig. 3 relate to pure Al after 4 pressings with $\Phi = 90°$ and Fig. 4 shows similar observations on the x plane for the same material pressed using route B_C but with internal angles of 112.5°, 135° and 157.5°, respectively. Since less strain is introduced when Φ is increased, the numbers of pressings were adjusted to give strains of the order of ~4 for each sample: namely, 6, 9 and 19 pressings for 112.5°, 135° and 157.5°, respectively. Inspection shows the SAED patterns in Fig. 4 exhibit a transition from the formation of rings at 112.5° to well-defined net patterns at the two larger angles. Thus, high angle grain boundaries are achieved most readily when Φ is near 90° [7].

By contrast, the pressing speed has a relatively minor influence except only that more time is available at slower speeds to attain an equilibrated microstructure. This is illustrated in Fig. 5 for samples of pure Al pressed through 4 passes in route B_C with $\Phi = 90°$ and pressing speeds of 8.5×10^{-3} mm s^{-1} and 7.6 mm s^{-1}, respectively [8].

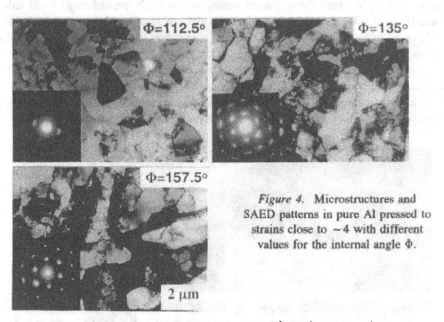

Figure 4. Microstructures and SAED patterns in pure Al pressed to strains close to ~4 with different values for the internal angle Φ.

Figure 5. Microstructures for pressing speeds of 8.5×10^{-3} mm s^{-1} and 7.6 mm s^{-1}.

4. Achieving superplasticity at very rapid strain rates

Although ECA pressing may give substantial grain refinement, these ultrafine grains are not necessarily stable at the high temperatures required for superplasticity. Figure 6 shows a plot of the measured grain size versus the annealing temperature for an Al-3% Mg alloy and an Al-3% Mg-0.2% Sc alloy where both materials were subjected to 8 pressings through route B_C using a die with $\Phi = 90°$ and with the samples subsequently annealed at each temperature for 1 hour [9]. A comparison of these two curves shows the ultrafine grain size is not stable in the Al-Mg solid solution alloy at temperatures above ~500 K whereas the Al-Mg-Sc alloy retains a grain size below ~1 μm up to annealing temperatures close to 700 K. The grain stability in the Al-Mg-Sc alloy is due to the presence of coherent Al₃Sc precipitates.

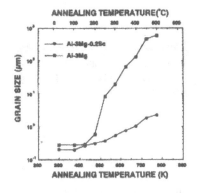

Figure 6. Grain size versus annealing temperature for two different alloys.

If the grains remain small at high temperatures, it is possible to achieve high tensile ductilities at very rapid strain rates [10-12]. An experimental example is shown in Fig. 7 for the Al-3% Mg-0.2% Sc alloy pressed through 8 passes with a die having Φ = 90°. The pressing was performed using routes A, B$_C$ and C and specimens were tested in tension at 673 K using a strain rate of 3.3 × 10⁻² s⁻¹. It is apparent that exceptional ductility is attained in this material when processing using route B$_C$ with an elongation to failure in excess of 2000%. This suggests the possibility of using the ECA pressing procedure to develop a superplastic forming capability at high strain rates.

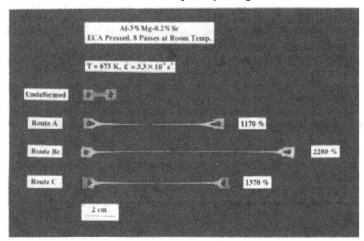

Figure 7. Exceptional ductility in an Al-Mg-Sc alloy after ECA pressing.

5. Summary and conclusions

1. Equal-channel angular (ECA) pressing gives substantial grain refinement but the nature of the microstructure depends upon the precise pressing conditions.
2. The development of a homogeneous microstructure of equiaxed grains separated by high angle boundaries requires a high angle between the two channels and rotation of

the sample by 90° in the same direction between each pass (route B_C). The pressing speed is relatively unimportant.

3. If the ultrafine grains introduced by ECA pressing are stable at high temperatures, it is possible to achieve superplasticity at very rapid strain rates.

6. Acknowledgements

This work was supported in part by the Light Metals Educational Foundation of Japan, in part by a Grant-in-Aid for Scientific Research from the Ministry of Education, Science, Sports and Culture of Japan and in part by the National Science Foundation of the United States under Grants No. DMR-9625969 and INT-9602919.

7. References

1. Segal, V.M., Reznikov, V.I., Drobyshevskiy, A.E. and Kopylov, V.I. (1981) Plastic working of metals by simple shear, *Russian Metall.* 1, 99-105.
2. Valiev, R.Z., Krasilnikov, N.A. and Tsenev, N.K. (1991) Plastic deformation of alloys with submicron-grained structure, *Mater. Sci. Eng.* A137, 35-40.
3. Iwahashi, Y., Wang, J., Horita, Z., Nemoto, M. and Langdon, T.G. (1996) Principle of equal-channel angular pressing for the processing of ultra-fine grained materials, *Scripta Mater.* 35, 143-46.
4. Furukawa, M., Iwahashi, Y., Horita, Z., Nemoto, M. and Langdon, T.G. (1998) The shearing characteristics associated with equal-channel angular pressing, *Mater. Sci. Eng.* A257, 328-32.
5. Iwahashi, Y., Horita, Z., Nemoto, M. and Langdon, T.G. (1998) The process of grain refinement in equal-channel angular pressing, *Acta Mater.* 46, 3317-31.
6. Oh-ishi, K., Horita, Z., Furukawa, M., Nemoto, M. and Langdon, T.G. (1998) Optimizing the rotation conditions for grain refinement in equal-channel angular pressing, *Metall. Mater. Trans.* 29A, 2011-13.
7. Nakashima, K., Horita, Z., Nemoto, M. and Langdon, T.G. (1998) Influence of channel angle on the development of ultrafine grains in equal-channel angular pressing, *Acta Mater.* 46, 1589-99.
8. Berbon, P.B., Furukawa, M., Horita, Z., Nemoto, M. and Langdon, T.G., (1999) Influence of pressing speed on microstructural development in equal-channel angular pressing, *Metall. Mater.Trans.* 30A, 1989-97.
9. Berbon, P.B., Komura, S., Utsunomiya, A., Horita, Z., Furukawa, M., Nemoto, M. and Langdon, T.G. (1999) An evaluation of superplasticity in aluminum-scandium alloys processed by equal-channel angular pressing, *Mater. Trans. JIM* 8, 772-78.
10. Valiev, R.Z., Salimonenko, D.A., Tsenev, N.K., Berbon, P.B. and Langdon, T.G. (1997) Observations of high strain rate superplasticity in commercial aluminum alloys with ultrafine grain sizes, *Scripta Mater.* 37, 1945-50.
11. Berbon, P.B., Furukawa, M., Horita, Z., Nemoto, M., Tsenev, N.K., Valiev, R.Z. and Langdon, T.G. (1998) Requirements for achieving high-strain-rate superplasticity in cast aluminium alloys, *Phil. Mag. Lett.* 78, 313-18.
12. Komura, S., Berbon, P.B., Furukawa, M., Horita, Z., Nemoto, M. and Langdon, T.G. (1998) High strain rate superplasticity in an Al-Mg alloy containing scandium, *Scripta Mater.* 38, 1851-56.

CHARACTERIZATION OF ULTRAFINE-GRAINED STRUCTURES PRODUCED BY SEVERE PLASTIC DEFORMATION

Z. HORITA[1], M. FURUKAWA[1], M. NEMOTO[1], R. Z. VALIEV[3], T. G. LANGDON[4]
[1]Department of Materials Science and Engineering
Faculty of Engineering, Kyushu University
Fukuoka 812-8581, Japan
[2]Department of Technology, Fukuoka University of Education
Munakata, Fukuoka 811-4192, Japan
[3]Institute of Physics of Advanced Materials
Ufa State Aviation Technical University
Ufa 450000, Russia.
[4]Departments of Materials Science and Mechanical Engineering
University of Southern California
Los Angeles, CA 90089-1453, U.S.A.

Abstract

Grain refinement of metallic materials may be achieved by imposing severe plastic deformation through procedures such as equal-channel angular pressing or high pressure torsion straining. The grain sizes produced by these techniques are generally below 1 μm and it is therefore necessary to make use of advanced analytical procedures in order to characterize the fine-grained microstructures. This report describes the application of high-resolution electron microscopy (HREM) to atomic-scale observations of the ultrafine-grained structures produced in Cu, Ni and alloys of Ni-Al-Cr and Al-Mg by torsion straining. It is demonstrated that the grain boundaries in these materials are in a high-energy nonequilibrium condition, with irregular arrangements of facets and steps at the interfaces. The thermal stability of these ultrafine-grained microstructures is examined by annealing the samples at elevated temperatures.

1. Introduction

It is now well established that procedures such as equal-channel angular (ECA) pressing and high pressure torsion (HPT) straining may be used to achieve substantial grain refinement in polycrystalline materials having large grain sizes, with the grain sizes reduced to the submicrometer or the nanometer level [1]. These procedures have a significant advantage over the standard methods for preparing materials with ultrafine grain sizes because they are capable of producing large bulk samples without the presence of any residual porosity. There have been several recent reports documenting

155

the evolution of the ultrafine microstructure produced in samples of high purity aluminum using ECA pressing [2-4].

The severe plastic deformation imposed during ECA pressing and HPT straining introduces a high dislocation density into the deformed samples and this leads to arrays of grains which are highly deformed and having grain boundaries which tend to be poorly delineated [5,6]. Thus, the microstructures produced using these processing methods are generally difficult to characterize using standard analytical methods and it is necessary to make use of alternative procedures such as high-resolution electron microscopy (HREM).

Several earlier reports described the application of HREM to characterize the microstructures produced in samples of pure Cu and pure Ni [7] and in alloys of Al-Mg [8,9] and Ni-Al-Cr [10] after subjecting samples to torsion straining. This report summarizes some of the important features of these observations.

2. Experimental materials and procedures

Detailed descriptions of the materials and the procedures were given in the earlier reports [7-10]. Briefly, the samples consisted of high purity (99.98%) Cu, high purity

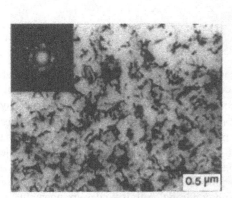

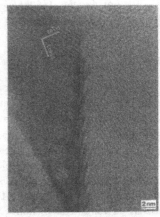

Figure 1. Low magnification image of Al-3% Mg alloy.

Figure 2 (right). Example of grain boundary faceting in Al-3% Mg alloy.

(99.99%) Ni, an Al-3% Mg solid solution alloy and a Ni-Al-Cr alloy containing, in at. %, 75.3% Ni, 16.5% Al and 8.2% Cr. The grain sizes of these four materials in the undeformed conditions were ~50 μm, ~50 μm, ~500 μm and ~60 μm, respectively. All of these materials were subjected to HPT straining by preparing small disks, with diameters of ~15 mm and thicknesses of ~0.3 mm, and then subjecting these disks to torsion straining at room temperature to a total strain of ~7 under an imposed pressure of the order of ~5 GPa. Following torsion straining, the disks were reduced by polishing to thicknesses of ~150-250 μm and then small disks with diameters of 3 mm were punched out. These discs were thinned to perforation and then examined using transmission electron microscopy (TEM) or high-resolution electron microscopy (HREM). For HREM, lattice images were recorded at close to the optimum defocus condition. To check on the thermal stability of the ultrafine-grained structures produced by torsion straining, some of these small disks were also annealed for 1 hour in an argon atmosphere at selected high temperatures.

3. Experimental results and discussion

Torsion straining leads to a very substantial refinement in the microstructure of polycrystalline materials. This is illustrated by the low magnification TEM image shown in Fig. 1 for the Al-3% Mg alloy, where the selected area electron diffraction (SAED) pattern was taken from a region having a diameter of 1.9 μm. Inspection shows that the SAED pattern exhibits rings so that the selected field of view contains many small grains having multiple orientations. It is also apparent that many of the grain boundaries are poorly defined and the visible boundaries tend to be curved or wavy. Measurements indicated that the grain size of the Al-3% Mg alloy in this condition, immediately following torsion straining, was of the order of ~0.09 μm. Thus, there has been a reduction in the grain size through severe plastic deformation by more than two orders of magnitude. It is also apparent that the contrast within the individual grains tends to be non-uniform. All of these observations suggest, therefore, that the microstructure produced by torsion straining is in a non-equilibrium condition.

The use of HREM provides an opportunity to observe the nature of this non-equilibrium structure in more detail. Figure 2 shows an example of a high resolution image of a grain boundary in the Al-3% Mg alloy after torsion straining. This grain boundary contains a regular array of facets which lie parallel to the (100) direction and with about 10 layers of (111) between each of the steps. Careful measurements suggested that the facet density for this boundary was ~5 x 10^8 m^{-1}. The microstructures shown in Figs 1 and 2 are representative after torsion straining.

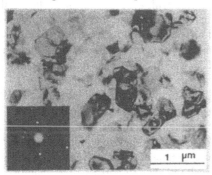

Figure 3. Low magnifica-tion image of Al-3% Mg alloy after annealing for 1 hour at 448 K.

The preceding results demonstrate the potential for achieving very substantial grain refinement using the torsion straining procedure. Similar grain refinements may be achieved also using the ECA pressing procedure. Experiments have shown that the equilibrium grain size introduced by ECA pressing is dependent upon the rate of recovery in the material. This means, for example, that the addition of magnesium. to an aluminum matrix in solid solution decreases the rate of recovery and leads to smaller equilibrium grain sizes by comparison with pure Al. Thus, the equilibrium grain sizes were recorded as ~1.3, ~0.45 and ~0.27 μm in pure Al, Al-1% Mg and Al-3% Mg after ECA pressing but the total strains required to establish reasonably homogeneous microstructures in these three materials increased from ~4 for pure Al to ~6 and ~8 for the Al-1% Mg and

158

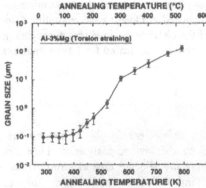

Figure 4. Variation of grain size with annealing temperature for the Al-3% Mg alloy.

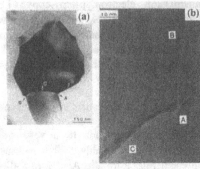

Figure 5. (a) Grain in Al-3% Mg after annealing, (b) enlargement of the triple point at A and grain boundaries B and C.

Al-3% Mg alloys, respectively [11]. Since a low stacking fault energy also leads to a low rate of recovery, it is possible to achieve a small stable grain size of ~0.27 μm in pure Cu by ECA pressing [12].

Although it is possible to achieve ultrafine grain sizes through severe plastic deformation, using either HPT straining or ECA pressing as the procedures for introducing very high strains into the materials, these very small grains may not be stable at high temperatures. The problem of grain growth becomes especially important if an attempt is made to achieve superplastic deformation in these fine-grained materials, since superplasticity is a diffusion-controlled process requiring a reasonably high temperature so that diffusion occurs fairly rapidly [13].

There have been several investigations of the thermal stability of the ultrafine grain sizes introduced by these processing techniques [14-16]. Figure 3 gives an example of the effect of annealing the Al-3% Mg alloy for 1 hour at a temperature of 448 K together with an SAED pattern taken from a region with a 1.3 μm diameter: this is the same alloy as in Fig. 1 and the ultrafine grain size was produced by HPT straining. A comparison with Fig. 1 shows two important differences. First, the grains have grown to an average size of ~0.3 μm as a result of the annealing treatment. Second, the grain boundaries are now essentially straight and well-defined. Thus, it is reasonable to conclude that the structure has evolved into a more equilibrated configuration.

Samples of the Al-3% Mg alloy were subjected to severe plastic deformation in HPT straining and then annealed for 1 hour at selected temperatures up to a maximum of 773 K. Figure 4 shows the variation of the average grain size with the annealing temperature. Thus, although HPT straining introduces a grain size within the nanometer range, these very small grain sizes are not stable at temperatures above ~400 K and the grains grow rapidly when annealed at temperatures above ~450 K. The inability to retain an ultrafine-grained structure at high temperatures is due to the absence of any precipitates that serve to restrict the grain boundary mobility. By contrast, it has been shown that an ultrafine grain size may be retained in the Al-3% Mg alloy at high temperatures by incorporating a dilute amount of scandium into the lattice, thereby introducing a distribution of fine coherent Al_3Sc particles which strengthen the matrix and serve also to inhibit the onset of recrystallization [17]. A similar retention of a very small grain size at high temperatures has been achieved also in an Al-5.5% Mg alloy containing 2.2% Li and 0.12% Zr [15].

Figure 5(a) shows an example, at a higher magnification, of a grain in the Al-3% Mg alloy after annealing for 1 hour at 443 K: it is apparent that the boundaries in this field of view are well-defined and they lie in arcs between the triple points. Some of the boundaries in Fig. 5(a) are inclined to the surface normal but the inclination angles remain essentially constant along the lengths of the boundaries. Thus, these boundaries are in a more equilibrated configuration than the boundaries visible in Figs 1 and 2 but the presence of the boundaries in the form of arcs rather than straight segments suggests that the equilibrium condition has not been fully achieved. The region around the triple point labeled A in Fig. 5(a) is shown at a higher magnification in Fig. 5(b) including the grain boundaries marked B and C. Inspection shows that boundary B is very smooth and straight but boundary C has a zigzag configuration along the entire length of the boundary. Results of this type demonstrate the need for the use of HREM in any evaluation of the grain boundary structure in these ultrafine-grained materials.

Torsion straining was conducted using a Ni-Cr-Al alloy with a composition, in at. %, of 75.3% Ni, 16.5% Al and 8.2% Cr. The initial grain size of this material was ~60 μm and the material was torsion strained and then annealed for 1 hour at selected temperatures up to a maximum of 1073 K. Figure 6 shows the microstructures after annealing at temperatures of (a) 473 K, (b) 673 K, (c) 873 K and (d) 1073 K, respectively, together with the SAED patterns taken from regions of 1.3 μm for (a) to (c) and 2.5 μm for (d).

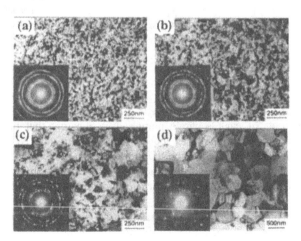

Figure 6. Microstructures and SAED patterns for a Ni-Al-Cr alloy after annealing for 1 hour at temperatures of (a) 473 K, (b) 673 K, (c) 873 K and (d) 1073 K.

Exceptionally small grains, of the order of ~34 nm, were achieved in the Ni-Al-Cr alloy by HPT straining. It is apparent from Fig. 6 that these grains gradually increase in size with increasing annealing temperature, although the rate of grain growth is very slow by comparison with the Al-3% Mg alloy shown in Fig. 4. Thus, at the highest temperature of 1073 K, the grain boundaries are well-defined and the average grain size remains below 1 μm. At the three highest temperatures shown in Fig. 6, the SAED patterns contain rings or spots which correspond to the [100] or [110] superlattice reflections from a Ni_3Al ordered phase, but this reflection is absent in Fig. 6(a) after annealing at 473 K and it was absent also in the as-strained condition.

The conclusion from these observations is that the Ni_3Al-based ordered phase is present prior to HPT straining but this ordered phase is lost during the intense plastic straining introduced by torsion straining at room temperature and the ordered phase only reforms again during an ordering of the structure which occurs at temperatures of the order of 673 K and above. The change in appearance of the SAED patterns is illustrated in Fig. 7 for (a) the as-strained alloy and after annealing at (b) 473 K, (c) 673 K, (d) 873 K and (e) 1073 K, respectively.

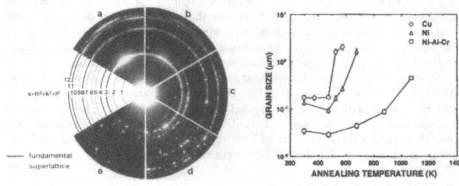

Figure 7. Appearance of SAED patterns for Ni-Al-Cr alloy (a) after straining, and after annealing at (b) 473 K, (c) 673 K, (d) 873 K and (e) 1073 K: shown on the left are rings for the fundamental and superlattice diffractions.

Figure 8. Variation of grain size with annealing temperature for Cu, Ni and the Ni-Al-Cr alloy.

Figure 8 shows the variation of the average grain size with the annealing temperature for pure Cu, pure Ni and the Ni-Al-Cr alloy, where the error bars denote the 95% confidence limit for each separate point. It is apparent that the grains grow significantly in pure Cu at temperatures above ~470 K and in pure Ni above ~500 K, whereas the growth is relatively slow and the grains remain below ~100 nm in the Ni-Al-Cr alloy even at temperatures as high as ~900 K. This stability of the ultrafine grains in the Ni-Al-Cr alloy is due to the formation of the Ni_3Al-based ordered phase with an $L1_2$-type structure which occurs at annealing temperatures in the vicinity of ~650 - 700 K. By contrast, the grains grow easily at the higher temperatures in the two pure metals in the absence of precipitates or an ordered structure.

4. Summary and conclusions

(1) Torsion straining and ECA pressing provide the capability of producing very substantial grain refinement in metals and metallic alloys. Examples are presented for several materials including for a Ni-Al-Cr alloy where the as-strained grain size is ~34 nm.

(2) Through the use of high-resolution electron microscopy, it is shown that the grain morphology in the as-strained materials is in a non-equilibrium condition. The ultrafine grains grow rapidly at high temperatures except when precipitates or ordered phases are present to inhibit boundary movement.

5. Acknowledgements

This work was supported in part by the Light Metals Educational Foundation of Japan, in part by a Grant-in-Aid for Scientific Research from the Ministry of Education, Science, Sports and Culture of Japan, in part by the National Science Foundation of the United States under Grants No. DMR-9625969 and INT-9602919 and in part by the U.S. Army Research Office under Grants 68171-98-M-5642 and DAAH04-96-1-0332.

6. References

1. Valiev, R.Z., Krasilnikov, N.A. and Tsenev, N.K. (1991) Plastic deformation of alloys with submicron-grained structure, *Mater. Sci. Eng.* **A137**, 35-40.
2. Iwahashi, Y., Horita, Z., Nemoto, M. and Langdon, T.G. (1997) An investigation of microstructural evolution during equal-channel angular pressing, *Acta Mater.* **45**, 4733-41.
3. Iwahashi, Y., Horita, Z., Nemoto, M. and Langdon, T.G. (1998) The process of grain refinement in equal-channel angular pressing, *Acta Mater.* **46**, 3317-31.
4. Iwahashi, Y., Furukawa, M., Horita, Z., Nemoto, M. and Langdon, T.G. (1998) Microstructural characteristics of ultrafine-grained aluminum produced using equal-channel angular pressing, *Metall. Mater. Trans.* **29A**, 2245-52.
5. Valiev, R.Z., Korznikov, A.V. and Mulyukov, R.R. (1993) Structure and properties of ultra-fine grained materials produced by severe plastic deformation, *Mater. Sci. Eng.* **A168**, 141-48.
6. Wang, J., Horita, Z., Furukawa, M., Nemoto, M., Tsenev, N.K., Valiev, R.Z., Ma, Y. and Langdon, T.G. (1993) An investigation of ductility and microstructural evolution in an Al-3% Mg alloy with submicron grain size, *J. Mater. Res.* **8**, 2810-18.
7. Horita, Z., Smith, D.J., Nemoto, M., Valiev, R.Z. and Langdon, T.G. (1998) Observations of grain boundary structure in submicrometer-grained Cu and Ni using high-resolution electron microscopy, *J. Mater. Res.* **13**, 448-50.
8. Horita, Z., Smith, D.J., Furukawa, M., Nemoto, M., Valiev, R.Z. and Langdon, T.G. (1996) An investigation of grain boundaries in submicrometer-grained Al-Mg solid solution alloys using high-resolution electron microscopy, *J. Mater. Res.* **11**, 1880-90.
9. Horita, Z., Smith, D.J., Furukawa, M., Nemoto, M., Valiev, R.Z. and Langdon, T.G. (1996) Evolution of grain boundary structure in submicrometer-grained Al-Mg alloy, *Mater. Character.* **37**, 285-94.
10. Oh-ishi, K., Horita, Z., Smith, D.J., Valiev, R.Z., Nemoto, M. and Langdon, T.G. (1999) Fabrication and thermal stability of a nanocrystalline Ni-Al-Cr alloy: comparison with pure Cu and Ni, *J. Mater. Res.* (submitted for publication).
11. Iwahashi, Y., Horita, Z., Nemoto, M. and Langdon, T.G. (1998) Factors influencing the equilibrium grain size in equal-channel angular pressing: role of Mg additions to aluminum, *Metall. Mater. Trans.* **29A**, 2503-10.
12. Komura, S., Horita, Z., Nemoto, M. and Langdon, T.G. (1999) Influence of stacking fault energy on microstructural development in equal-channel angular pressing, *J. Mater. Res.* (submitted for publication).
13. Berbon, P.B., Furukawa, M., Horita, Z., Nemoto, M., Tsenev, N.K., Valiev, R.Z. and Langdon T.G. (1998) Requirements for achieving high-strain-rate superplasticity in cast aluminum alloys, *Phil. Mag. Lett.* **78**, 313-18.
14. Wang, J., Furukawa, M., Horita, Z., Nemoto, M., Valiev, R.Z. and Langdon, T.G. (1996) Enhanced grain growth in an Al-Mg alloy with ultrafine grain size, *Mater. Sci. Eng.* **A216**, 41-46.

162

15. Furukawa, M., Iwahashi, Y., Horita, Z., Nemoto, M., Tsenev, N.K., Valiev, R.Z. and Langdon, T.G. (1997) Structural evolution and the Hall-Petch relationship in an Al-Mg-Li-Zr alloy with ultra-fine grain size, *Acta Mater.* **45**, 4751-57.
16. Hasegawa, H., Komura, S., Utsunomiya, A., Horita, Z., Furukawa, M., Nemoto, M. and Langdon, T.G. (1999) Thermal stability of ultrafine-grained aluminum in the presence of Mg and Zr additions, *Mater. Sci. Eng.* **A265**, 188-96.
17. Berbon, P.B., Komura, S., Utsunomiya, A., Horita, Z., Furukawa, M., Nemoto, M. and Langdon, T.G. (1999) An evaluation of superplasticity in aluminum-scandium alloys processed by equal-channel angular pressing, *Mater. Trans. JIM* (in press).

FRAGMENTATION IN LARGE STRAIN COLD ROLLED ALUMINIUM AS OBSERVED BY SYNCHROTRON X-RAY BRAGG PEAK PROFILE ANALYSIS (SXPA), ELECTRON BACK SCATTER PATTERNING (EBSP) AND TRANSMISSION ELECTRON MICROSCOPY (TEM)

E. SCHAFLER[1], M. ZEHETBAUER[1], P. HANAK[3], T. UNGAR[3],
T. HEBESBERGER[2], R. PIPPAN[2], B. MINGLER[1],
H.P. KARNTHALER[1], H. AMENITSCH[4], S. BERNSTORFF[5]
[1]Institute of Materials Physics, University Vienna, A-1090 Vienna
Strudlhofgasse 4
[2]Institute for Material Science, Austrian Academy of Science, A-8700
Leoben
[3]Institute for General Physics, Eötvös University Budapest, H-1518 P.B,
Hungary
[4]Institute of Biophysics and X-ray Structure Research, Austria Academy
of Sciences, A-8020 Graz; [5] Sincrotrone ELETTRA, Basovizza, I-34012
Trieste

1. Introduction

In the last decade investigations of plastic deformation have focussed on the large strain ranges, *i.e.* stage IV and V of deformation [1-5]. Although several models have been developed to explain the work hardening behavior in these stages [2, 6-8], the existing experimental findings are not sufficient to identify the real microstructural processes governing stage IV and V hardening. This situation arises mainly from two facts: (i) traditional methods were not convenient to measure dislocation densities, local internal stresses and misorientation of the substructure, and (ii) most of microstructural investigations were done without relation to the specific mechanical properties. Two new methods, X-ray Bragg Profile Analysis (XPA) [9-12] using a rotating anode and/or synchrotron radiation, and Electron Back Scatter Patterning (EBSP) [13, 14], are effectively for studying microstructures induced by large strains. The XPA-method implemented with a rotating anode generator allows investigation of microstructural evolution within one or more grains by using a focal spot size of several tenth of a mm [15]. Using synchrotron radiation with intensities up to 10^{12} photons/mm/s allows a reduction in the footprint of the beam on the sample to an order of a few tens of microns. This allows the investigation of the microstructural evolution within a single grain [16, 17]. The EBSP method evaluates the Kikuchi-line pattern from back scattered electrons in a Scanning Electron Microscope (SEM) with a spatial resolution down to 0.5µm. This is done by computer support in a very efficient way so that a very large number of different lattice sites can be studied in short time.

The aim of the present work is to study large deformation substructures in Al (representing a fcc metal with a high stacking fault energy) applying three techniques:

T.C. Lowe and R.Z. Valiev (eds.), Investigations and Applications of Severe Plastic Deformation, 163–171.
© 2000 *Kluwer Academic Publishers. Printed in the Netherlands.*

SXPA, EBSP and TEM, anticipating stage-specific information on the evolution of dislocation density and arrangement, of local internal stresses, and of size and orientation of tilted lattice areas, during large strain plastic deformation.

2. Experimental

2.1. SAMPLES

Rectangular blocks (15 x 15 x 2.2-15 mm) were cut from a Al 99.99% material by spark erosion. The samples were annealed for 6 h at 650 K in high vacuum in order to obtain a relatively large grain size of about 400 µm. After this treatment the specimen were cold-rolled at room temperature to true strains $\varepsilon = \ln(d_0/d)$ between 0.11 and 2.07 in steps $\Delta\varepsilon = 0.1$ (d_0 and d denote the initial and the final sample thickness, respectively). Before the experiments the surface layer being 50 µm in thickness was removed by etching. For the investigation by the EBSP-method the samples were electro-chemical-polished. The preparation of the TEM samples was done by electrolytic thinning using a Tenupol 2 (70% CH_4O + 30% HNO_3, 35V, at 243K). TEM was the final investigation method and was performed on the same samples, which were used for SXPA and EBSP measurements.

2.2. METHODS

For the Synchrotron Bragg Peak Profile Analysis (SXPA) experiments the 8keV radiation at the SAXS beamline at ELETTRA Trieste was used [18]. It was possible to reduce the beam size to 40 x 50 μm^2. The diffraction profiles of the {200} Bragg reflection was recorded by a position sensitive gas flow detector (PSD-50M, Braun, Munich) with a linear spatial resolution of 80µm. The measurements were performed in steps of about 50µm from grain boundary to grain boundary [16, 17]. From the resulting profile the formal dislocation density ρ^*, the dislocation arrangement parameter M and the long range internal stresses were evaluated [11, 20, 21].

The basis of the Electron Back Scatter Patterning (EBSP) equipment was a Leica Stereoscan 440 scanning electron microscope with a crystal orientation system (COS) for recording the pattern. Orientation image microscopy (OIM) managed the external beam control and the display of the data.

For the Transmission Electron Microscopy (TEM) investigations both a Philips CM 300 and a CM 30 were used with maximum acceleration voltages of 200 kV and 300 kV, respectively. For the analysis of the microstructure bright-field images of large areas were taken.

3. Results and Discussion

Fig. 1 shows the long range internal stresses as a function of ε, with the averaged values of each deformation degree. After an increase up to about $\varepsilon \approx 0.4$ – the transition range of stage III and IV – there is a slight decrease followed by reincrease at higher deformation. This wavy behavior confirms the transformation of the cell walls from PDW (polarized dipolar wall, consisting of dislocation dipoles) [22] to PTW (polarized tilt wall, similar to subboundary) structure, as found by quantitative contrast analysis of XPA [23, 24].

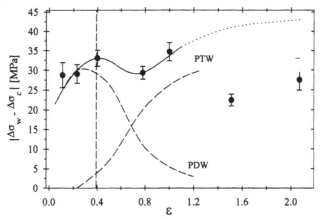

Figure 1: The internal stresses as function of strain. The dashed curves representate the supposed contribution of the different cell wall structures to the evolution of the internal stresses

The TEM image related to stage III in Fig. 2a, distinctly show large cells/cell blocks without misorientation representing the typical PDW-structure. Nearly dislocation free areas are surrounded by rough cell walls and the cells have same contrast, because of no misorientation. As shown in [16], the driving force for the transformation of the cell wall structure are the increasing internal stresses. Entering the transition range to stage IV the PDWs transform into dislocation arrays similar to sub-grain walls (PTWs), which account for the tilt of the lattice between adjacent cells and leads to a reduction of the internal stresses. These PTWs can comparably incorporate more dislocations than the PDWs during further deformation, because the internal stresses are lowered. Fig. 2b shows the cell structure at $\varepsilon = 0.78$, where the cell walls become sharper and exhibit a tilt between whole cell blocks. In Fig. 2c at $\varepsilon = 1$ the propagation of PTWs has expanded and led to an further fragmentation of the cell blocks. This evolution can also be observed in the line-scans of EBSP-measurements. At small deformation (Fig. 3a) the misorientation is only up to about 3° between the measured points. At higher deformation the fluctuations of the misorientation angle increase and also values of the misorientation angle up to about 6° at $\varepsilon=0.78$ (Fig. 3b) and up to 10° at $\varepsilon =1$ (Fig. 3c).

Fig. 1 also shows a reduction of the internal stresses for deformations over $\varepsilon =1$. This is in fact a consequence of the iterative character of rolling deformation, because of the occurrence of recovery processes in between single rolling passes. This has been shown in previous work [25] and for rolled Ni [17]. The effect starts in stage V and even increases with further deformation.

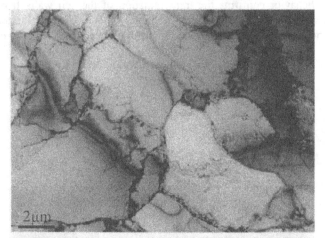

Figure 2a.
Transmission electron microscope image of Aluminium cold-rolled to ε=0.11. Dislocation free large cells/cell blocks are surrounded by thick cell walls called polarized dipolar walls PDWs.

Figure 2b.
Transmission electron microscope image of Aluminium cold-rolled to ε=0.78. The cell blocks become smaller and the cell walls sharper. These walls are called polarized tilt walls (PTWs) and cause misorientation between adjacent cells/cell walls.

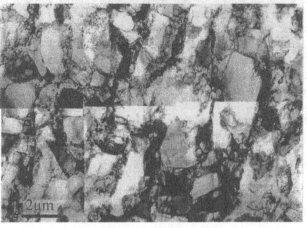

Figure 2c.
Transmission electron microscope image of Aluminium cold-rolled to ε = 1. The fragmentation of cell blocks has propagated, the regions with the same orientation are very small.

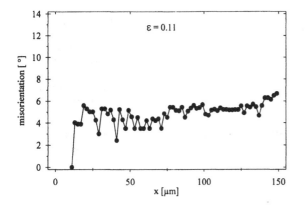

Figure 3a. EBSP line-scan across a grain of an Al sample cold-rolled to ε=0.11. The stepwidth was 2μm. The misorientation keeps rather low up to 3°. One can also see the distance of orientation changes as about 20μm.

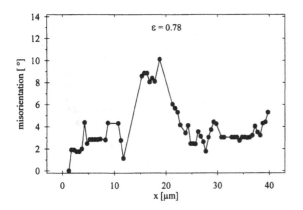

Figure 3b. EBSP line-scan across a grain of an Al sample cold-rolled to ε=0.78. The stepwidth was 0.5μm. The misorientation angles has increased up to about 6° and the distance of orientation changes are smaller than 5μm.

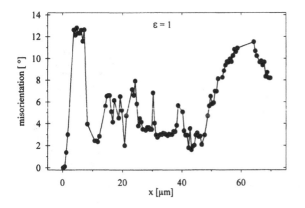

Figure 3c. EBSP line-scan across a grain of an Al sample cold-rolled to ε = 1. The stepwidth was 0.5μm. The misorientation angles again has increased up to about 10°. The distance of orientation changes are now about 5μm.

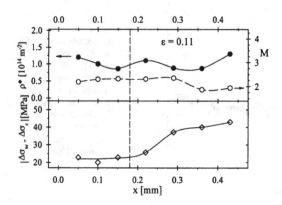

Figure 4a. Formal dislocation density ρ*, dislocation arrangement parameter M and local internal stresses |Δσ_w − Δσ_c| measured along a spatial grain scan in Aluminium cold rolled to ε = 0.11. The dashed lines indicate the location of grain boundaries.

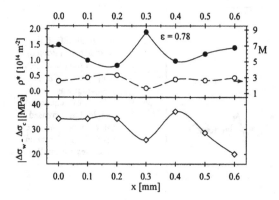

Figure 4b. Formal dislocation density ρ*, dislocation arrangement parameter M and local internal stresses |Δσ_w − Δσ_c| measured along a spatial grain scan in Aluminium cold rolled to ε = 0.78.

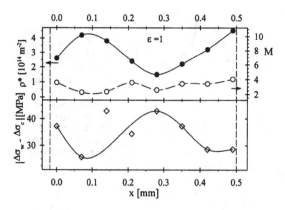

Figure 4c. Formal dislocation density ρ*, dislocation arrangement parameter M and local internal stresses |Δσ_w − Δσ_c| measured along a spatial grain scan in Aluminium cold rolled to ε = 1. The dashed lines indicate the location of grain boundaries.

Three typical grain scans of the formal dislocation density ρ^*, the arrangement parameter M and the long range internal stresses $|\Delta\sigma_w - \Delta\sigma_c|$, are shown in Fig. 4(a)-(c) at three deformation values: a small deformation in stages II-III, ε =0.11; a deformation in stage IV, ε = 0.78; and well in stage IV, ε =1.

In Fig. 4a it can be seen that all three parameters, formal dislocation density ρ^*, arrangement parameter M and long range internal stresses $|\Delta\sigma_w - \Delta\sigma_c|$ remain rather constant within the two single grains (the dashed lines represent the grain boundaries). At larger deformations the three quantities increasingly fluctuate and in Fig. 4b, ε = 0.78, at x=0.3 mm the value of ρ^* markedly exceeds the neighboring values, but $|\Delta\sigma_w - \Delta\sigma_c|$ and M are lowered. This fact indicates the formation of PTWs inside the PDW-structure: The PTW structure can store many more dislocations than the PDW since the long range internal stresses are quite low. Arriving at higher strains ε = 1 (stage V of deformation) the situation reverses as seen in Fig. 4c at x=0.28 mm, where a local minimum of ρ^* appears with a local maximum of $|\Delta\sigma_w - \Delta\sigma_c|$. These local values correspond to the last remainders of a few local PDW regions with relatively low dislocation density.

Due to the higher mobility of dislocation in Al, there is no apparent tendency for forming pile ups at grain boundaries as it is the case in Cu [16].

At higher strains corresponding to the decrease of the internal stresses in Fig. 1, correlations of ρ^* with $|\Delta\sigma_w - \Delta\sigma_c|$ typical of the PTW could not be found; rather, PDW-like correlations were present. This sudden absence of PTW formation at the highest strains can be ascribed to the static recovery processes becoming effective in stage V, as seen during rolling deformation of Al. If the rolling pass chosen is too small, the level of local internal stress reached is too low to launch a transformation from a PDW to a PTW structure.

4. Summary and Conclusions

Large strain cold worked polycrystalline Al has been investigated by X-ray diffraction profile analysis using Synchrotron radiation, Electron Back Scatter Patterning and Transmission Electron Microscopy. The results can be summarized as follows:

1. Similar to previous findings in polycrystalline Cu, Fe and Ni, Al exhibits a marked change of dislocation cell walls when the plastic deformation proceeds from stage III to stage IV. The cell wall transforms from a polarized dipolar dislocation arrangement to a polarized tilt one.

2. The microstructural evolution in cold working of Al at very large strains differs from that of Cu because of higher dislocation mobility, caused by (i) the higher stacking fault energy and (ii) the higher homologous deformation temperature.

3. The consequence of reason (i) is the absence of partial dislocations and stacking faults in Al as compared with Cu, and thus no pile-ups at grain boundaries are formed in the lower deformation range of Al since the dislocations will be incorporated into the cell walls. Reason (ii) seems to be responsible for the enhancement of recovery processes in Al as compared to Cu, both of dynamic recovery limiting the extension of stage IV, and of static recovery which occurs at very high strains during the single pass–breaks of iterative rolling deformation. The latter will cause both the degeneration of the

dislocation density as well as of the long range internal stresses which even prevents the formation of the PTW structure.

5. Acknowledgements

The Science Foundations of Austria and Hungary are acknowledged for financial support under projects 12944/12945-OPY (FWF, Austria), OTKA T022968/ T017609 and FKFP 0116/1997OTKA (Hungary), respectively. Costs for external stays of authors at the Synchrotron ELETTRA Trieste have been provided within the Scientific-Technical Co-operation between Austria and Italy, under project 2/97-98.

6. References

1. Gil Sevillano, J., van Houtte, P, and Aernoudt, E. (1980) Large strain work hardening and textures, *Progr. Mater. Sci.* **25**, 69-412.
2. Zehetbauer, M., and Les, P. (1998) Micromechanisms of plastic deformation in metals, *Kovove Materialy (Metallic Materials)* **36**, 153-161.
3. Kuhlmann-Wilsdorf, D. and Hansen, N. (1991) Geometrically necessary, incidental and subgrain boundaries, *Scripta metall. mater.* **25**, 1557-1562.
4. Hughes D.A. (1995) The evolution of deformation microstructures and local orientations, in: Proc.16th Int.Symp.Mater.Sci., ed. Hansen, N. et al., Riso National Laboratory, Roskilde (Denmark), p. 63-85.
5. Hansen, N. (1990) Cold deformation microstructures, *Mater. Sci. Technol.* **6**, 1039-1047.
6. Rollett, A.D., Kocks, U.F., Embury, J.D., Stout, M.G., Doherty, R.D. (1987) Strain hardening at large strains, Proc. 8th Int.Conf.Strength Met. & Alloys (ICSMA 8), Tampere, Finland, ed. Kettunen, P.O. et al., Pergamon, Oxford, p.433-438.
7. Argon, A.S., Haasen, P. (1993) A new mechanism of work hardening in the late stages of large strain plastic flow in fcc and diamond cubic crystals, *Acta metall. mater.* **41**, 3289-3306.
8. Zehetbauer, M. (1993) Cold work hardening in stages IV and V of fcc metals-II. model fits and physical results, *Acta metall.mater.* **41**, 589- 599.
9. Wilkens, M.and Bargouth, M.O. (1968) Die Bestimmung der Versetzungsdichte verformter Kupfer-Einkristalle aus verbreiterten Röntgenbeugungsprofilen, *Acta metall.* **16**, 465-468.
10. Oettel, H. (1971) X-ray analysis of dislocation densities and resistivity changes of plastically deformed fcc Ni-Co alloys, *phys. stat. sol. (a)* **6**, 265-272.
11. Ungár, T., Mughrabi, H. and Wilkens, M. (1982) An X-ray line-broadening study of dislocations near the surface and in the bulk of deformed copper single crystals, *Acta metall.* **30**, 1861-1867.
12. Ungár, T., Mughrabi, H., Rönnpagel D. and Wilkens, M. (1984) X-ray line broadening study of the dislocation cell structure in deformed [001]-orientated copper single crystals, *Acta metall.* **32.**, 333-342
13. Juul Jensen, D. (1993) Automatic EBSP analysis for recrystallization studies, *Textures and Microstructures* **20**, 55-65.
14. Hjelen, J., Oresund, R., Hoel, E., Runde, P., Furu, T. and Nes, E. (1993) EBSP, progress in technique and applications, *Textures and microstructures* **20**, 29-40.
15. Ungàr, T.and Zehetbauer, M. (1996) Stage IV work hardening in cell forming materials, pt. II: A new mechanism, *Scripta mater.* **35**, 1467-1473.
16. Zehetbauer, M., Ungar, T., Kral, R., Borbely, A., Schafler, E., Ortner, B., Amenitsch, H., and Bernstorff, S. (1999) Scanning X-ray diffraction peak profile analysis in deformed Cu-polycrystals by synchrotron radiation, *Acta mater.* **47**, 1053-1061.
17. Schafler, E.,Zehetbauer, M., Kopacz, I., Ungar, T., Hanak, P., Amenitsch, H., and Bernstorff, S. (1999) Microstructural parameters in large strain deformed Ni-polycrystals as investigated by synchrotron radiation, *phys. stat. sol. (a)*, **175**, No.2.
18. Amenitsch, H., Bernstorff S. and Laggner P. (1995), High Flux Beamline for Small Angle X-ray Scattering at ELETTRA, *Rev. Sci. Instrum.* **66**, 1624-1626.

19. Ungár, T., I. Groma and M. Wilkens, (1989) Asymmetric X-ray line broadening of plastically deformed crystals.II. Evaluation procedure and application to [001]-Cu Crystals, *J. Appl. Cryst.* **22**, 26-34.
20. Groma, I., Ungár, T. and Wilkens, M. (1988) Asymmetric X-ray line broadening of plastically deformed crystals. Theory, *J. Appl. Cryst.* **21**, 47-53.
21. Wilkens, M.(1970) The determination of density and distribution of dislocations in deformed single crystals form broadened X-ray diffraction profiles, *phys. stat. sol. (a)* **2**, 359-370.
22. Mughrabi, H. (1983) Dislocation wall and cell structures and long range internal stresses in deformed metal crystals, *Acta metall.* **31**, 1367-1379.
23. Schafler, E. (1998) Investigation of the microstructural evolution in large strain deformed metals by X-ray Bragg Peak Profile Analysis, Thesis, University Vienna, Austria.
24. Zehetbauer, M. (1999) Strengthening processes of metals at severe plàstic deformation, Proc. NATO Adv.Res.Workshop "Investigations & Applications of Severe Plastic Deformation", August 2-6 (Moscow, Russia), ed. Lowe, T.C. and Valiev, R.Z., Kluwer Acad. Publ., The Netherlands, this issue.
25. Zehetbauer, M., and Trattner, D. (1987) Effects of stress-aided static recovery in iteratively cold-worked aluminium and copper, *Mater. Sci. Eng.* **89**, 93-101.

INFLUENCE OF THERMAL TREATMENT AND CYCLIC PLASTIC DEFORMATION ON THE DEFECT STRUCTURE IN ULTRAFINE-GRAINED NICKEL

E. THIELE[1], J. BRETSCHNEIDER[1], L. HOLLANG[1], N. SCHELL[2], C. HOLSTE[1]

[1]*Institut für Physikalische Metallkunde, Technische Universität Dresden, 01062 Dresden, Germany*
[2]*Institut für Ionenstrahlphysik und Materialforschung, FZ Rossendorf, 01314 Dresden, Germany*

Abstract

Ultrafine-grained (UFG) high purity nickel samples produced by equichannel angular pressing were submitted to thermal treatment and cyclic plastic deformation at different temperatures in order to investigate the stability of the defect structure. Recrystallization was observed already after annealing above 425K. Cyclic plastic deformation at 300K and 425K leads to a coarsening of grains and to a dynamic recrystallization, respectively. Investigations performed by means of synchrotron radiation diffraction revealed that the mean volume expansion, long-range and short-range internal strains are diminished in consequence of the cyclic plastic deformation.

1. Introduction

Compared with conventional polycrystalline materials, ultrafine-grained (UFG) metals produced by equichannel angular pressing (EAP) show particular physical properties, such as high microhardness and high yield stress combined with a sufficient ductility [1]. However, for the application of the material, information is needed concerning the stability of the defect structure. Therefore, the purpose of the present paper is to characterize the structure in the original UFG material and to analyze the change of the grain structure and of the internal strains due to thermal treatment and due to cyclic deformation.

2. Experimental Procedures

2.1. SAMPLE PREPARATION, CYCLIC DEFORMATION AND ANNEALING

UFG nickel billets of 99.992% purity with a rectangular cross section were produced by ECAP using a 90 degree channel angle. After the first pass, the billets were turned and pressed through the die in the opposite direction. This procedure was repeated for four

173

T.C. Lowe and R.Z. Valiev (eds.), Investigations and Applications of Severe Plastic Deformation, 173–178.
© *2000 Kluwer Academic Publishers. Printed in the Netherlands.*

additional times rotating the billet 90 degrees clockwise about the billet pressing axis in between, so that the total number of passes was ten.

The loading axis of the fatigue samples was chosen parallel to the billet pressing axis. The samples were cyclically deformed at constant plastic strain amplitudes $\varepsilon_{pa} = 5 \times 10^{-4}$ and 1×10^{-3} with a plastic strain rate of $\Delta\varepsilon_p/\Delta t \approx 1 \times 10^{-3} s^{-1}$ at 300K and 425K. The X-ray diffraction investigations were carried out on a sample side face lying perpendicular to the shear plane of the last pass.

For the annealing experiments small blocks were prepared with surfaces parallel to the billet faces. The thermal treatment was applied for 10 minutes at different temperatures T_{an} (cf. Tab. 1). The heating and cooling rates were $15 Ks^{-1}$ and $-4 Kmin^{-1}$, respectively.

2.2. INVESTIGATIONS OF THE DEFECT STRUCTURE

The grain structure was observed using the orientation contrast of back scattered electrons in a scanning electron microscope (SEM). Measurements of X-ray diffraction profiles were carried out at the beamline ROBL of the Forschungszentrum Rossendorf at the European Synchrotron Radiation Facility in Grenoble in order to characterize the distribution of internal strains and the size of coherently scattering regions in the samples investigated. The linear polarized synchrotron radiation was collimated, vertical focused and monochromized at a wavelength $\lambda = 0.1539 nm$ ($E = 8.05 keV$) to achieve an energy resolution of $\Delta E/E \approx 1.5 \times 10^{-4}$ with an integrated flux of 6×10^{11} photons/s for 20keV and 200mA. The instrumental profile broadening was negligibly small in comparison with the strain and particle size induced broadening observed at the UFG samples. The profiles of the {111}-, {200}-, {220}-, {311}-, {222}- and {400}-Bragg-reflections were measured using a ϑ-2ϑ-mode. The Bragg-angles ϑ_p and ϑ_c of the peak and of the centre of gravity of the profile, respectively, the corresponding asymmetry parameter $\vartheta_a = \vartheta_p - \vartheta_c$, the integral breadth (IB) and the profile Fourier coefficients were determined.

3. Experimental Results and Discussion

3.1. CYCLIC STRESS-STRAIN RESPONSE

During the fatigue tests at room temperature the stress amplitude σ_a remains nearly constant until macrocracks occur, whereas a remarkable decrease of σ_a with increasing cycle number was observed for deformation at 425K (Fig. 1). In all deformation experiments it was observed that at the beginning of the cyclic deformation the maximum stress σ_{max} was higher than the absolute value of the minimum stress σ_{min} in a loading cycle. The mean stress $\sigma_m = (\sigma_{max} + \sigma_{min})/2$ is reduced to zero from its initial value σ_{m0} with increasing cycle number. This is probably a consequence of the reduction of long-range internal stresses (cf. 3.3). In Fig. 2, the constant stress amplitude of the UFG material is plotted against the deformation temperature. This curve is drawn together with the temperature dependence of the saturation stress amplitude for conventional polycrystals (grain size 40 µm). As in the case of polycrystals, at a given plastic strain amplitude ε_{pa} the stress amplitude σ_a decreases with increasing deformation temperature.

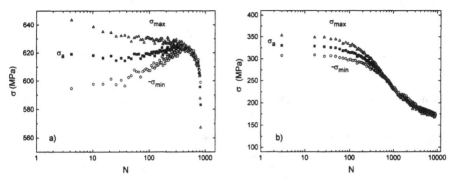

Figure 1. The stress amplitude σ_a, the maximum stress σ_{max} and the minimum stress $-\sigma_{min}$ in a deformation cycle in dependence of the cycle number N: a) at 300K with $\varepsilon_{pa} = 1 \times 10^{-3}$, b) at 425K with $\varepsilon_{pa} = 5 \times 10^{-4}$.

As described in 3.2 the softening at 425 K is accompanied by a dynamic recrystallization enlarging the mean grain diameter from 260 nm in the original UFG material to 3.2 µm after 10000 cycles. In Fig. 2 this structural change is reflected by a transition from the "UFG-curve" towards the curve for conventional polycrystals. The data point for the UFG material at 475 K results from an experiment at a sample with a higher content of impurities in which the UFG grain structure remained stable even at this temperature [2]. From a general point of view, Fig. 2 confirms the experience that stresses increase with decreasing grain size.

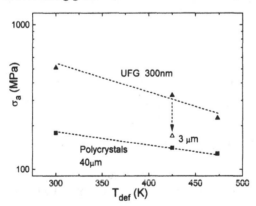

Figure 2. The stress amplitude σ_a for cyclic deformation at $\varepsilon_{pa} = 5 \times 10^{-4}$ in dependence of the deformation temperature T_{def} for nickel samples with different grain size.

3.2. GRAIN STRUCTURE

Together with the figures of the mean grain diameter listed in Tab. 1, Fig. 3 illustrates the influence of thermal treatment and cyclic plastic deformation on the grain structure.

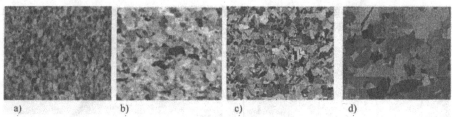

Figure 3. The grain structure of the original UFG material (a), after cyclic deformation with $\varepsilon_{pa} = 5 \times 10^{-4}$ at 300K (b) and at 425K (c) and after annealing at 525K (d). The micrographs show areas of 5 x 4 μm² (a and b) and of 50 x 40 μm² (c and d).

Taking into account the results of the X-ray diffraction experiments (cf. 3.3) it seems to be reasonable to classify the structure types into four groups.

Annealing at temperatures T_{an} slightly above room temperature doesn't effect significant changes of the original UFG defect structure (group I). After cyclic deformation at 300 K a coarsening of the grains was observed (group II). This coarsening is the more pronounced the higher the degree of deformation is, in particular near the cracks. Cyclic deformation at 425 K induces a dynamic recrystallization (group III) while static recrystallization occurs after annealing at $T \geq 475$ K (group IV). It should be noted that annealing at 425 K for the same time as the cycling proceeded (8h) leads to a static recrystallization, too, similar to that observed after annealing over 10 min at 475 K.

TABLE 1. Overview of the annealing and deformation conditions; the mean grain diameter d. Group numbers for the classification of the defect structure are added.

sample	T_{an} (K)	T_{def} (K)	grain structure	ε_{pa} (10^{-4})	σ_a (MPa)	σ_{m0} (MPa)	d (nm)	group
N2_3	-	-	UFG original	-	-	-	260	I
N2_6	375	-	UFG	-	-	-	350	I
N2_5	425	-	UFG	-	-	-	350	I
N1_1	-	300	UFG coarsened	10	620	48	530	II
N1_2	-	300	UFG coarsened	5	510	94	450	II
N1_5	-	425	UFG	5	325	45	-	-
N1_5	-	425	recrystallized	5	170	0	3200	III
N2_4	475	-	recrystallized	-	-	-	6200	IV
N2_2	525	-	recrystallized	-	-	-	6100	IV
N2_7	425 (8h)	-	recrystallized	-	-	-	5900	IV

3.3   ANALYSIS OF X-RAY DIFFRACTION PROFILES

Using the Williamson-Hall (WH) plot according to [3] the root mean square (rms) strain $<\varepsilon^2>^{1/2}$ was determined for the four groups classified in Tab. 1. As shown in Fig. 4 the value $<\varepsilon^2>^{1/2}$, which is proportional to the slope of the dashed lines, is reduced by cyclic deformation, in particular at elevated temperatures. However, at 300 K no significant influence of the strain amplitude on the decrease of $<\varepsilon^2>^{1/2}$ was found. As expected, the recrystallized samples (group IV) are nearly free of internal strains. These results are supported by the behaviour of the mean dislocation density ρ estimated from a Krivoglaz-Wilkens plot of the Fourier coefficients [4] for the {311}-diffraction profiles (cf. Tab. 2).

For group I and II the size D_{WH} of coherent scattering regions was estimated from the integral breadth of profiles measured at {111}-planes which are insensitive for long-range strains caused by heterogeneous shear strains [2]. D_{WH} is found to be somewhat smaller than the mean grain size d determined by SEM.

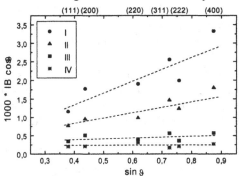

Figure 4. Williamson-Hall plot of the normalized integral breadth *IB* of the diffraction profiles in dependence of the Bragg angle ϑ for the different groups of samples *(cf.* Tab. 1).

The Warren-Averbach (WA) procedure [5] was employed to get information concerning the distributions of particle size and internal strains by an analysis of the Fourier coefficients of the profiles. From Fig. 5a it can be derived that the particle size D_{WA} given by the reciprocal value of the initial slope of the curves increases due to cyclic deformation already at 300 K. The effect is more pronounced for the specimen deformed at 425 K. The WA procedure yields a smaller particle size than the WH analysis for all samples investigated here. As stated in [6], D_{WA} and D_{WH} can be interpreted as the minimum and the maximum value of the particle size distribution, respectively.

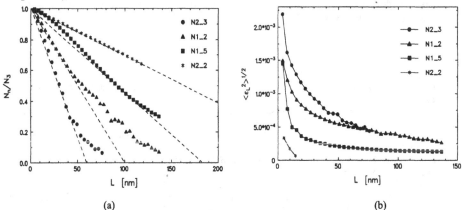

(a)   (b)

Figure 5. The size Fourier coefficient N_n/N_3 (a) and the rms $<\varepsilon_L{}^2>^{1/2}$ (b) in dependence of the Fourier length L [5] for different samples representing a group of samples (cf. Tab. 1).

As shown in Fig. 5b, the rms strains $<\varepsilon_L{}^2>^{1/2}$ obtained using the WA procedure decrease during cycling, particularly at elevated temperature. This is in agreement with the data $<\varepsilon^2>^{1/2}$ resulting from the WH analysis. From the behaviour of $<\varepsilon_L^2>^{1/2}$ vs. the Fourier length L it can be concluded that mainly the short-range strains are reduced during cyclic deformation at room temperature. Additionally, the long-range strains vanish at higher deformation temperatures.

The mean volume expansion $\Delta V/V$ was estimated from the shift of ϑ_c compared to ϑ_{c0} of profiles measured at the reference sample N2_2 [2]. As a peculiarity, only for the original UFG material a mean volume expansion $\Delta V/V = 5 \times 10^{-4}$ was found whereas for all the other samples $\Delta V/V \approx 0$.

Up to now the reasons for the occurrence of the mean stress σ_m and its reduction during cyclic deformation have not been clear. A more detailed investigation of the internal long-range stresses could be useful to solve this problem taking into account the asymmetry of the diffraction profiles observed for the UFG material.

TABLE 2. Defect structure parameters derived from the X-ray diffraction profiles for the different sample groups

group	$<\varepsilon^2>^{1/2}$ (10^{-4})	$\rho\,(10^{14}\mathrm{m}^{-2})$	D_{WH} (nm) from {111}	D_{WA} (nm) from all {hkl}
I	17 ± 4	≈ 4.0	≈ 220	60 ± 5
II	9 ± 2	≈ 2.8	≈ 240	110 ± 10
III	1 ± 1	≈ 0.5	-	180 ± 10
IV	0 ± 1	-	-	330 ± 30

Summarizing, these investigations of high purity UFG nickel samples have shown that their defect structure is relatively unstable against cyclic plastic deformation and thermal treatment. This is indicated by the increase of the grain size as well as by the reduction of the internal strains. In the case of cyclic deformation this occurs already at 300 K, while thermal treatment slightly above room temperature leads to recrystallization and vanishing of internal strains. Compared with the results of previous experiments [2] it can be speculated that a higher content of impurities increases the stability of the UFG defect structure.

4. References

1. Valiev, R. (1997) Structure and mechanical properties of ultrafine-grained metals, *Mat. Sci. Eng.* A234-236, 59-66.
2. Thiele, E., Hecker, M., Schell, N. (1998) Change of internal strains in ultrafine-grained nickel due to cyclic plastic deformation, *Mat. Sci. Forum* (in press).
3. Williamson, G. K., Hall, W. H. (1953) X-ray line broadening from filed aluminium and wolfram, *Acta Metall.* 1, 22-31.
4. Klimanek, P. in: Hasek, J. (ed.) (1988) *X-ray and Neutron Structure Analysis in Materials Science*, Plenum Press, New York, 125-137.
5. Warren, B. E. (1990) *X-Ray Diffraction*, Dover Publications Inc., New York.
6. Ungár, T., Borbély, A. (1996) The effect of dislocation contrast on X-ray line broadening: A new approach to line profile analysis, *Appl. Phys. Lett.* 69 (21) 3173-3175.

NANOSTRUCTURE STATE AS NONEQUILIBRIUM TRANSITION IN GRAIN BOUNDARY DEFECTS IN SPD CONDITION

O.B. NAIMARK

Insitute of Mechanics of Continuum Media, Urals Division, Russian Academy of Sciences, Acad. Koroleva st. 1, Perm, 614013 Russia

1. Introduction

A promising way of generating novel physical properties of solids is possible by the modification of their microstructures. In nanostructured solids (crystalline size *ca.*10 nm) the physical properties are strongly modified by the disordered structure of grain boundaries (GB). Nanocrystals are produced under severe (high pressures and heavy deformations) or nonequilibrium (growth from an amorphous state) conditions. Much attention is presently given to studies of nanocrystalline solid as nonequilibrium systems whose mechanical and physical properties are related to collective phenomena in the system of GB defects. The GB defects are thermodynamically nonequilibrium, but can be mechanically stable at ambient temperatures. The nonequilibrium nature of GBs may originate from the intrinsic properties of the dislocation structures that form these mesoscopic defects, as well as from the interactions in the system of GB defects. Since the effect of GB defects on the mechanical properties of polycrystals should be considered as a cooperative phenomenon (*i.e.*, as the effect of the whole ensemble of GB defects), a statistical approach seems to be most relevant in accounting for both the nonequilibrium nature of a separate mesoscopic defect (a grain boundary) and the interactions of GB defects in the ensemble. In this work, we attempt to construct a statistical theory of the polycrystalline state that explicitly allows for the presence of GB defects and defect structures that develop in the ensemble of such GB defects. One result of the theory is the development of a new approach to the appearance of the nanocrystalline state, which is considered as a consequence of a structural (topological) transition that takes place in the system of GB defects with decreasing grain size.

2. Statistical Model

2.1. ENERGY OF GRAIN BOUNDARY DEFECTS

The structure of GBs and its model representations are the subject of continuous interest [1, 2]. Any boundary in a crystal is a nonequilibrium defect. Considering the GB defects, we can define the equilibrium GB structure by requiring that long-range stress fields associated with the boundaries are absent and the GB structure itself is mechanically stable. A qualitatively acceptable GB model that satisfies the equilibrium conditions is the model according to which the boundary is a region between two

179

T.C. Lowe and R.Z. Valiev (eds.), Investigations and Applications of Severe Plastic Deformation, 179–187.
© 2000 *Kluwer Academic Publishers. Printed in the Netherlands.*

crystallites misoriented due to the presence of a dislocation pile-up [3]. It is known that, contrary to a dislocation pile-up, a dislocation wall has a minimum self-energy per dislocation. Considering each dislocation in the wall as a generalized Volterra dislocation with Burgers vector $\vec{B}$, which is defined as a full displacement in a given shear zone, we can write the energy density of the dislocation wall in the form

$$\varepsilon = C_1 a_{xx}^2, \qquad (C_1 = \frac{\mu}{4\pi} hD \ln\frac{h}{D}), \qquad (1)$$

where μ is the shear modulus, h and D are characteristic sizes of grains, and $B/hD = a_{xx}$ is the dislocation density tensor. In [4], we represented the defects as superdislocations formed by dislocation pileups. In such a representation, the energy of short-range stress fields that develop in the regions of this type of incompatibility. To account for the collective effects in the system of GB defects, we must consider deformation defects responsible for the total strain and introduce notations describing their orientation and the intensity. The translation defects can be introduced by the following "microscopic" parameters:

$$s_{ik}^n = s v_i v_k, \qquad s_{ik}^\tau = \frac{1}{2} s^\tau (v_i l_k + l_i v_k).$$

The tensor s_{ik}^n describes a microcrack of "normal" mode; the microcrack volume is $S = S_0 B$, where S_0 is the area of the base of a disk-like defect with the normal $\vec{v}$ to the base parallel to the direction of the Burgers vector ($\vec{B} = B\vec{v}$). Another parameter, S_{ik}^τ, is the microscopic is the microshear tensor; its intensity is S^τ, and $\vec{l}$ is the slip vector.

2.2. SELF-CONSISTENCY FIELD METHOD

At the developed stages of deformation, the GB defects exhibit clear features of collective behavior. The defect concentration can be as high as $10^{13} - 10^{14} cm^{-3}$, so that, as was first noted in [5, 6], we can speak about statistical collective effects. However, we must always keep in mind that each of the GB defects is a thermodynamically nonequilibrium system. Evolution of the ensemble of GB defects is governed by the statistical distribution of defect nuclei, interactions between defects, and the effects of external fields. The corresponding equation for the microscopic parameter S_{ik} is the Langevin equation [7]

$$\dot{s}_{ik} = K_{ik}(s_{lm}) - F_{ik}(t), \qquad (2)$$

where $K_{ik}(s_{lm})$ and $F_{ik}(t)$ are the deterministic and stochastic contributions to the interaction forces. The stochastic or fluctuation part usually satisfies the following conditions:

$$\langle F_{ik}(t) \rangle = 0, \qquad \langle F_{ik}(t')F_{ik}(t) \rangle = Q\delta(t - t'),$$

where the averages are taken over the stochastic process, and Q is the correlator of the fluctuating force. In our case, however, it is more convenient to consider the Fokker-Plank equation for the distribution function $W(s,t)$

$$\frac{\partial}{\partial t}W = -\frac{\partial}{\partial s_{ik}}(K_{ik}(s_{lm})W) + \frac{1}{2}Q\frac{\partial}{\partial s_{ik}}\left(\frac{\partial}{\partial s_{ik}}W\right). \tag{3}$$

The distribution function $W(s_{ik}, t)$ gives the probability of finding the system in the state $(s_{ik}, \dot{s}_{ik})$ of the phase space. As it was shown in [8] the distributions of defects are characterized by statistical self-similarity. The statistically self-similar solutions are the stationary solutions to the Fokker-Plank equation. The latter can be obtained from (3), assuming the boundary conditions $W(s_{ik}) \to 0$ for $s_{ik} \to \pm\infty$. The result is $W = Z^{-1}\exp\left(\int_0^s 2K_{ik}(s'_{lm})ds_{ik} / Q\right)$, where Z is the normalization constant that is determined by the condition $\int W(s_{ik})ds_{ik} = Z$.

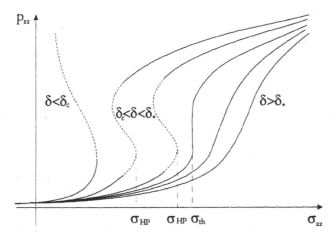

Figure 1. Solution of self-consistency equations for various values of parameter $\delta = 2\alpha / \lambda n$

The parameter Q characterizes the initial energy relief of the structure (the nonequilibrium potential) [9]. To account for the appearance of long-range stress fields in the ensemble of interacting GB defects, we introduce an additional term $H_{ik}s_{ik}$ in (2)

$$E = E_0 - H_{ik}s_{ik} + \alpha s_{ik}^2 . \qquad (4)$$

Here, we assume that the force acting on a GB defect consists of two contributions: the first term is proportional to the external field σ_{ik}, and the second term, $\lambda n\langle s_{ik}\rangle$, is the macroscopic (mean) long-range field created by other defects. The effective field in this expression is $H_{ik} = \gamma\sigma_{ik} + \lambda n\langle s_{ik}\rangle$. The striking feature of nanocrystalline materials prepared by different techniques is often the existence of very high internal elastic stresses [10, 11]. This occurs due to the fact that dislocations are unstable in ultra-fine grains making they structure a nonequilibrium one. Averaging S_{ik}, using the distribution function W, we obtain the self-consistency equation for $p_{ik} = n\langle s_{ik}\rangle$:

$$p_{ik} = n \int s_{ik} Z^{-1} \exp(-E/Q) ds d^3\vec{v} , \qquad (5)$$

where $Z = \int \exp(-E/Q) ds d^3\vec{v}$. It could be shown that p_{ik} is a macroscopic strain arising from the presence of defects. The self-consistency equation was solved earlier [12] for the particular case of uni-axial tension with a constant stress σ_{zz}. Figure 1 shows the dependence of the tensor component p_{zz} on the stress σ_{zz} for various values of the dimensionless parameter $\delta = 2\alpha/\lambda n$. This parameter δ characterizes the "ability" of GB defects to grow from the pre-existing nuclei and the condition of the interaction between defects arising from the internal stress. The dislocation model of mesoscopic defect [3] gives $\alpha \approx G/V_0$, where G is the shear modulus and $V_0 \approx l_n^3$ is the initial "free" volume (volume of the nucleus). The estimation of the mean-field constant as $\lambda \approx G$ gives $\delta \approx l_c/l_n$, where $l_c \approx n^{-1/3}$. Parameter δ is determined by two characteristic scales in a heterogeneous media: the characteristic size l_n of the nucleus of a mesoscopic defect and the characteristic scale of the internal stresses produced by the GB defects l_c. Three regions separated by the asymptotic curves δ_c and δ_* in Fig. 1 correspond to qualitatively different solutions, which reflect three qualitatively different responses of a polycrystal to the growth of defects under the load. The actual behavior depends on the initial grain size and on the characteristic scale of the internal stress. The values of $\delta \sim (\delta_c, \delta_*)$ for the asymptotic curves are the bifurcation points of the solution for the self-consistency equation and for the mentioned uni-axial case $\delta_c = 1$ and $\delta_* = 1.3$.

3. Spatial-Temporal Structures of Grain Boundary Defects

The curves in Fig. 1 correspond to the characteristic responses of a material to a change in the basic modes describing an ensemble of the GB defects, and phenomenologically this behavior can be reflected in the form of the Ginzburg-Landau expansion for the free energy F [13, 15]:

$$F = \frac{1}{2} A \left(1 - \frac{\delta}{\delta_*}\right) p_{ik}^2 + \frac{1}{4} B p_{ik}^4 - \frac{1}{6} C \left(1 - \frac{\delta}{\delta_c}\right) p_{ik}^6 - \sigma_{ik} p_{ik} + \frac{1}{2} \chi (\nabla p_{ik})^2, \quad (6)$$

where A, B and C are expansion parameters; χ is the non-locality parameter. The motion equation for the defect density tensor follows from the evolution inequality and has the form [12]

$$\frac{dp_{ik}}{dt} = -L \frac{\delta F}{\delta p_{ik}}, \quad (7)$$

where $\delta \big/ \delta p_{ik}$ is the variation derivative, L is the kinetic coefficient.

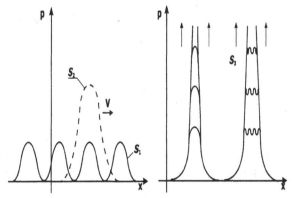

Figure 2. Spatial-time structures in the ensembles of grain boundary defects

Transitions through the bifurcation points δ_c and δ_* lead to a sharp change in the symmetry of the GB defects distribution. In the region $\delta > \delta_*$ this equation is of the elliptic type with periodic solutions $p_{xx} = P \exp(i\phi)$ with spatial period Λ given mainly by the internal stress. As $\delta \to \delta_*$ Eq. (7) changes locally from elliptic to hyperbolic and the periodic solution transforms into a solitary-wave solution. This transition is accompanied by divergence of the inner scale Λ: $\Lambda \approx -\ln(\delta - \delta_*)$. In this case the solution has the form of the solitary wave solution $p_{xz}(\xi) = p_{xz}(x - Vt)$. A transition through the bifurcation point δ_c is accompanied by the appearance of spatial-temporal structures of a qualitatively new type characterized by explosive accumulation of defects as $t \to t_c$ in the spectrum of spatial scales (peaking regimes). The transition from spatially periodic distribution of GB defect density to dissipative structures such as localized waves of dislocation instability is accompanied by changes in the topological invariants of the nonlinear system [3]. This may cause abrupt changes in the mechanisms of deformation and macroscopic properties of the polycrystals in the course of the grain change. The transition at $\delta = \delta_*$ can be considered as a topological transition in the system of GB defects, because the connectivity parameters, correlation length, and other scaling characteristics of the system abruptly change, as it is observed in the case of usual phase transitions. At $\delta = \delta_c$, another topological transition takes place to the state characterized by a quasi-brittle fracture behavior, i.e., by appearance of local regions with explosive kinetics of the GB defects growth. The localization effects in the range of $\delta < \delta_c$ were analyzed in [14].

4. Violation of the Hall-Petch Law in Nanocrystalline Materials

The constitutive equations which describe the influence of the defects kinetics on relaxation and deformation properties of polycrystals include the motion equation (7) for p_{ik} and the state equation for the solid with GB defects [15]

$$\varepsilon_{ik} = \frac{1}{2\mu}(\sigma_{ik} - \frac{1}{3}\sigma_{ll}\delta_{ik}) + \frac{1}{9K}\sigma_{ll}\delta_{ik} + \gamma p_{ik}, \tag{8}$$

where ε_{ik} is the total deformation, μ and K are the shear and bulk modulus. According to the Hall-Petch relationship, the strength of polycrystalline materials increases as the crystallite size l decreases and the yield stress of polycrystals obeys the well-known law $\sigma_y = \sigma_0 + kl^{-\frac{1}{2}}$, where σ_0 is the friction stress required to move dislocations in single crystals. According to this law, one might expect a hundredfold increase in the yield stress for nanocrystals in comparison with conventional polycrystals ($100\,\mu m$). However, recent studies on the microhardness of nanocrystals have shown [16–18] that although there is an approximate $l^{-\frac{1}{2}}$ dependence (with a large experimental scatter than in coarse-grained samples). The yield stress is usually no more than several times that for conventional polycrystals: Some workers have noted a decrease in strength with decreasing grain size in ultrafine-grain materials (UFG) [10]. Attempts have been made to calculate an approximate value for the grain size at which the Hall-Petch relationship breaks down. Several derivations of the Hall- Petch equation are based on the concept of dislocation pile-ups at grain boundaries [19]. A number of important questions could be posed, for instance. At what point does dislocation generation and mobility become so restricted that the Hall-Petch relationship no longer applies? How does the presence of a significant volume fraction of grain boundaries (30% or higher) affect the mechanical behavior of nanocrystalline material? According to results given by the developed statistical approach, the conventional plasticity area corresponds to the range $\delta_c < \delta < \delta_*$, where the ensemble of the GB defects reveals the pronounced collective effect caused by the interaction of the orientation modes of the defects. The evolution of the deformation caused by the defects occurs in this range of δ in the course of the so-called orientation transition. The orientation transition area includes the stress metastability area where the threshold character of the transition from the bottom to the upper thermodynamic branch proceeds (Figure 1). Taking in view the scale dependence of $\delta \sim l_c / l_n$ it is evident that the threshold stresses increases with the decrease of the GB defect nucleus l_n which is linked with a grain size l: $l_n \sim l$. The size $l_*(\delta_*)$ corresponds to the critical grain size when the collective effects caused by the orientation mode interaction are degenerated. As the consequence the conventional threshold mechanism of the plasticity for the polycrystals disappears. Morphologically it means that the plastic flow with pronounced strain localization effects (adiabatic shear bands, Portevin- Le Chatelier effect), when few strain directions are involved in deformation, is observed only in the range $\delta_c < \delta < \delta_*$ as the consequence of the sharp ordering of the defects. In the range $\delta > \delta_*$ the plastic flow involves a number of slip directions. Weertman et al. [20] reported the results of MD vizualization of a 5.2 nm grain size copper material, before and after 3% deformation. The sharp change of the symmetry (topological) properties in the GB defect system under the pass of the critical grain size l_* is the reason of the violation of the Hall-Petch law. Qualitatively the dependence of the threshold stress σ_{th} (Fig.1) on the grain size l (the size of the GB defect nucleus l_n) has the similar form given by the Hall-Petch law.

There is another reason for the δ_*-transition, taking in view the second spatial scale l_c that determines the scale of the internal stress induced by the GB defects. It is natural to link this scale with the state of the dislocation pile-ups representing the GB defects. A larger l_c-scale could be identified with a more non-equilibrium state of the GB defects. It provides deeper penetration in the nanoscale area with pronounced disorder of the internal stress field. As a consequence, the deformation response of the UFG materials with non-equilibrium GBs reveals higher hardening work. Valiev et al. [11] studied the Hall-Petch dependence in the UFG aluminum alloy with different degrees of a non-equilibrium state of GBs and established three typical stress-strain curves for the alloy which correspond to non-equilibrium and relatively equilibrium states. The flow stresses for these curves are quite similar but there is a sharp difference in the elongation to failure. However, the curves show no signs of pronounced strain hardening and this is what principally distinguishes the UFG alloy behavior with the grain size 10 μm. The concentration of the grain boundaries (grain boundary phase [11]) determines the average scale l_c as the measure of the "physical width" of the GBs and l_c could be considered as the characteristics of the non-equilibrium state of the GBs and the intensity of the internal stress in the UFG materials.

5. Severe Plastic Deformation in Ultrafine-Grained Materials

Severe plastic deformation (SPD) can be considered as an effective method of producing UFG materials. The study of highly strained materials behavior indicates that the rotation modes of deformation play an active role and result in the formation of a severe misoriented fragmented structure. The main specfic feature of the structure of UFG materials is the existence of non-equilibrium (unrelaxed) GBs. TEM photographs of the UFG alloys [11] reveal specific "diffusive" diffraction contrast of GBs in the alloy after SPD. This contrast is a typical feature of non-equilibrium GBs with a high energy and long-range stresses and has been observed in different UFG materials. HREM investigation of UFG materials [11] showed two main peculiarities of the GB structure. First, the crystallographic width of the GBs, as in usual materials, is narrow and does not exceed 1 nm. Secondly, near a GB there is an elastically distorted layer, about 10 nm in width, where the value of the elastic strains is rather high. This layer was named in [11] as the "grain boundary phase (GBP)" that coexists with a grain phase. The observed difference between these phase parameters (electric and magnetic properties) is associated first with a higher dynamic activity of GBP atoms. This fact is related with the atom behavior in the elastically distorted layer. The thickness of the GBP (the physical width of a GB) is determined by the degree of the non-equilibrium state of GB structure.

The analysis of thermodynamics of the plasticity represented in Fig.1 shows that the deformation curves for the conventional polycrystals ($\delta_c < \delta < \delta_*$) and the UFG materials have the close asymptotics for $\sigma > \sigma_{th}^*$. As it was shown in [3] the irreversible deformation (plastic flow) occurs for both above cases under excitation of more large scale GB defects. This mechanism can be described in terms of δ as the δ-change induced by the current state of the GB defect system. The natural tendency of the δ-decrease. This process can be considered as the effective growth of the scale of the defect nuclei l_n. The multi-axial stress (strain) field that is realized under the SPD provides the decrease of the size of the structural elements (new grains) due to the formation of the power defect net with a high degree of the disorder. Mathematically speaking it means that multi axial stress field produced by SPD can form macroscopically isotropic net of the GB defects. This leads finally to the fine grains with a size close to the net size and to high internal stresses. Both of these factors lead to the increase of the δ-parameter, that according to the considered property of the polycrystals can provide the UFG material responses.

6. Conclusion

1. Transition to the nanocrystalline state can be considered as a topological transition in the ensemble of GB defects that consists of abrupt changes in its connectivity (changes in fractal dimension) and critical exponents in the scaling relationships.
2. Changes in the mechanism of deformation (deviation from the Hall-Petch law) at the transition to the submicrocrystalline state are the consequence of sharp change of the symmetry properties in the GB defects ensemble and the creation of the disordered GB defect system.
3. UFG structure formation under severe plastic deformation is the consequence of the creation of the GB defect network which produces new grain structure and long range internal stress. The nanocrystalline structure is the limiting case of UFG structure produced by extremely multi-axial straining of conventional polycrystals.

7. References

1. Gleiter, H. and Chalmers, B. (1972) High-Angle Grain Boundaries, *Progress in Materials Science,* **16,** Chalmers, B., Christian, J.W., and Massalski, T.B, Eds., Oxford: Pergamon.
2. Orlov, A.N., Perevezentsev, V.N., and Rybin, V.V. (1980) *Granitsy zeren v metallakh* (Grain Boundaries in Metals), Moscow: Metallurgiya.
3. Naimark O.B. Nanocrystalline state as a topological transition in an ensemble of grain-boundary defects, *The Physics of Metals and Methallogrophy* **84,** No.4, 327–337.
4. Naimark, O.B. (1982) *Thermodynamics of Deformation and Fracture of Solids with Microcracks,* Preprint of Inst. of Mechanics of Continuum Media, Ural Sci. Center, USSR Acad. Sci., Sverdlovsk, no.22.
5. Vladimirov, V.I. (1984) *Fizicheskaya priroda razrusheniya metallov (Physics of the Fracture of Metals),* Moscow: Metallurgiya.
6. Gleiter, H. (1969) *Philos. Mag.* **20,** 821–827.
7. Landau, L.D. and Lifshitz, E.M. (1976) Statistical Physics, Moscow: Nauka.
8. Botvina, L.R. and Barenblatt, G.I., (1985) Self-Similarity in Accumulation of Damage, *Probl. Prochn.* **12,** 17–24.
9. Haken, H. (1983) *Advances Synergetics. Instability Hierarchies of Self-Organizing Systems and Devices,* Heidelberg: Springer.
10. Weertman, J.R. (1993) Hall-Petch strenghening in nanocrystalline metals, *Materials Science and Engineering* **A166,** 161–167.
11. Valiev, R.Z, Korznikov, A.V., and Mulyukov, R.R. (1993) Structure and properties of ultrafine-grained materials produced by severe plastic deformation, *Material Science and Engineering* **A168,** 141–148.
12. Naimark, O.B., (1995) *Structural Transitions in Solids and Mechanics of Plasticity and Failure.* Preprint of Inst. of Mechanics of Continuum Media, Urals Division, Russ. Acad. Sci., Perm.
13. Naimark, O.B. (1998) Defect induced instabilities in condensed media, *JETP Letters* **67,** 751–757.
14. Naimark, O.B., Davydova, M.M., and Plekhov, O.A. (1998) *Failure scaling as multiscale instability in defects ensemble, PROBAMAT – 21 Century: Probabilities and Materials,* (G.N.Frantziskonis, ed.), Kluwer, 127–142.
15. Naimark, O.B. and Silbershmidt, V.V. (1991) On fracture of solids with microcracks, *Eur. J. Mech. A: Solids,* **10,** No. 6, 607–619.
16. Valiev, R.Z, Korznikov, A.K. and Mulyukov, R.R. (1992) *Met.Phys.Metall.* **4,** 4.
17. Valiev, R.Z, Krasilnikov, N.A., and Tzenev, N.K. (1991)*Mater.Sci.Eng.* **A137,** 35.
18. Valiev, R.Z., Chmelik, R., Bordeaux, F., Kapelski, G., and Baudelet, B. (1992) *Scr.Metall.Mater.* **27,** 855.
19. Meyers, M.A. and Chawla K.K. (1984) Mechanical Metallurgy, Prentice Hall, Englewood Cliffts, NJ.
20. Weertman, J.R, Farkas, D., Hemker, K., Kung, H., Mayo, M., Mitra, R., and Van Swygenhowen, H. (1999) Structure and mechanical behavior of bulk nanocrystalline materials, *MRS Bull.* **24,** 44–50.

TEXTURE, STRUCTURAL EVOLUTION AND MECHANICAL PROPERTIES IN AA5083 PROCESSED BY ECAE

L. DUPUY, E.F RAUCH, J.J. BLANDIN
Génie Physique et Mécanique des Matériaux (ESA CNRS n°5010)
Institut National Polytechnique de Grenoble – ENSPG
BP 46 - 38402 Saint Martin d'Hères CEDEX - France

1. Introduction

A reduction of the mean grain size in metallic alloys is expected to increase their yield stress at room temperature and to promote superplastic properties at higher strain rates and/or lower temperatures than those conventionally used in superplastic forming. By conventional thermomechanical treatments, grain sizes of about 10 µm are developed for aluminium-magnesium alloys and superplastic deformation is thus typically obtained at a temperature close to 500°C and a strain rate in the range 10^{-4}-10^{-3} s^{-1} [1,2]. Very fine microstructures can be produced by severe plastic deformation. Equal channel angular extrusion (ECAE) is one of the most popular techniques and has been applied to aluminium alloys [3], particularly in the case of Al-Mg alloys due to their large industrial use, especially in terms of superplastic forming [1,2].

In the ECAE process, the number of extrusions is a key parameter. An important refinement of the microstructure is obtained after the first pass, associated with a sharp increase of the hardness [3]. For further passes, a relative constancy of the mean structural size is frequently observed [3,4] but recent works suggest that the knowledge of the mean structural size is not sufficient to predict the mechanical behavior of the alloy. Indeed, despite similar structural size, some variations of the hardness [3] and very different superplastic behaviors [4] are obtained. Nevertheless, it must be kept in mind that the structural size is generally measured by transmission electron microscopy (TEM) micrographs and takes into account grains as well as subgrains.

The nature of the boundaries is also an important parameter. It has been reported that the average misorientation angle, estimated from selected area electron diffraction, increases with the number of extrusions [3]. However, only limited data are available and, consequently, a procedure to measure misorientations appears fruitful to improve the microstructural characterisation of severely deformed materials.

2. Experimental Material and Procedures

The chemical composition of the commercial Al-Mg alloy under investigation is given in Table 1. The AA5083 was received as a 10 mm thick hot-rolled plate with a grain diameter of about 40 µm. Billets with dimensions of 10x10x60 mm^3 were deformed by ECAE up to 8 times at 150°C via Route C [6]. The angle between the two channels

189

T.C. Lowe and R.Z. Valiev (eds.), Investigations and Applications of Severe Plastic Deformation, 189–195.
© 2000 *Kluwer Academic Publishers. Printed in the Netherlands.*

being 90°, each pass promotes a strain of $\varepsilon = 1.15$ [5]. For the first extrusion, the billet was introduced in the first channel in the rolling direction and with the normal direction parallel to the second channel.

TABLE 1. Chemical composition of AA5083 (weight%)

Mg	Mn	Cr	Fe	Si	Cu	Al
4.69	0.80	0.08	0.20	0.10	0.05	Bal

Room temperature compression tests were performed on cylinders 12 mm high and 8 mm in diameter with the compression axis parallel to the extrusion direction. The cross-head displacement rate was 1 mm/min and gave an initial strain rate of 1.4×10^{-3} s^{-1}. Pole figures were extracted from X-ray measurements using the Berkeley Texture Package (BEARTEX [7]). The samples for texture measurements were cut from the middle of each billet, polished to a metallographic finish and observed in the mirror plane of the process which contains both the entrance and the exit channels direction. TEM observations were performed on thin foil discs taken parallel to the mirror plane and perforated in a twinjet polisher at 15V using a mixture of 25% HNO_3 and 75% methanol at $-15°C$. They were examined using a TEM operating at 200kV. Selected area electron diffraction (SAED) patterns were recorded from regions having diameters of 2.7 µm. The measurement of the grain size was made directly from TEM micrographs by the line intercept method.

To measure accurately the orientations of each dislocation cell, a procedure has been developed using the electron microscope nano or convergent beam technique and computer analyses of the diffraction patterns. The samples were scanned in a JEOL-3010FX electron microscope with a very fine electron beam (~ 5 nm) in straight lines over a distance of 18 µm with steps of 50 nm, parallel and orthogonal to the shear direction. The diffraction patterns were directly extracted with a digital camera and analyzed with semi-automatic computation software. The accuracy of the measurements was about 1° while no Kikuchi lines are required.

3. Results and Discussion

3.1. MECHANICAL BEHAVIOR

The room temperature mechanical properties of the extruded alloy were investigated through hardness measurements and compression tests (Figure 1). As expected, severe plastic deformation leads to a sharp increase of the strength. The hardness of the material saturates at a value of HV130 after few passes. Compression tests appear more sensitive. Indeed, both the yield stress and the ultimate stress increase up to 8 passes; a 25 MPa hardening is obtained between the two last records. This work-hardening may be related to stage IV. The value is lower than room temperature stage IV hardening reported for Al-Mg alloys [8] by a factor of three. However, extrusion was performed at 150°C and it is known that stage IV hardening depends on temperature [9]. Whatever the origin of the stress increase, the results suggest that structural evolutions did not reach a steady state.

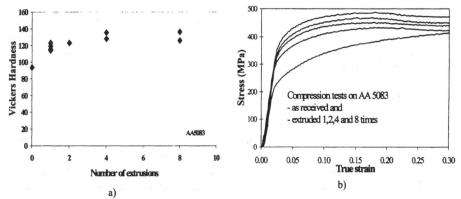

a)

Figure 1. Results of a) hardness and a) compression tests on AA5083 for increasing number of extrusions.

During the compression tests, softening is recorded after 20% strain. The stress decrease is correlated with the development of strain localization. The latter is promoted by the strongly anisotropic shape of the samples due to a sharp texture of the material.

3.2. TEXTURAL EVOLUTION

The extrusion process implying simple shear, the crystallographic rotations should be similar to the ones observed during planar simple shear or torsional deformation (*e.g.* [10]). The as received material was mainly produced by hot rolling. The resulting texture is composed of a sharp volume percentage of the brass ({110}<112>) and S ({123}<634>) components, respectively 50% and 23% (within 15° from the ideal orientations). To facilitate comparisons, the texture of the as received material is shown with a (111) pole figure containing both the rolling and the normal direction. With this unusual projection, an intense peak is observed in the centre (*i.e.* the transverse direction) and corresponds merely to the brass component (Figure 2.a).

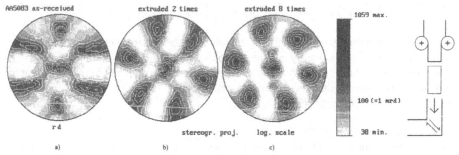

Figure 2. (111) pole figures of a) as received, b-c) extruded samples (respectively 2 and 8 passes). The same texture is produces for the different levels of strain.

The textures after extrusion are similar for the different number of passes (Figure 2.b-c). It consists mainly of the as received texture rotated around the transverse direction up to get the [110] axis aligned with the shear direction. The resulting orientations are stable and correspond to grains in which collinear slip systems are strongly activated, the common burgers vector being placed on the [110] axis considered above. Remembering that route C consists in changing the sign but not the direction of the strain, the constancy of the texture is not surprising but means that the strengthening observed in compression between 1 and 8 passes is not related to crystallographic orientations.

3.3. STRUCTURAL EVOLUTIONS

3.3.1. TEM observations

The pictures in Figures 3 a-b contain the principal axes of deformation and show typical morphology and size of the microstructure for 1 and 8 passes, respectively. The microstructure is severely affected by extrusion. Even after one pass, the initial grain boundaries are hard to trace. The intragranular structure consists of more or less well defined cells bounded by walls containing a high density of defects. Internal stresses are revealed by the perturbed contrast within cells which makes the observation of individual dislocations difficult. The microstructure is elongated in the direction of shear after one pass. For a higher number of passes the microstructure is more isotropic. The average structural size is roughly constant in the course of processing (d≈0.25 μm).

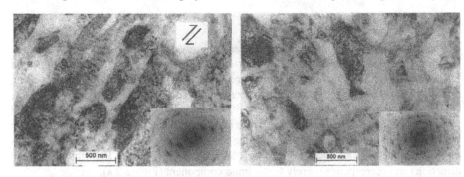

Figure 3 .TEM micrographs of typical structures of extruded AA5083 : a) one pass, b) 8 passes. Selected diffraction patterns are given for the two states of deformation.

Typical selected area diffraction patterns are added to the pictures. The dispersion of the spots in the pattern is frequently considered to determine the nature (subgrain or grain) of the observed features. Following the usual procedure, the present observations would lead to the conclusion that the observed small volumes are separated by high angle boundaries. This conclusion will be discussed in more details in the following section.

3.3.2. Misorientation measurements

The size and morphology of the microstructure being nearly independent of the number of passes, another structural property must be responsible for the slight but definite evolution of the mechanical properties. The nature of the interface is a possible candidate to explain the observation. The degree of misorientations between adjacent cells being poorly characterized by selected area diffractions, the orientations of cells were individually determined in indexing spots diffraction patterns.

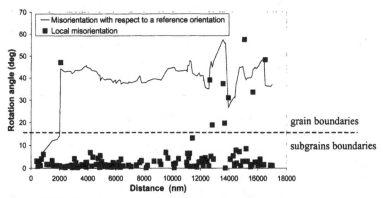

Figure 4. Misorientation along the shear direction after one extrusion.

Typical line measurements are shown on Figures 4 and 5. Full lines denote misorientations with respect to an arbitrary reference grain. Dots correspond to local misorientations *i.e.* between adjacent volumes. To analyze the structure, it is proposed to separated the boundaries in two groups : low angles and high angles boundaries with a limit arbitrarily fixed to 15° for local misorientations. The first class corresponds merely to cells. High angle boundaries are observed even after one pass but their distribution is inhomogeneous (Fig. 4). At large levels of strain, the number of high angle boundaries has increased and their spacing is more regular. The first conclusion which may be drawn from these measurements is that the microstructure observed is not uniquely composed of grains. Rather a mixture of low and high angle boundaries is denoted. As a consequence, the substructure size cannot be directly interpreted as a grain size.

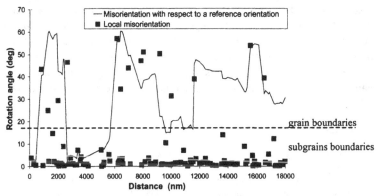

Figure 5. Misorientation along the shear direction after eight passes.

Of importance is the increase of the number of grain boundaries with strain. The grain size is estimated by dividing the length of the measured lines by the number of high angle boundaries. The calculated grain size decreases in the course of extrusion. The initial value of 40 μm is reduced to 3 ± 0.5 μm after the first extrusion but still diminishes down to 1 ± 0.3 µm after 8 passes (Table 2). Care was taken to perform the measurements on grains whose orientation was representative of the strong texture.

TABLE 2. Structural sizes in µm estimated on TEM micrographs or from misorientation measurements

	one extrusion	eight extrusions
TEM	0.25 ± 0.15	0.25 ± 0.15
Misorientations	3.0 ± 0.5	1.0 ± 0.3

This is attested by the partial pole figure reconstructed from the recorded orientations which exhibits roughly the same peaks as the X-ray pole figures. It should be noted that the high angle boundaries are not periodically placed along the measured lines. The grain splitting is inhomogeneous which means in turn that the deformation was heterogeneous. Consequently the calculated grain size is an average. The striking feature pointed out by Table 2 is the decrease of the grain size while the structural size, including high and low angle boundaries, is nearly constant. Measurements at intermediate number of passes are under progress. The available results strongly suggest that, despite its apparent stability, the structure did not reach a steady state. The recorded hardening is believed to result from the grain diameter evolution.

4. Conclusions

Samples of AA5083 were subjected to equal channel angular extrusion. Up to eight passes were performed using the so-called route C at 150°C. The mechanical behavior, as measured by room temperature compression tests, exhibits a small hardening which does not saturate within the investigated levels of strain. By contrast, both the overall texture and the structural size deduced from TEM pictures are stable after the first extrusion. The real nature of the microstructure was analyzed in terms of misorientations between adjacent cells. A dedicated experimental procedure based on TEM diffraction patterns and computer indexings was developed. Measurements along straight lines show that the substructure is a mixture of grains and subgrains even at large strains. The average grain diameter appears to decrease down to 1 µm while the apparent substructure size takes rapidly a constant value lower than 0.2 µm. It is claimed that the recorded hardening is related to the evolution of the grain size.

5. References

1. Vetrano, J.S., Lavender, C.A., Hamilton, C.H., Smith, M.T., and Bruemmer, S.M. (1994) *Scripta Metall. Mater.* 30, 565.
2. Friedman, P.A., Ghosh, A.K. (1996) *Metall. Mater. Trans.* 27A, 3827.
3. Iwasashi, Y., Horita, Z., Nemoto, M., and Langdon, T.G. (1997) *Acta Mater.* 45, 4733.
4. Berbon, P.B., Tsenev, N.K., Valiev, R.Z., Furukawa, M., Horita, Z., Nemoto, M., and Langdon, T.G. (1998) *Metall. Mater. Trans.* 29A, 2237.
5. Y. Iwahashi, J. Wang, Z. Horita, M. Nemoto, and Langdon, T.G. (1996) *Scripta Mater.*, 35, 143.
6. Segal, V.M. (1995) *Materials Science and Engineering* A197, 157.
7. Wenk, H.R., Matthies, S., Donovan, J., and Chateigner, D.J. (1998) *J. Appl. Crystallogr.* 31, 262.
8. Rollet, A.D. (1988) Ph.D. Thesis, Drexel University, Department of Mater. Sci. Eng. Philadelphia. Edited By Los Alamos National Laboratory, LA-11202-T. Los Alamos, NM.

9.   Sevillano, J.V. (1993) chapter 2, in Mughrabi, H. (ed.), *Plastic Deformation and Fracture of Materials*, VCH (Germany), pp.19-88.
10. U.F. Kocks, in *Texture and Anisotropy : preferred orientations in polycrystals and their effect on materials properties*, U.F. Kocks, C.N. Tomé and H.-R. Wenk (Authors and Editors), Cambridge University Press, 1998, p.86

A TEM-BASED DISCLINATION MODEL FOR THE SUBSTRUCTURE EVOLUTION UNDER SEVERE PLASTIC DEFORMATION

M. SEEFELDT*, V. KLEMM, P. KLIMANEK
Freiberg University of Mining and Technology
Institute of Physical Metallurgy
D-09596 Freiberg, Germany
Now at K.U. Leuven, Departement MTM,
de Croylaan 2, B-3001 Heverlee, Belgium

1. Introduction

Physical modelling of the macroscopic mechanical properties of metallic materials requires a coupling between these properties and the underlying micro- and substructural features. The substructure development under plastic deformation up to large strains at low and intermediate temperatures is characterized by the *coexistence of two substructures on different size and misorientation scales*, namely of a *cell structure* and a *fragment (cell block) structure* [1,2]. While the cell structure saturates with respect to size and misorientation, the mean fragment size decreases and the mean misorientation between fragments increases monotonously. Long-range stresses are present in the cell as well as in the fragment interiors. The present model describes the cell and the fragment structure development, that means the substructure development on the *microscopic* and the *mesoscopic* length scales, through separate, but coupled evolution equations for dislocations carrying deformation and work-hardening on the microscopic scale and for disclinations carrying them on the mesoscopic scale. The authors propose *non-compensated nodes of fragment (cell block) boundaries*, that means triple junctions with an orientational mismatch around them, to have a *disclination character*, and, thereby, to be *sources of long-range stresses*.

2. Disclinations

Volterra introduced disclinations into the theory of elasticity in 1907 as the rotational twin-sisters of the translational dislocations. Fundamentals and applications of disclination theory can be found in the review papers by deWit [3] and Romanov and Vladimirov [4]. Considering self-screening configurations of partial disclinations (PD) [4] instead of single perfect ones allows the "translation" of PDs into dislocation groups [3,4] [1] (fig. 1).

Using the above "translation", Romanov and Vladimirov showed that PD dipoles (PDD) can *propagate* by capturing and incorporating dislocations [4]. Such propagating PDDs are *immobilized* resulting in a network of sessile PDDs or - "translated" into

---

[1] "Translation" means that PDs and dislocation groups have the same long-range stress fields.

T.C. Lowe and R.Z. Valiev (eds.), Investigations and Applications of Severe Plastic Deformation, 197–202.
© *2000 Kluwer Academic Publishers. Printed in the Netherlands.*

198

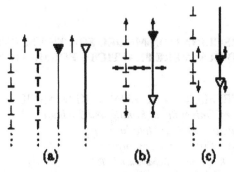

Figure 1. Types of wedge PDDs (black and white triangles) and their "translations" into dislocation "language": (a) two parallel semi-infinite excess edge dislocation boundaries, (b) a both-side terminated excess edge dislocation boundary, (c) an excess dislocation boundary segment carrying an enlarged misorientation.

dislocation "language" - in a mosaic of excess dislocation boundaries. If arising boundaries grow into an existing one by PDD propagation and if the misorientations across all three boundaries are preserved, then *orientational mismatches* are left behind around the new triple junctions or --- "translated" into disclination "language" --- an immobile PDD with the Frank vectors of the mobile one locked at the existing boundary. Since such *non-compensated nodes* are preserved after load relaxation, they are suitable objects for an experimental confirmation of a disclination interpretation of large strain substructures (fig. 2).

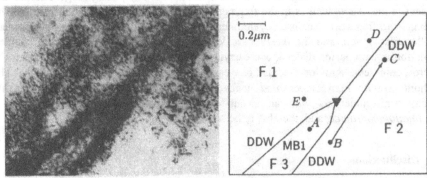

Figure 2. PD in the splitting line of a dense dislocation wall (DDW) into a first generation microband (MB1) in a Cu polycrystal rolled down to 50% at room temperature. the bullets A-E indicate where local orientation measurements were taken.

In local orientation measurements by TEM microdiffraction at the indicated spots, misorientation matrices were found whose product does not give the identity, but an orientational mismatch matrix resulting into a Frank vector [1,3,4] of

$$\omega = [-0.0046 \ -0.0316 \ -0.0004] \quad \text{with} \quad |\omega| = 1.83° \tag{1}$$

This PD power is in good agreement with the theoretical expectation of $|\omega| \approx 1\text{-}3°$ given by Romanov and Vladimirov in [4]. A *continuous* variation of the local orientation was observed under shift of the beam from spot B to spot C – indicating an elastic distortion. The corresponding stress field has a marked influence on defect dynamics. For more details and measurements see [5].

3. Model

In the model, the substructure develops according to the following scenario: Mobile dislocations (density ρ_m) are generated multiplicatively and immobilized at or annihilated with redundant cell wall dislocations (ρ_r). *Incidental non-balanced trapping of neighboring excess dislocations generates the first propagating PDDs (θ_b) along cell walls. By capturing further mobile dislocations, a PDD spreads out misorientation from a nucleus over the whole cell wall and superposes the cell wall with a fragment boundary.* Incidental trapping of mobile dislocations with "proper" Burgers vector into existing fragment boundaries generates propagating PDDs enlarging misorientation (fig. 1, (c)). Propagating PDDs are immobilized in the network of other propagating and immobile PDDs (θ_i). For details, see [6].

3.1. REACTION MECHANISMS INCLUDING DISCLINATION DENSITIES[2]

Since cell walls are obstacles for glide dislocations, interface dislocations (giving tensile and compressive stresses) are generated in reactions between two mobile dislocations approaching the wall on intersecting slip planes from the same side (fig. 3, (a), cf. [7]), whereas excess dislocations (giving misorientations) are formed as a result of reactions between two mobile dislocations approaching it from different sides (fig. 3, (c), cf. [8]). A PDD is generated, if several (say three) neighboring mobiles are trapped as excess dislocations on the one side and are not balanced by opposite excess dislocations on the other side (fig. 3, (d, e)). Thus, the probability P_c^{PDD} of a combined mobile dislocation generating a PDD when being trapped into a cell wall can be expressed by the probability P_c of being trapped times the probability $(1-P_c)$ of the corresponding combined mobile dislocation on the other side *not* being trapped, raised to the third power to take into account the required collective behavior, $P_c^{PDD} = (P_c(1-P_c))^3$. The loss rate of mobile dislocations due to PDD-generating trapping into cell walls is approximated as (w - cell wall width, f_g - geometric factor for non-equiaxed cells, ξ - cell wall volume fraction, K - Holt's constant, m_S - Schmid factor, using similitude and Holt's law)

$$\left(\frac{d\rho_m}{d\varepsilon}\right)_{K_c}^{-} \approx -4\xi \frac{P_c^{PDD}}{w}\frac{1}{m_s b} = -4\frac{\left(P_c(1-P_c)\right)^3 f_g}{K}\frac{1}{m_s b}\sqrt{\rho_r} =: -K_c\sqrt{\rho_r}. \quad (2)$$

Trapping into a *fragment boundary* gets easier with increasing misorientation. The product $P_f(1-P_f)$ (P_f - probability of a mobile dislocation being trapped by a fragment boundary) takes its maximum for $P_f = 1/2$. Two factors of 1/2 take into account that the

---

[2] For reaction terms including only dislocation densities, we refer to the literature.

trapped dislocation must have the same sign and the same character as the excess dislocations in the boundary to generate a PDD, whereas collective behavior is not required, $P_f^{PDD} = \frac{1}{4} P_f \left(1 - P_f\right)$.

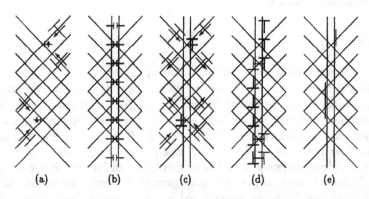

Figure 3. Generation of PDDs at cell walls. (a) formation of a cell wall, (b) cell wall of the Mughrabi type, (c) cell wall as a glide obstacle triggering the formation of excess dislocations, (d) incidental irregularities in the dipolar character of the excess dislocation walls, (e) translation of (d) into disclination "language". Large edge symbols denote mobile, small ones redundant cell wall, bold ones non-redundant excess edges. See [7,8].

The fragment boundary width is assumed to be small compared to the mean fragment diameter, so that the loss rate of mobile dislocations due to PDD-generating trapping into fragment boundaries scales with $1/d_f$ instead of with $1/w_f$. A Poisson-Voronoi geometry of the fragment structure gives $d_f \approx 1.66/\sqrt{\theta_i}$ [6] and

$$\left(\frac{d\rho_m}{d\varepsilon}\right)_{K_f}^{-} \approx -4 \frac{P_f^{PDD}}{d_f} \frac{1}{m_s b} \approx -0.602 P_f \left(1 - P_f\right) \frac{1}{m_s b} \sqrt{\theta_i} =: -K_f \sqrt{\theta_i} . \tag{3}$$

According to Romanov and Vladimirov [4], a partial disclination dipole (PDD) of width a propagates by capturing and incorporating mobile dislocations into its backward excess dislocation boundaries. The Peach-Koehler force exerted by the PDD [4] drives a mobile dislocation of appropriate Burgers vector into the dipole plane. To estimate the capture length of the dipole, the stress $\tau_{xy}^{PDD} + \tau_{ext}$ is compared to the resistance τ^* due to a statistical dislocation distribution giving $y_{cl} \approx 10a$. In contrast to [4], the present model "feeds" the dipole not by a sessile dislocation density but by a dislocation current. One finds a loss rate of

$$\left(\frac{d\rho_m}{d\varepsilon}\right)_E^{-} = -y_{cl} \frac{1}{m_s b} \theta_p \approx -10a \frac{1}{m_s b} \theta_p =: -E\theta_p . \tag{4}$$

This mechanism also relates the average velocities $\bar{v}$ and $\bar{V}$ of the mobile dislocations and the propagating PDDs to each other. The time within which the dislocation current provides enough dislocations to allow the PDD to propagate by its capturing length can be estimated as $t_c \approx 2\omega/b\bar{v}\rho_m$ giving $\bar{V} \approx y_{cl}/t_c \approx \left(5ab/\omega\right)\bar{v}\rho_m$. Due to the force exerted by a sessile on a propagating PDD [4], each of the partials of a propagating PDD extending along a cell wall stops at the next fragment boundary, *i.e.* in the next mesh of the immobile PDD network. Each of the partials of a propagating PDD extending on a fragment boundary will undergo piling-up or partial annihilation and stop at the next triple junction, *i.e.* in the next node of the immobile PDD network. Besides, immobilization at mixed junctions between fragment boundaries and cell walls is possible. This effect is roughly taken into account by reducing the mean free path by a factor of 5. With a Poisson-Voronoi geometry of the fragment structure and the above relationship between $\bar{v}$ and $\bar{V}$, the propagating PDD loss rate then is

$$\left(\frac{d\theta_p}{d\varepsilon}\right)_J^- \approx -\frac{5}{d_f}\frac{1}{m_s b}\frac{5ab}{\omega}\theta_p \approx -15.1\frac{a}{\omega}\frac{1}{m_s}\sqrt{\theta_i}\,\theta_p =: -J\sqrt{\theta_i}\,\theta_p . \tag{5}$$

Since the average value of the dipole width a keeps roughly constant and the disclination power ω is slowly varying compared to the defect velocities and densities, they are regarded as constants when evaluating coefficients E and J.

3.2. EVOLUTION EQUATIONS

The resulting system of evolution equations reads (M, I, R - constants from dislocation dynamics, see [6], N-number of excess dislocations forming a new PDD)

$$\frac{d\rho_m}{d\varepsilon} = M - I\sqrt{\rho_r} - R\rho_r - K_c\sqrt{\rho_r} - K_f\sqrt{\theta_i} - E\theta_p \tag{6}$$

$$\frac{d\rho_r}{d\varepsilon} = I\sqrt{\rho_r} - R\rho_r \tag{7}$$

$$\frac{d\theta_p}{d\varepsilon} = \frac{K_c}{N}\sqrt{\rho_r} + \frac{K_f}{N}\sqrt{\theta_i} - J\sqrt{\theta_i}\,\theta_p \tag{8}$$

$$\frac{d\theta_i}{d\varepsilon} = J\sqrt{\theta_i}\,\theta_p \tag{9}$$

3.3. COUPLING TO THE MECHANICAL BEHAVIOR

One flow stress contribution, σ_ρ, is coupled to the (global average) redundant cell wall dislocation density ρ_r, whereas the other one, σ_θ, is coupled to the immobile PDD density θ_i. The mean disclination power $|\omega|$, can - according to section 2 - be approximated with the mean misorientation ϕ which can be calculated from the excess dislocation density and the PDD density for a Poisson-Voronoi geometry of the fragment structure, $\phi \approx b\rho_{exc}/1.21\sqrt{\theta_i}$ [6]. Romanov's and Vladimirov's law $\Delta\sigma_\theta = \beta G|\omega|$ gives

$$\sigma_{tot} \approx \sigma_\theta + \sigma_\rho = \beta G b \frac{\rho_{exc}}{1.21\sqrt{\theta_i}} + \frac{\xi \alpha G b}{m_s} \sqrt{\frac{\rho_r}{\xi}}. \tag{10}$$

Thus, a dislocation or cell contribution and a disclination or fragment contribution to the total flow stress can clearly be distinguished.

4. Results and Discussion

The model is able to reproduce the saturation of the cell structure, a hyperbolic decrease of the mean fragment size and a linear increase of the mean misorientation between fragments [6]. Figure 4 shows a flow curve evaluated for $P_c=1/4$, $P_f=1/2$, $\xi=0.20$, $f_g=2$, $K=16$, $a=0.1$ μm, $\alpha=0.3$, $\beta=0.1$, $R=2y_e/m^Sb=12/m^S$. Stage III and stage IV can clearly be distinguished. Stage III is due to the cell structure or dislocation contribution and stage IV is due to the fragment structure or disclination contribution.

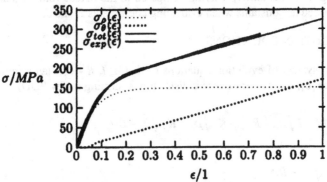

Figure 4. Calculated flow curve and its cell and fragment structure contributions, experimental flow curve from Cu single crystal compression at room temperature.

5. References

1. Rybin, V.V. (1986) *Bolshie plasticheskie deformatsii i razrushenie metallov*, Metallurgiya, Moscow (in Russian).
2. Bay, B., Hansen, N., Hughes, D.A., Kuhlmann-Wilsdorf, D. (1992) Evolution of f.c.c. deformation structures in polyslip, *Acta metallurgica et materialia* **40**, 205-219.
3. deWit, R. (1973) Theory of Disclinations, *Journal of Research of the National Bureau of Standards A* **77A**, 49-100, 359-368 and 607-658.
4. Romanov, A.E., Vladimirov, V.I. (1992) Disclinations in Crystalline Solids, in F.R.N. Nabarro (ed.), *Dislocations in Solids, vol. 9*, North Holland, Amsterdam, 191-422.
5. Klemm, V., Klimanek, P., Seefeldt, M. (1999) A tem microdiffraction method for the characterization of partial disclinations in plastically deformed metals, *physica status solidi (a)* **175**, 569-576.
6. Seefeldt, M. (1999*) Modellierung der Substrukturentwicklung bei Kaltumformung mit Hilfe von Disklinationen*, accepted as Ph.D. thesis, Freiberg University of Mining and Technology, Freiberg (in German).
7. Mughrabi, H. (1983) Dislocation wall and cell structures and long-range internal stresses in deformed metal crystals, *Acta metallurgica* **31**, 1367-1379.
8. Ungar, T., and Zehetbauer, M. (1996) Stage IV work hardening in cell forming materials, part II: A new mechanism, *Scripta materialia* **35**, 1467-1473.

PHYSICAL MESOMECHANICS OF ULTRAFINE-GRAINED METALS

V.E. PANIN
Institute of Strength Physics and Materials Science, SB, RAS
Tomsk, 634021, Russia

1. Introduction

In the last two decades, a new scientific trend, the physical mesomecanics of materials [1–6], which links continuum mechanics, physics of plasticity and strength (dislocation theory), and physical materials science has been developed in Tomsk.

Conventional approaches to the description of a deformable solid in continuum mechanics (macrolevel) and dislocation theory (microlevel) are based on force models. It is assumed that, on reaching the yield point of a macroparticle in some local region of a specimen, the mean applied stress produces plastic flow in the region that occurs along with strain hardening.

In physical mesomechanics, there is a qualitatively different synergetic methodology based on the scale levels of shear-stability loss in local stress-concentrator zones. The mean applied stress is fundamentally unable to produce plastic flow of the macrovolume of a loaded solid, which is macroscopically shear-stable. Plastic shear is localized in certain stress-concentrator zones, where the original crystal loses its shear stability and takes on a different structure. The evolution of the shear-stability loss of a loaded solid at different scale levels culminates in its failure. The basic principles of physical mesomechanics and their application to analysis of the mechanical behavior of ultrafine-grained metals are presented.

2. Fundamental principles of physical mesomechanics

The following fundamental principles of physical mesomechanics of materials are put forward [5].

1. A loaded solid is a self-organizing multilevel system with a high degree of non-equilibrium, where the plastic flow develops as synergetic evolution of the shear-stability loss at micro-, meso-, and macrolevels (Fig. 1).

2. At the microlevel, shear-stability loss of the crystal lattice occurs in local stress-microconcentrator zones. The basic crystal-lattice defect is a dislocation resulting from local structural transformation of the crystal lattice within the stress-microconcentrator gradient field. The dislocation core is a fragment of a new structure, which is of higher energy than the original crystal lattice (Fig. 1(a)).

3. At the mesolevel, shear-stability loss occurs in local zones of the loaded specimen as a whole. The basic mesodefect is a mesoband generated in local zones of stress me-

T.C. Lowe and R.Z. Valiev (eds.), Investigations and Applications of Severe Plastic Deformation, 203–209.
© *2000 Kluwer Academic Publishers. Printed in the Netherlands.*

soconcentrators and propagating in the directions of the maximum tangential stress τ_{max}, regardless of the crystallographic orientation of the lattice (Fig. 1(c)).

4. The basic plastic-flow carrier at the mesolevel is a three-dimensional mesovolume: cells of dislocation substructures (Fig. 1(b)), deformation domains, subgrains, grains, their conglomerates, etc. Their motion is realized by the scheme "shear + rotation".

5. A crystalline material capable only of translational shear forms a hierarchy of dissipative substructures at the mesolevel whose deformation follows the scheme "shear + rotation". This process leads to fragmentation of the material at the mesolevel.

6. Failure is the global shear stability loss of a loaded specimen. A stress macroconcentrator appears in a deformed solid and transition of the specimen fragmentation from the mesolevel to the

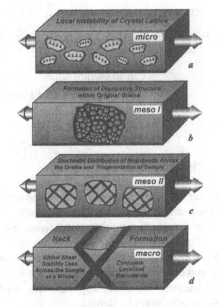

Figure 1. Schematic of scale levels of shear stability loss in deformed solid.

macrolevel takes place. Two macrobands of localized deformation (parallel to τ_{max} or in conjugate directions) propagate through the whole specimen cross-section, culminating in the limiting case of macrolevel fragmentation: separation of the specimen into two parts (Fig. 1(d)).

7. The plastic flow mechanisms, their carriers and the corresponding stages of the stress-strain curve are scale-invariant (the scaling principle).

8. Within the framework of the methods of physical mesomechanics all the data on a loaded material, including the data on its internal structure, may be entered in a computer, and computer-aided design of materials with specified mechanical properties (the inverse problem) can be realized.

The principle of scale invariance leads to the fundamental conclusion: any type of deformation may be represented by a certain combination of standard characteristic elements (plastic shears) of different scale levels [4]. This provides the basis for construction of a clear algorithm of mathematical representation as to the plastic flow of material with any internal structure and for any loading conditions. In each case, a certain scale level (and its corresponding elements) is dominant. The elements of other scale levels are either accommodational or absent.

The basic works on the dislocation theory that use a transmission electron microscope are related to the conditions, where the microlevel is dominant. The corresponding deformation mechanisms are exclusively interpreted in terms of the nucleation, motion, and interaction of dislocations. Attempts to extend this interpretation to all possible cases of plastic deformation encounter insurmountable difficulties. We now understand that the meso- and macrolevels may also be dominant. In these cases, the interpretation must be based on strain-induced defects at the meso- and macrolevels (deformation meso- and macrobands, disclinations, etc.), and dislocation motion must be regarded as an accommodational process. Typical situations when the mesolevel is dominant include: deformation of high-strength alloys [6] or surface-hardened materials [5], tension of specimens after preliminary drawing or rolling [1], and all types of fatigue fracture [4]. The macrolevel is dominant in the case of deformation of specimens with welded joints, holes or grooves, etc. [1].

Within the limits of physical mesomechanics, systematic efforts are being made to compile standard elements of the scale levels of deformation for a wide range of materials and various loads. Such data are invaluable for designers and specialists in the field of computer-aided design of advanced materials.

3. Relationships governing the mechanical behavior of ultrafine-grained copper at different temperatures

The mechanical properties of ultrafine-grained (UFG) metals at room temperature are characterized by high yield point and strength, but such metals exhibit very low plasticity at tension. It is typical for plastic deformation of solids at the mesoscale level. There are good grounds to believe that the mechanical behaviour of UFG metals should be analyzed on the basis of physical mesomechanics

The present work is devoted to the overview of results [5, 7, 8] related to investigations into

Figure 2. Microstructure of UFG copper as received. × 41 000.

the mechanisms of plastic deformation in UFG copper at different temperatures. The UFG copper with the grain size ~0.2–0.5 μm (Fig. 2) was produced through intensive preliminary deformation by equal-channel angular extrusion up to the strain $\varepsilon_{real} = 3$, *i.e.*, the copper was in an extremely non-equilibrium state. After thorough mechanical polishing a number of flat specimens were subjected to tension at room temperature at a rate of 9.6 mm/h, for which an IMASH-2078 testing machine equipped by the Television-Optical Meter for Surface Characterization (TOMSC) was employed. The remainder of the specimens was annealed at 473 K for 30 min (without visible changes in internal structure, but the σ–ε curve was gently descending) and subjected to tension at different temperatures: 373, 423, and 453 K. For comparison similar specimens of the coarse-grained copper with the grain size 40 μm were tested at room temperature and at 453 K. The mesoscale deformation patterns of the specimen were registered in situ every 7 seconds and they were used to calculate the displacement vector fields. TEM studies of the original grain structure and strain-induced defects were performed using an M-125 K microscope by the methods of self-shading replica and foils.

206

3.1. RESULTS OF MECHANICAL TESTS

Shown in Fig. 3 are the tensile curves for UFG and coarse-grained copper specimens at different temperatures. It is evident that under the conditions of equal-channel angular extrusion and formation of the UFG structure the mechanical properties of the copper are essentially dependent on temperature.

At room temperature the yield point of UFG copper increases by a factor of 7 and plasticity drops by a factor of 4 as compared to the corresponding characteristics of coarse-grained copper. The stage of quasi-uniform flow for UFG copper appears to be abnormally short, 1–2% only, and is characterized by the absence of any strain hardening, whereas the neck formation stage for UFG copper is much longer than that of coarse-grained copper.

The onset of plastic deformation of UFG copper specimens is readily detected by the appearance of odd pronounced mesobands of localized deformation (~10 μm wide) propagating at 54° to the axis of tension and almost instantaneously transversing the specimen's cross-section. This short stage of quasi-uniform flow is characteristic of plastic deformation at the mesoscale level and ends up by the formation of two localized

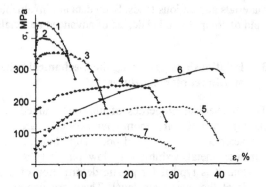

Figure 3. Stress-strain curves for copper specimens under tension at different temperatures. UFG copper: T_{test} = 293 K, as received (1); tension after preliminary annealing at 473 K: T_{test} = 293 K (2), 373 K (3), 423 K (4), 453 K (5). Coarse-grained copper: T_{test} = 293 K (6), 353 K (7).

conjugate macrobands (Fig. 4(a)). The microstructure of a typical localized macroband is shown in Fig. 4 (b). It can be classified as a substructure with continuum disclination density, where the crystal-lattice curvature is up to 40 deg/μm. Their evolution defines the neck formation corresponding to the macroscale level of shear stability loss of a loaded specimen. A very long process of the neck formation culminates in ductile fracture along one of the conjugate macrobands of localized deformation.

An increasing test temperature results in extension of quasi-uniform flow of the UFG specimen and in appearance of strain hardening (Fig. 3). Comparison of the σ–ε curves at 453 K for UFG and coarse-grained copper shows that the yield point of UFG copper is higher than that of coarse-grained copper by a factor of 2.9. But contrary to the behavior of UFG copper at room temperature the plasticity of UFG copper at elevated temperature as compared to coarse-grained copper is increased by a factor of 1.3.

Electron-microscope data (Fig. 4(c)) show that the plastic flow of UFG copper at 453 K are realized by the mechanisms inherent in the microscale level: motion of dislocations, deformation twinning, dynamic recrystallization being accompanied by the grain growth up to 3–5 μm.

The transition of plastic deformation of UFG copper from the mesoscale level to the microscale level develops gradually as testing temperature is increased.

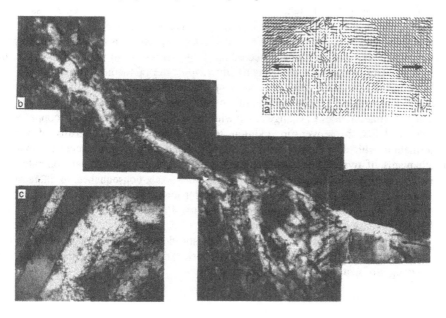

Figure 4. The displacement vector field in the neck of the UFG copper specimen (a) and the light-field image of the localized macroband in the neck; 293 K, ε= 9 %, × 7 500 (b); microstructure of UFG copper after annealing at 473 K and subsequent tension at 453 K, ε= 33 %, × 15 500 (c).

3.2. DISCUSSION OF UFG COPPER MECHANICAL BEHAVIOR

UFG copper exhibits its shear instability at all testing temperatures. But the mechanisms of the shear stability loss and corresponding mechanical behavior of UFG copper are essentially dependent on temperature.

There is reason to think that low plasticity, clearly defined disclination mechanisms of plastic flow and absence of strain hardening are attributable to the fact that dislocations can not originate in UFG copper under tension at room temperature. Formation of the UFG structure in copper leads to a qualitative alternation of the material type as compared to coarse-grained copper. In essence, UFG copper is a composite material, with 90 % of its volume being occupied by hard inclusions of a crystalline phase and 10 % by a shear-unstable interlayer of the defect "phase". The collective effects of the strain-induced defects behavior in the defect interlayer are responsible for the behavior of the latter as a hydrodynamic medium. The relaxation times of collective atomic redistribution in the defect interlayer are extremely short. It is, therefore, unlikely for the shear stress components to occur in it. The latter are necessary to cause stress microconcentrators capable of generating dislocations in the course of tension. Hence, the crystalline inclusions show no dislocation accumulation and, therefore, no strain hardening. When stress mesoconcentrators become active in the UFG copper under loading, the loss of shear stability occurs along the τ_{max} direction. Mesobands propagate throughout the cross-section of the specimen regardless of its crystallographic structure. This stage is, however, very short on the σ–ε plot.

A global loss of shear stability of the UFG copper specimen results in two conjugate macrobands. Their development according to the scheme of the phase wave of switching ensures high local plasticity of UFG copper during the neck formation [7].

Contrary to the deformation at room temperature, the plastic flow of UFG copper at elevated temperatures proceeds at the microscale level. That is the scale level at which plastic deformation doesn't change along with the reduction in grain size in copper from coarse to UFG. Moreover, in addition to the dislocation mechanisms of plastic deformation, which are characteristic of coarse-grained copper, very effective relaxation mechanisms of dynamic recrystallization and deformation twinning are operative in UFG copper under loading at elevated temperatures. As a consequence, an increase of the yield point and strength of UFG copper under loading at elevated temperatures as compared to the corresponding properties of coarse-grained copper is accompanied by sufficient increase of material plasticity.

Thus, the retention of the scale level of plastic deformation at transition from the coarse-grained structure to the UFG one is a necessary condition for increase of not only strength of materials but also their plasticity.

4. Conclusion

The mechanical behavior of UFG copper at different temperatures is defined by the plastic flow mechanisms being characteristic of different scale levels: meso- and macrolevels at room temperature and microlevel at elevated temperatures. The familiar statement that added strength of a solid occurs along with decreased plasticity is true for the conditions when the solid strengthening causes the elevation of the scale level of plastic flow. UFG copper at elevated temperatures is deformed by the mechanisms of the microscale level similar to coarse-grained copper. Therefore, the transition in these conditions from coarse-grained copper to UFG copper leads to increase of the both strength and plasticity of a solid. This regulation should be taken into account for industrial applications of UFG materials.

5. Acknowledgments

The author would like to thank Prof. R.Z. Valiev, Prof. A.N. Tyumentsev, Dr. L.S. Derevyagina, and Dr. N.A. Dubovic for support of this overview. This work has been financially supported by the Russian Fund for Basic Research, Project No. 99-01-00583.

6. References

1. Panin, V.E. (ed.) (1998) *Physical Mesomechanics of Heterogeneous Media and Computer-Aided Design of Materials*, Cambridge Interscience Publishing, Cambridge.

2. Panin, V.E. (1994) Physical mesomechanics of plastic deformation and fracture of solids, *in Proc. of the 10-th Int. Conf. on the Strength of Materials, Japanese Inst. of Metals*, Sendai, pp. 415–418.
3. Panin, V.E. (1997) Plastic deformation and fracture of solids at the mesoscale level, *Mat. Sci. Eng. A* **234–236**, 944–948.
4. Panin, V.E. (1998) Overview on mesomechanics of plastic deformation and fracture of solids, *Theor. Appl. Fracture Mech.* **30**, 1–11.
5. Panin, V.E. (1998) Foundations of physical mesomechanics, *Phys. Mesomech.* **1**, No. 1, 5–20.
6. Panin, V.E., Korotaev, A.D., Makarov, P.V., and Kuznetsov, V.M. (1998) Physical mesomechanics of materials, *Rus. Phys. J.* **41**, 856–884.
7. Panin, V.E., Derevyagina, L.S., and Valiev, R.Z. (1999) Mechanism of shear band formation in submicrocrystalline copper under tension, *Phys. Mesomech.* **2**, No. 1–2, 89–94.
8. Panin, V.E., Derevyagina, L.S., Tyumentsev, A.N., Valiev, R.Z., and Dubovic, N.A. (in press) Temperature dependence of ultrafine-grained copper mechanical properties and mechanisms of plastic deformation at tension.

MICROSTRUCTURE EVOLUTION IN TI-ALLOYS DURING SEVERE DEFORMATION BY ELECTRIC UPSETTING AND IMPACT FUSED-FORGING MODELING

New power-saving technologies

B.N. KODESS[1,2], L.A. KOMMEL[3], G.P.TETERIN[4], V.K.OVCHAROV[1]
[1]*VNIIMS, 46 Ozernaya St., Moscow 119361*
[2]*ICS&E at Denver*
21277 E. Aberdeen Pl., Aurora, CO 80015
[3]*TTU, 5 Enitajate Tallin EE-0026, Estonia*
[4]*MSOU, 22 P. Korchagina St., Moscow 129805, Russia*

Abstract

The microstructure and micro-strain level during severe plastic deformation (SPD) of 30 heat-resistant Ti-based billets has been characterized using of OM, TEM, EDS and XRD methods. Samples of Ti-based alloys was produced using SPD with simultaneous high speed heating (up to 250 C/sec), and titanium aluminide with rapid solidification. Bragg reflections analysis for different order of diffraction shows metastable state of phases after SPD. The micro-strain make major contribution to broadening of the peaks. The subfine-grained microstructure and texture change have been found in a zone of maximum deformation. The synergetic approach has been used for explanation of the peculiarities of the system of micro- and macro- strains and microstructure evolution during SPD with simultaneous supply of energy.

1. Introduction

Replacement of Ni-based alloys with more light weight Ti-based alloys with preservation of strength characteristics at high temperatures is a long-standing objective of modern engine-building. Among the diversity of the bimetallic metallic compositions, the Ti-based dual alloys are of a high importance for both creation and improvement of engine parts and medical equipment, and other commercial applications.

The microstructure of Ti-based alloys is determined mainly by technological parameters during hot deformation and cannot be completely changed by subsequent thermal treatment. At the same time it is possible to improve a number of mechanical and service characteristics when there is an appropriate favorable initial microstructure.

For manufacturing of parts made of Ti-based alloys with regulated and predetermined macro- and microstructure it was proposed to use severe plastic deformation, SPD, with simultaneous intensive and high speed heating [1-3] or rapid

T.C. Lowe and R.Z. Valiev (eds.), Investigations and Applications of Severe Plastic Deformation, 211–218.
© *2000 Kluwer Academic Publishers. Printed in the Netherlands.*

solidification [4-5]. Combination of the various methods allows to control the conditions more flexibly and to reduce number of stages and time of parts finishing. In the article [4] the authors show the feasibility of reliably bonding dual alloy components along the whole length of contact, using the IFFM method. Disks of automotive valves were made of titanium-aluminide with superalpha 2 orthorombic aluminide and a stem was made of high-strength VT-titanium alloys with conventional additions. It is important that process modeling with finite elements [3] and results of first experiments showed that Ti-alloys may take high strength properties as a result of preliminary severe deformation of billets. The purpose of the present article is to report on the results of microstructure and x-ray examination of one of such VT-alloys. Instant heating combined with deformation was used as the first stage of SPD, and it allowed to reach sub-fine grains (of 500-1000 nm) already at this first stage. We also discuss usage of the preliminary processed billets in different applications including manufacturing of bimetallic parts and in the process of multiple forging. The synergetic approach has been used to explain the peculiarities of the system of micro- and macro-strains and microstructure evolution during SPD. This approach has us to successfully explain temporal evolution of the structural parameters and formation of similar defects hierarchy in multicomponent metalloceramics [7-10].

2. Results

Scheme and technology features of samples production are described in [1-6] and choisingce of thermal-physical parameters for deformation with instant electric heating is in [1-3]. The axial plane macrostructure examination has shown that for all samples (more than 30) and heating speeds between 30-170°C/sec microstructure can be divided into five zones. In this technique zone of the deformation and heating of billets is shifted sequentially to nondeformed parts of billets. Logarithmic degree of deformation has reached a value of $\cong 1.4$ and relative deformation - of 69% correspondingly at the temperature of 1010-1020°C. If the speed of heating increases higher than 170°C/sec (j=8-26 MA/m^2), microstructure distinctions between zones disappear smoothly and for maximum speed of heating of 250°C/sec only two zones remain distinctive.

Both X-ray diffraction (XRD) and energy-dispersive analysis (EDS), and also optic and electron microscopy techniques (OM, SEM, TEM) were employed to characterize the crystal structure, phase composition and microstructural features of the samples. The diffraction pattern was determined by XRD techniques using Cu-K α radiation in X-ray diffractometers. Bragg angles and the half-width of Bragg reflections profiles of high and low orders for the α- and β-phases, and two intermetallic phases were measured. Microhardness measurements were undertaken using 15-g, 50-g and 100-g loads with Micrometer-2001 and PMT-3 testers.

The structure under observation is obtained as a result of a processes in course of deformation of billets which has been instantly heated (EHF) or undercooled (IFFM) during their transfer through the conduit, which connects the upper chamber with the evacuated die cavity (see also Fig. 1 in [4]). Analysis of X-ray pictures exposes two major sets of Bragg reflections. Each set for intermetallics corresponds to γ-phase

(TiAl) and α_2-phase (Ti$_3$Al) respectively, or for α- and β-phases for Ti-based VT alloys. The part of α_2-phase is in the range of 8-15% of the total content, and β-phase (for VT alloys) is in the range of 20-35%. Fig. 1 shows comparison of x-ray results for three billets of VT-18 alloys (TiAl$_{6.8}$Zr$_{3.5}$Sn$_2$MoNb) with different degree of deformation. Fig. 2a and 2b are typical for the zone of 69% degree of deformation at 1010°C and for a zone with 15% deformation at 950°C, respectively. Fig. 3a shows a general (X 500) microstructure of VT -18 of the stem in a zone of greatest deformation at the temperature of 1010°C.

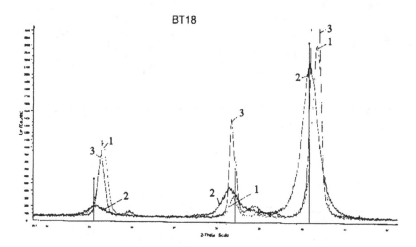

Figure 1. VT-18 billets X-ray difractograms (1-initial, 2-deformation degree 69%, 3- deformation degree 15%).

It is seen from these figures that the block boundary dies out and evolves into an equiaxial and subfine-grained microstructure (with grain size of up to 1000 nm). The microstructure has an appearance of bands drawn in the direction of metal flow with axial-symmetrical kind of flows.

X-ray data show that the texture and micro-strain level has been changed during deformation too. It is of interest that after deformation the intensity of the peaks (11.0)α and (200)β was higher than normal for all samples. This phenomenon was the most pronounced in a zone of maximal deformation. As it is noticeable on the x-ray diffractograms, there is a remarkable increase of background, appearance of additional peaks of Bragg reflections and shift of angular position of main phases peaks after deformation.

For qualitative estimation of the possible contributions to observed broadening, measurements of half-widths (FHWW) of Bragg's lines of different order were made for α- and β-phases of alloy and respectively γ- and α_2-phases of intermetallic. Ratios of FHWW of low and high angles peaks were within the range of 1.25 and 1.40.

214

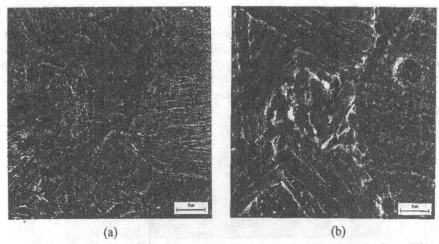

(a)   (b)

Figure 2. Micrograph of the VT-18 billets, (a)- deformation degree 69%, (b)- deformation degree 15%.

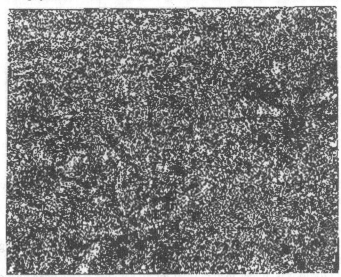

Figure 3. Micrograph of the VT-18 billets -deformation degree 69% (zoom factor is x 500).

Measurements of microhardness (Hc) value have been done as a function of cross-section diameter. The results for zone with maximal deformation show that microhardness is as low as 520-600 kg/mm^2 even for the most sub-fine fraction of samples deformed with simultaneous heating. Similar results have been obtained for other Ti-based and Ni-based alloys (VT-3, 25, *etc.*).

Blades of high-pressure double-contour compressors of turbo-jet engines have also been made of this techniques. Comparison of the fatigue strength testing results have been done for similar blades, which were manufactured in a horizontal upsetting machine with conventional preliminary heating and hot extrusion, (testing conditions were: number of cycles $N=2 \bullet 10^7$, frequency $f_{1,0} = 1007 - 1079$ Hgc). The results show that preliminary SPD with electric heating cause better homogeneous distribution of grain sizes and narrower interval of fatigue strength. As a result engine operating life lengthen several times.

3. Discussion

The results show, that preliminary SPD with electric heating allows to achieve subfine microstructure (of 500-1000 nm) what is surprisingly exceptional for this temperatures. The structure have been obtained already at the first stage. At the same time, x-ray analysis shows, that texture, phases composition and micro-strain level change remarkably in zone of deformation. Similar data obtained for other Ti-based and Ni-based alloys which have remarkably different combination of alloying elements.

Broadening ratios data show that all phases are metastable. The fact, that high-angle peaks experience more noticeable broadening shows, that micro-strain put in the major contribution. Besides shift of high-angle Bragg reflections after SPD deformation is usually connected with a static shift of atoms of crystal lattice and stacking faults caused by deformation.

Let's pay attention to process of developing of defect substructure taking into account existence of an additional source of nonequilibrium when a remarkable energy is supplied/taken out from outside. In technologies under consideration the intense change of energy is accompanied with simultaneous SPD which cause certain hierarchy of defect structure, texture and grain size.

The self-organization of multilevel system of defects in multicomponent samples is determined by various barriers (obstacles, which are surrounded and concentrating fields of strain). We observed similar interrelation between external sources of energy and formation of multilevel system of defects of a various kind in four-component metalloceramics of Y-Ba-Cu-O system, so-called 1:2:3 phase or YBCO [7-10].

In Ti-alloys and in its intermetallics which combining properties of metals and ceramics, the situation differs some from YBCO, where a powerful and continuous source of energy is one of three weak-bound atoms of oxygen supplied from micro-pore and environment. Its anomaly high diffusion on vacant sublattice provides for spontaneous complex-periodic formations of fractals [12] and complex-periodic relaxation of this structure of defects in time even at room temperature. In development of process, associates of vacancies do not play the main role for materials based on titan as in above mentioned metal-ceramics, though the occurrence of vacancies cumulations is noticed in separate places of a zone of intermetallics and alloy connection (see also Fig 3 in [4]).

Extend defects (linear, flat or 3-dimensional) play prevailing role in the process. Genetically, from the beginning of deformation, they are thermodynamically non-equilibrium formations and keep the information about both previous external influences and about interaction of defects with each other.

The influence of defects distribution on both substructure and relaxation processes varies for a wide range of space organization of materials (different ranges of the length scale). At a subatomic level the distribution of electronic density is defined by both transfer of atomic charges of components and formation of essential share of collective electrons. The data of x-ray spectral researches and direct diffraction measurements of distribution of electronic density show that the redistribution of electronic density when temperature decreases or for the account of alloying (including stabilizers) determines both character of phase transitions in titan and its intermetallics and ability to brittle-ductile transition [11,13].

At an atomic level the degree of disordering is controlled with change of defects of nonstoichiometry and concentration of impurity elements (Zr, Nb- for TiAl; Al, Sn, Cr, Mo *etc.* for Ti). Concentration of impurity elements and vacancies determine a range of temperatures and pressure ensuring optimum structure.

The accumulation and interactions of defects of a linear type create local disordering in a crystal lattice, which is transmitted to submicroscopic scale. The x-ray data show, that at a microlevel the strong plastic deformation imposes texture and size of grains, the boundaries between which are determined by flat defects. The decomposition results in local heterogeneity on structure. It can include formation of dispersoids with sharply distinguished crystal structure. The discrepancy between parameters of elementary cells of these inclusions and matrix results in occurrence of additional macroscopic concentrators of strain.

Changing reactionary ability of crystals, increase of concentration of defects determine kinetics of substance carrying. Let's notice, for example, that for all technological regimes [4-6] we observed the wavy characteristic of the distribution of phase structure and microhardness. In multicomponent Ti-based alloys the periodic distribution was more appreciable than for pure metals (Cu and Nb [5]). This is possibly due to the additional effects of martensitic transformation.

As was already noted for Ti-alloys, it is very difficult to change this microstructure by the subsequent heat treatment. Therefore fine-grained microstructures were created during the first stage during intense plastic deformation. Simultaneous high-speed heating technology [1,2] results in processes of dynamic recovery in the grains, thus providing some relaxation of internal stresses. This allows one to conduct additional severe deformation following the initial stage of SPD, which is to obtain the final shape of parts.

The tests data show that for rather large products such as compressor blades, for example, SPD plastic deformation before giving a necessary form allows to improve mechanical properties and increase their life resource [2]. At the same time, when manufacturing details of a small size (nozzles, hollow balloons, low-temperature adapters etc.) [5], formed globular microstructure and high quality of a surface allow to use these products without additional expenses for a subsequent processing of details.

4. Conclusions

Two technologies have been examined from the common synergetic point of view of their ability to produce optimal microstructures in Ti-based alloys and TiAl-based dual alloys.

For Ti-based alloys with different composition subject to SPD and high speed heating, one obtains textured grains with sizes between 500 nm and 1500 nm. The shape of microstructure reflects the axial-symmetry of flow and also partial relaxation of strains within of the grain volume. Changes of microstructure and diffraction patterns for samples with different degrees of deformation reflects the appearance and growth of the number of concentrators - sources of micro-strains of different scale: from static atoms displacement of the position in crystal lattice to micro-strains connected with dispersoids.

For the samples obtained in result of impulse processing during supercooling the wave type distribution of the strain fields and micro-hardness also reflects specifics of interaction and multi-level hierarchy of defects caused by additional energy take-off. Activation in the feedback system of the deformed sample determines optimal mechanical features of the parts produce with the help of such self-organized technologies and ability of this alloys use in different applications, including commercial. A remarkable growth of mechanical properties of parts, their life-time and manufacturing time reduction of complex forms parts, including bimetallic have been achieve.

5. Acknowledgments

We would like to express our gratitude to Mrs. M.I. Ermolova and Mr. N.G. Kisel for assistance in the process characterization and preparation of materials. This work is in part supported by ICS&E.

6. References

1. Kommel, L.A., Teterin, G.P. and Kodess, B.N., (1998) Influence of thermophysical properties of Titanium alloys on electric upset forging parameters, *Material Working by Pressure* (Kuznechn.-Shtampov. Proizvodstvo)**7**, 29-34.
2. Kommel, L.A. and Teterin, G.P., (1998) Electric upset forging of preforms for compressor short blade material working by pressure (Kuznechn.-Shtampov. Proizvodstvo) **10**, 18-22.
3. Biba, N.V., Lyshnig, F.I. and Vlasov, A.V. (1998) Simulation of couple problem of electric upsetting, *Proceed. of the VI Int. Conf. on numerical methods in industrial forming processes, NUMIFORM*, Netherlands 22-25 VI.98, 523-528.
4. Kodess, B.N., Teterin, G.P., Kommel, L.A., and Ovcharov, V.K. (1999) Structure and mechanical properties of the engine valves with intermetallic disk, *Proceed. of MRS. Fall Meeting 1998: High-Ordered Intermetalic Compounds, VIII,* **487**, 685-692.
5. Teterin, G.P. and Volkov, A.E., (1994) Impulse stamping of difficult-deformable parts, *Forging and Stamping Production. Material Working by Pressure* (Kuznechn.- Shtampov. Proizvodstvo), **7**, 2-6.
6. Kodess B.N. and Medetbekov, M.T. (1996) The distribution of phases, composition and properties of Ti-Al alloys after various perturbation action, VNIIMS, GosStandart of Russia 14.05.96, Depozit VINITI No. 3137-B96, 1-45.
7. Kodess B.N. (1995), Auto-oscillations processes in a crystalline substances, *Nature* **K09016**, 04.09.95. 1-5.
8. Kodess B.N. and Medetbekov M.T. (1996) The processes self-organization of structural parameters investigation in HTSP materials. VNIIMS, 25 December, 1995, Moscow, Depozit VINITI, 1996, **48A**, N **2075-B96**, 1-44.
9. Gubayidulin Z.K., Kodess, B.N. and Medetbekov M.T. (1996) Electron paramagmetic resonance spectra and structural parameters of the KDP Family. VNIIMS, Dep.VINITI 05.01.96, N41, 1-22.
10. Avduchina, V.M., Kolesova, N.S., Kodess, B.N., Kaznelson, A.A., Revkevich, G.P. and Zivotov, S.M. (1997) Structure changes in crystals with various nonstoichiometries under long aging and exciting influence, Surface Investigation. *Physics, Chemistry, Mechanics* **12**, 695-709.
11. Kodess, B.N. (1991) Electron density distribution, lattice dynamic parameters of compounds based on X-ray Investigations, Moscow State University.
12. Mosolov, A.B. (1986) Fractal Geometry of HTSP, *Letters JEP* **15**, 64-68.

13. Eberhart, M, E., Clougherty, D.P. and MacLaren, J.M. (1993) Bonding-property relationships in intermetallic alloys, *J. Material Research Society* 3, 438-448.

IV. PHYSICAL AND MECHANICAL PROPERTIES OF SEVERE PLASTIC DEFORMATION MATERIALS

SPD PROCESSING AND ENHANCED PROPERTIES IN METALLIC MATERIALS

R.Z. VALIEV
Institute of Physics of Advanced Materials, Ufa State Aviation Technical University, K. Marks Str., 12, Ufa, 450000, Russia

Abstract

When severe plastic deformation (SPD) is applied to metals and alloys, the processed materials can possess ultrafine-grained nanostructures having highly non-equilibrium grain boundaries and a distorted crystal lattice. These new states can lead to novel properties of SPD materials. The present paper considers the effects of nanostructures on the unusual mechanical properties in several metals and alloys subjected to severe plastic deformation.

1. Introduction

A strong refinement of microstructure and the formation of nanostructures in bulk metallic materials by severe plastic deformation (SPD), *i.e.* intense plastic straining under high imposed pressure, provides a potential to achieve their new and extraordinary properties [1-3]. However, attaining such properties is a complex problem, which depends upon multiple processing and microstructural parameters. It has been shown that for nanostructured SPD materials there is typically present not only very small grain sizes, but also specific defect structures, high internal stresses, crystallographic texture, and often a change of phase composition [2]. Moreover, SPD leads to highly metastable states due to formation of supersaturated solid solutions and disordering or amorphization of intermetallic phases in multi-phase alloys [4]. On the other hand, the resulting microstructural parameters are related to details of SPD processing (*e.g.* processing routes, temperature, strain and strain rates). Therefore, it is important to analyze the relationship: processing-nanostructures-new properties in SPD-fabricated materials. This paper is devoted to investigations of structural features and enhanced properties in several pure metals and alloys processed by SPD: pure Cu, Ni, Ti, Fe, Al-based alloys and Ni_3Al intermetallics.

2. Materials and Experimental Procedures

Two techniques of SPD were used for processing nanostructures, as illustrated in Fig. 1. Equal-channel angular (ECA) pressing, shown in Fig. 1a is a procedure in which the sample is pressed under a load P through two channels of equal cross-section

T.C. Lowe and R.Z. Valiev (eds.), Investigations and Applications of Severe Plastic Deformation, 221–230.
© *2000 Kluwer Academic Publishers. Printed in the Netherlands.*

intersecting at an angle ϕ and the pressing are repeated to attain the required level of strain. This technique was developed several years ago in order to introduce an intense plastic strain into materials with no change in the cross-sectional area [5]. Later it was developed as a method for attaining a submicrometer or nanometer grain size [2, 6]. Torsion straining under high pressure, shown in Figure 1b, is a procedure in which disk samples are subjected to large plastic deformation at room temperature by torsion under imposed pressure of about 5 GPa. The samples processed by this technique are in the form of disks with a diameter of 12 mm or 20 mm and thickness of 0.2 mm or 1 mm.

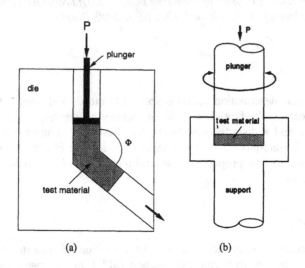

Figure 1. Principles of SPD methods: (a) ECA pressing , (b) torsion under high pressure.

Transmission electron microscopy (TEM) studies of the foils were conducted using a JEM–200 SX and a JEOL ARM 1000 (NCEM, Berkley) electron microscopes. X-ray diffraction was used for structural analysis and for estimating the long range order parameter. Microhardness was measured using a Vickers diamond pyramidal indenter under a load of 200 g. Mechanical compression and tensile tests were conducted at various strain rates using an INSTRON machine. Due to their small sizes, titanium samples were tested in bending. Cyclic loading experiments on SPD copper were carried out under fully-reversed tension/compression at constant plastic strain amplitudes, $\Delta\varepsilon_{pl}$ in the range 4×10^{-4} – 1×10^{-3} [7]. Details of these experiments and the results of investigations of the Bauschinger effect are described in [7]. Further details on processing and tests are also given elsewhere [2].

3. Results

3.1. THE AS-PROCESSED MATERIALS AND EFFECT OF ANNEALING

Formation of ultrafine-grained nanostructures with high angle grain boundaries requires application of large strains, usually with a true strain greater than 10, during severe plastic deformation [1]. There are also other processing parameters which affect fabrication of ultrafine-grained structures by SPD, namely, temperature and strain rate, imposed pressure, processing routes, lubrication and others. These parameters were controlled in order to process homogeneous nanostructures and a grain size as small as possible. The structural macrohomogeneity of the resulting samples was evidenced by a uniform distribution of microhardness values measured along each specimen.

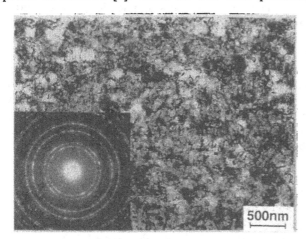

Figure 2. Typical microstructure observed by TEM in Ni after severe torsion straining

Figure 2 shows the typical microstructure observed in Ni after severe torsion straining (5 turns, P = 6 GPa) at room temperature, together with a selected area electron diffraction (SAED) pattern taken from a 1 μm diameter region. The pattern exhibits rings, indicating that there are many small grains with random misorientations in the selected area of view. The average grain size measured from the dark field image was about 120 nm. In the structure, some grain boundaries are visible but many grain boundaries are poorly defined. The contrast observed is not uniform and changes in a complex way. Such observations are typical for metals subjected to severe plastic straining and suggest a highly distorted state of the material. This fits well with the X-ray results demonstrating high internal stresses after torsion straining under high pressure [2, 6]. Similar microstructures were observed in the as-processed Cu [4, 8], Ti [9] and Fe [10], although in SPD Ti and Fe the mean grain size was less than 100 nm, but crystal lattice distortions were higher.

Inspection showed that the microstructure of copper after ECA pressing was also homogeneous with a mean grain size about 210 nm [8]. This is a granular type microstructure and recently the presence of high angle grain boundaries in the structure of ECA-pressed Cu was also confirmed by direct measurements of separate grain boundary misorientations [11].

In alloys, SPD leads to microstructure refinement and often to change of phase composition due to formation of supersaturated solution and disordering of intermetallic phases. For example, the processed nanostructure in boron-doped $Ni_3Al(Cr)$ was completely disordered and quite stable up to 500° C [3, 12]. However, as the annealing temperature is increased, the grain boundaries become better defined, suggesting some recovery process of the nanostructure. After heating to 750° C and holding for 5 min, the mean grain size is still very small, averaging 70 nm. HREM was used to examine the nature of the grain boundaries in SPD processed and annealed samples. These observations showed that in the as-processed and annealed Ni_3Al at

650° C grain boundaries are mostly wavy, curved, or corrugated and their large portions are inclined to the specimen. Figure 3 shows a lattice image of one grain boundary in the specimen, annealed at 650^0 C for 30 min. It appears that the grain boundary is very narrow, having a width less than ~ 0.5 nm, equivalent to two atomic spacings. However, again, the boundary is not atomistically smooth but has small steps and the lattice fringes are continuous up to the boundary. Some lattice fringes are terminated, suggesting the presence of grain boundary dislocations. An examination of a large number of grain boundaries in the SPD-processed Ni_3Al annealed at 650^0 C showed that, even after this annealing, boundaries are in high-energy and non-equilibrium configurations, suggesting only partial relaxation of the grain boundary structure. Similar HREM observations were performed in other metals after torsion straining [13].

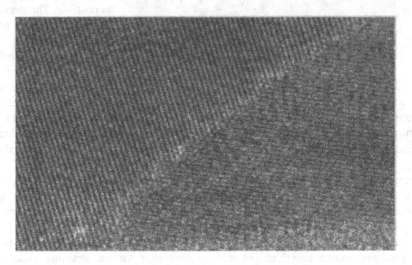

Figure 3. HREM micrograph of typical grain boundary in SPD Ni_3Al

These results fit well with the structural model of SPD materials [14] suggesting the formation of ultrafine-grained nanostructures having high density of defects at grain boundaries and crystal lattice distortions near interfaces.

3.2. MECHANICAL BEHAVIOR

3.2.1. Tests at Room Temperature

First let us consider the data on the mechanical tests of SPD copper [8, 15]. Figure 4 shows "true stress-strain" curves of the specimens tested in compression and tension. For comparison, similar curves are given for annealed specimens. It is seen (Figure 4a) that Cu specimens with nanostructure obtained by ECA pressing have yield strength that is several times higher and strain hardening at the stage of plastic flow that is much smaller than for coarse-grained annealed Cu.

The stress-strain curve for a specimen subjected to additional annealing at 473 K for 3 min is shown on Figure 4b. The annealing does not cause noticeable grain growth but

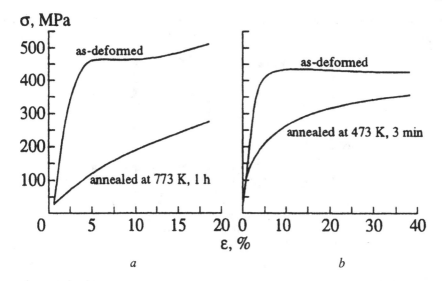

Figure 4. "True stress strain" curves of SPD Cu at room temperature a) tension, b) compression

leads to an abrupt decrease in the level of internal stresses [2, 15]. It is seen that, despite similar grain sizes, the specimens in these two states substantially differ in deformation behavior. The curve for the specimen subjected to such short-term annealing is similar to that of coarse-grained copper. This result is of great importance, because it illustrates that the strength of SPD materials is affected not only by grain size, but also by grain-boundary defects.

The mechanical properties of titanium specimens with a grain size of about 100 nm subjected to deformation by torsion are also of interest for understanding the deformation behavior of nanomaterials [9]. Specimens about 10 mm in diameter were tested by bending to determine yield strength σ_y, ultimate tensile strength σ_u and the value of maximum bend in the as-deformed condition and after annealing at various temperatures. According to the TEM data, changes in structure of SPD Ti on heating begin at 250° C, a decrease in elastic lattice distortions due to recovery was clearly seen from dark-field images of the specimens annealed at 250° C. The average grain size measured from these images somewhat increases after annealing, although no migration of grain boundaries occurs at this temperature. Grain growth in titanium subjected to SPD starts at 350° C.

The results of microhardness and bending tests of SPD Ti show (Figure 5) that the H_v values remain almost constant with increasing annealing temperature to 300–350° C and sharply decrease at higher temperatures. At the same time, the yield strength substantially decreases after annealing at 250° C. The plasticity measured from the maximum deflection is small for the initial specimens, but substantially increases after annealing and reaches 0.35 mm in the specimen annealed at 250° C, at which temperature both Hv and the yield strength are maximum.

As was already mentioned, structure refinement in the alloys during SPD may be accompanied by substantial changes in phase composition. The latter phenomenon is

226

also important for mechanical properties. For example, the aging behavior of aluminum alloys Al 1420 [2] and Al-11% Fe [16] after severe torsion straining is considered below.

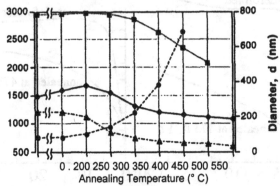

Figure 5. Average grain size and mechanical properties of SPD titanium as a function of annealing temperature

To obtain a high-strength state, the alloy Al 1420 (Al-5.5% Mg-2.2% Li-0.12% Zr) was sequentially quenched, deformed by torsion, and aged at 120° C. This aging temperature was selected to ensure additional strengthening at the expense of precipitation of a second phase while retaining a small grain size. The microhardness of the as-quenched alloy was 540 MPa (Table 1). Severe torsion plastic deformation of this alloy caused the formation of a nanostructure with a mean grain size of 70 nm.

TABLE 1. Microhardness of aluminium alloys (MPa)

Treatment	Alloy	
	Al 1420	Al-11wt.%Fe
After quenching	540	-
After intense deformation	1750	1700
After quenching + aging	2300	3020

SPD increases microhardness to 1750 MPa. This level of microhardness is retained with increasing annealing temperature up to 150° C. Aging at 120° C retains the average grain size of 70 nm but causes the precipitation of second-phase particles with a size of 20 nm and increases the microhardness to 2300 MPa.

Similar effects were seen in the Al-11wt%Fe alloy whose structure and aging behavior was studied in detail in [16]. The microhardness of the alloy is 1700 MPa after SPD and substantially increased after annealing at 100°C for 5 h (Table 1) due to intense aging of the alloy. This shows that the formation of nanostructures in Al alloys is accompanied by the formation of new metastable states whose decomposition during aging causes a substantial strengthening of the alloys.

3.2.2. Fatigue Behavior
The formation of nanostructures by SPD processing may have a substantial effect on

fatigue behavior of metals and alloys. For example, the cyclic strengthening of four copper specimens was studied in [7]. Two specimens were taken in the as-deformed state, and the other two specimens were additionally subjected to short-term annealing (473 K, 3 min), which retains the grain size, or to annealing at 773 K, which causes substantial grain growth (up to 50 μm). All specimens exhibited saturation after several cycles, but the value of the saturation stress σ_s was greatly affected by the preliminary treatment of the copper.

A substantial difference in the fatigue behavior of ultrafine-grained and coarse-grained copper was established from the dependence of the hysteresis loss parameters (Bauschinger-energy parameter) B_E as a function of accumulated plastic deformation. The B_E value was estimated from the shape of the hysteresis loop. The larger the value of B_E, the more pronounced is the Bauschinger effect. As it was shown the B_E value for all states is weakly affected by the accumulated deformation, but the maximum B_E value is observed for the SPD copper subjected to short-term annealing. Thus, the results of [7] indicate that SPD materials can have unusual fatigue properties. A higher fatigue strength of SPD Cu is seen from the fact that its B_E value is smaller than that of the specimen additionally subjected to short-term annealing.

3.2.3. Superplasticity

In recent years, several works were devoted to the unusual superplastic properties of SPD nanostructured alloys. In [17], grain size of 150 nm was obtained in the Al-4% Cu-0.5% Zr alloy by ECA pressing. This material contains ultrafine Al_3Zr particles of 30 nm in size. These specimens, tested by tension at 250° C at strain rates ranging from 2.8×10^{-5} to 1.4×10^{-3} s^{-1}, exhibit very high elongation to failure, despite a relatively low test temperature. The maximum elongation (850%) was obtained at a starting strain rate of 1.4×10^{-3} s^{-1}. The strain-rate sensitivity m in this case was 0.46. Note that the same alloy with a grain size of 8 μm exhibits similar superplastic behavior only at temperatures as high as 500°C. This example illustrates the superplasticity in SPD materials at relatively low temperatures. This effect was observed in several other SPD materials [3,12].

Another unusual effect is seen for high strain rate superplasticity, which was recently demonstrated in an Al-based alloy 1420 (Al-5.5% Mg-2.2% Li-0.12% Zr) subjected to ECA pressing [18]. The grain size of this material was about 1 μm and grain boundaries were mostly high-angle. Mechanical tests showed very high superplastic properties of this material. Thus, the elongation to failure of the specimens tested at 350° C at a strain rate of 10^{-2} s^{-1} was 1180%. Moreover, the specimens demonstrated superplastic behavior even at a deformation rate of 10^{-1} s^{-1}, exhibiting elongation of about 1000% [19]. These results are the first evidence of high-strain-rate superplasticity of SPD Al alloys.

4. Discussion

The results considered in the present study demonstrate that nanostructures with high-angle grain boundaries may be successfully processed by the methods of SPD. At the same time, the grain boundaries are usually nonequilibrium because they have dislocation distorted structure. Moreover, nanostructured SPD materials are often characterized by considerable lattice elastic distortions, high density of lattice dislocations, and strong crystallographic texture. All these structural parameters should be taken into account when considering mechanical properties of SPD materials.

The above experimental data on ultrafine-grained copper demonstrate that the room-temperature mechanical behavior of this material has some specific features, such as high yield strength, weak strain hardening, and quite high ductility. This effect is attributed [8] to enhancement of the recovery processes at grain boundaries and grain boundary sliding during the deformation of nanostructured metals. For example, it was shown that the activation energy Q for grain-boundary diffusion in the ECA-pressed copper is 78 kJ/mol [8]. This value is substantially lower than that for ordinary coarse-grained copper (107 kJ/mol). These data and the results of the other studies [20] demonstrate the enhanced diffusivity along grain boundaries in SPD materials. The enhanced diffusion, which assists the recovery processes at grain boundaries and grain boundary sliding, may be responsible for the absence of noticeable strain hardening and the ductility observed in the ECA-pressed metals. On the other hand, the high strength may be attributed to the difficulty of generation and motion of dislocations at non-equilibrium grain boundaries in SPD nanostructured metals, which have high internal stresses [2].

The effect of non-equilibrium grain boundaries was also used [21] to explain the unusual fatigue behavior of the ECA-pressed copper, whose short-term annealing also decreases yield strength but increases the Bauschinger energy parameter [7]. These boundaries also may explain the mechanical behavior of SPD titanium, which exhibits strong dependence of yield strength not only on the grain size, but also on internal stresses (Figure 5).

Two important conclusions were reached from these data [21]. First, it becomes apparent that the well-known empirical equation $H_v = 3\sigma_y$ is not valid for nanostructured SPD materials in the initial and annealed states. Second, it is necessary to take into account the role of the defect structure of grain boundaries when analyzing the Hall-Petch relationship. The yield strength of SPD materials may be strongly affected by grain-boundary defects or precipitation during aging of metastable alloys.

In the context of the mechanical behavior of SPD materials at elevated temperatures, the most important phenomena revealed are low temperature and high strain rate superplasticity. The nature of these effects was discussed in [17], and it was suggested that these effects are, obviously, also associated with non-equilibrium grain boundaries in SPD materials.

5. Conclusions

(1) Ultrafine-grained nanostructures having many high-angle grain boundaries may successfully be produced in bulk specimens of various metals and alloys by severe plastic deformation. The resulting materials have specific structural features, resulting from highly distorted non-equilibrium grain boundaries leading to high internal stresses and lattice distortions.

(2) Ultrafine-grained SPD materials often display extraordinary mechanical behaviors and properties, such as high strength and ductility, unusual fatigue properties, and enhanced superplasticity. However, these properties originate from very small grain sizes, grain size distributions, internal stresses, texture and other structural features, which depends on SPD processing methods and regimes.

(3) Establishing SPD processing guidelines enables new advanced properties in metallic materials to be obtained.

6. References

1. Valiev, R.Z. (1996) Ultrafine-grained materials produced by severe plastic deformation: a thematical issue, *Ed.,Ann. Chim.* (Paris) **21**, 369–480.
2. Valiev, R.Z., Alexandrov, I.V., and Islamgaliev, R.K. (1998) In: Nanocrystalline Materials, Science and Technology. NATO ASI, Ed.s G.M. Chow and N.I. Noskova, Kluwer Publ, The Netherlands, 121-142.
3. McFadden, S.X., Mishra, R.S., Valiev, R.Z., Zhilyaev, A.P., Mukherjee, A.K., Low-temperature superplasticity in nanostructured nickel and metal alloys, *Nature* **398** (22 Apr. 1999), p.684.
4. Stolyarov, V.V. and Valiev, R.Z., (1999) Bulk nanostructured metastable alloys prepared by severe plastic deformation, *J. of Metastable and Nanocrystalline Materials* **1**, 185.
5. Segal, V.M., Reznikov, V.I., Drobyshevskiy, A.E., and Kopylov, V.I (1981) plastic working of metals by simple shear, *Russian Metallurgy (Metally)* **1**, 115.
6. Valiev, R.Z., Korznikov, A.V, and Mulyukov, R.R. (1993) Structure and properties of ultrafine-grained materials produced by severe plastic deformation, *Mater. Sci. Eng.* **168**, 141–148.
7. Vinogradov, A., Kaneko, Y., Kitagawa, K., Hashimoto, S., Stolyarov, V.V., Valiev, R.Z. (1997) Fatigue behaviour of ultrafine-grained copper, *Scr. Metall. Mater.* **36**, 1345–1349.
8. Valiev, R.Z., Kozlov, E.V., Ivanov,Yu.I., Lian, J., Nazarov, A.A., Baudelet B. (1994) Deformation behaviour of ultrafine-grained copper, *Acta Metall. Mater.* **42**, 2467–2475.
9. Popov, A.A., Pyshimintsev, I.Yu., Demakov, S.L., Illarionov, A.G., Lowe, T.G., Sergeyeva, A.V., Valiev, R.Z., (1997) Structural and mechanical properties of nanocrystalline titanium processed by severe plastic deformation, *Scripta Mater.* **37**, 1089–1095.
10. Valiev, R.Z., Ivanisenko, Yu.V., Rauch, E.F., Baudelet, B., (1996) Structure and deformation behaviour of armco iron subjected to severe plastic deformation, *Acta Mater.* **44**, 4705–4712.
11. Mishin, O.V, Gertsman, V.Yu., Valiev, R.Z., and Gottstein, G. (1996) Grain-boundary distributions and texture in ultrafine-grained copper produced by severe plastic deformation, *Scr. Metall. Mater.* **35**, 873–878.
12. Mishra, R.S., Valiev, R.Z., McFadden, S.X., Mukherjee, A.K. (1998) Tensile superplasticity in nanostructured nickel aluminide, *Mater. Sci. Eng.* **A252**, 174.
13. Horita, Z., Smith, D.J., Nemoto, M., Valiev, R.Z., Langdon, T.G., (1998) Observations of grain boundary structure in submicrometer Cu and Ni using high-resolution electron microscopy, *J. Mater. Res.* **13**, 446–450.
14. Valiev, R.Z., (1995) Approach to nanostructured solids through the studies of submicron grained polycrystals, *Nano-struct. Mater.* **6**, 73–82.
15. Gertsman, V.Yu., Valiev, R.Z., Akhmadeev, N.A, and Mishin, O.V. (1996) Mechanical properties of ultrafine-grained metals, *Mater. Sci. Forum* **223**, 80–90.
16. Senkov, O.N., Froes, F.N., Stolyarov, V.V., Valiev, R.Z., and Liu, J. (1997) Nonequilibrium structure in aluminium alloys subjected to severe plastic deformation, *Proc. 5th Conf. on Advanced Particulate Materials and Processes*, West Palm Beach.
17. Valiev, R.Z. (1997) Superplasticity in nanocrystalline metallic materials, *Mater. Sci. Forum* **243-245**, 207–216.
18. Valiev, R.Z., Salimonenko, D.A., Tsenev, N.K., et al. (1997) Observations of High Strain Rate Superplasticity in commercial aluminium alloys with ultrafine grain sizes, *Scr. Metall. Mater.* **37**, 724–729.
19. Berbon, P.B., Furukava, M., Horita, Z., Nemoto, M., Tsenev, N.K., Valiev, R.Z., Langdon, T.G. (1998) Requirements for achieving high-strain superplasticity in cast aluminium alloys, *Phil. Mag. Let.* **78**, 313–318.

230

20. Kolobov, Yu. R., Grabovetskaya, G.P., Ratochka, I.V., Kabanova, E.V., Naidenkin, E.V., Lowe, T.C. (1996) Effect of grain boundary diffusion fluxes of copper on the acceleration of creep in submicrocrystalline nickel, *Ann. Chim. Fr.* **21**, 369–378.
21. Valiev, R.Z. (1997) Structure and mechanical properties of ultrafine-grained metals, *Mater. Sci. Eng.* **A234-236**, 59–66.

TENSILE SUPERPLASTICITY IN NANOCRYSTALLINE MATERIALS PRODUCED BY SEVERE PLASTIC DEFORMATION

R.S. MISHRA, S.X. MCFADDEN, A.K. MUKHERJEE
Department of Chemical Engineering and Materials Science
University of California, One Shield Avenue, Davis, CA 95616

Abstract

Tensile superplasticity has been observed in a number of severe plastic deformation (SePD) processed alloys with nanocrystalline microstructure. The observations of superplasticity in nanocrystalline materials are briefly reviewed with emphasis on the aspects that are different from superplasticity in microcrystalline materials. The temperature for onset of superplastic elongation coincides with microstructural instability. The important features include, high strain rate superplasticity in an aluminum alloy, low temperature superplasticity, extensive strain hardening and high flow stresses. A comparison of the experimental results with existing models shows the difference in superplastic deformation kinetics. The deformation mechanisms for microcrystalline materials are not simply scaleable to nanocrystalline range. It is difficult to establish the parameters for deformation mechanism because of grain growth. The observations of low temperature and high strain rate superplasticity in nanocrystalline materials with some unique features opens up new possibilities for scientific and technological advancements.

1. Introduction

Severe plastic deformation produces ultrafine grained microstructure (grain size <1 μm) in several bulk metallic materials [1,2]. Unlike powder processed materials, bulk SePD processed materials do not have problems associated with surface contamination and residual porosity. This makes SePD one of the ideal processing techniques to produce specimens for investigation of grain size dependent phenomenon in bulk materials. A number of SePD techniques have been explored to produce ultrafine grained materials. The most promising SePD techniques are equal channel angular extrusion (ECAE) and torsion straining (TS). TS-SePD is particularly effective in producing nanocrystalline microstructure.

Bulk nanocrystalline materials provide an opportunity to investigate the scalability of grain size dependent phenomenon to a much finer scale than was previously possible. Superplasticity is a well established grain size dependent phenomenon that is exhibited by fine grained materials at elevated temperatures. The flow behavior during superplasticity can be represented by a generalized constitutive relationship [3]

T.C. Lowe and R.Z. Valiev (eds.), Investigations and Applications of Severe Plastic Deformation, 231–240.
© *2000 Kluwer Academic Publishers. Printed in the Netherlands.*

$$\dot{\varepsilon} = A\frac{DGb}{kT}\left(\frac{b}{d}\right)^{p}\left(\frac{\sigma}{E}\right)^{2} \tag{1}$$

where $\dot{\varepsilon}$ is the strain rate, D is the appropriate diffusivity (lattice or grain boundary), G is the shear modulus, b is the Burger's vector, k is the Boltzmann's constant, T is the test temperature, d is the grain size, p is the grain size exponent and σ is the applied stress. It is well accepted that grain boundary sliding is the dominant mechanism for structural superplasticity [4]. The grain boundary sliding leads to stress concentration at the grain boundary triple junctions. The grain boundary sliding is accommodated by dislocation or diffusional processes. Usually the grain size exponent has a value of 2 or 3 depending on the micromechanism and the dominant diffusion process (i.e. grain boundary diffusion or lattice diffusion). This equation has been validated for metals, intermetallics and ceramics materials. A common issue for investigation of superplasticity in nanocrystalline materials is that of scalability or transition. It is possible that equation (1) is valid down to the smallest grain size. On the other hand, if some of the micromechanisms involved during superplasticity are influenced in a fundamental manner, that can result in a change of parametric dependencies or in the kinetics of deformation.

We briefly review some of the experimental observations of superplasticity in SePD nanocrystalline materials. A comparison with the microcrystalline materials leads to some general understanding. The nanocrystalline materials show superplasticity at significantly lower temperatures as well as at enhanced strain rates. However, on the normalized basis the kinetics of superplastic deformation in nanocrystalline materials is slower. This highlights the need for a closer examination of the influence of grain size on the micromechanisms involved in superplasticity.

2. Experimental Trends and Discussion

Figure 1 shows a transmission electron micrograph of a TS-SePD processed Ni$_3$Al-Cr alloy. This is a typical microstructure of materials processed by TS-SePD, particularly materials with high melting points. The dynamic recovery at ambient temperature is not significant and that leads to microstructure with high residual stress and ill-defined grains. The defect density is very high in the as processed state. The microstructure is relatively less distorted for aluminum alloys. But even in aluminum alloys the boundaries have poor contrast. Valiev et al. [2] have suggested that it is an indication of non-equilibrium boundaries.

A number of nanocrystalline materials produced by severe plastic deformation processing have exhibited superplasticity [5,11]. Table 1 summarizes some of the observations. The general trends of superplasticity in SePD materials include the following features: significant reduction in minimum temperature for superplasticity, increase in optimum strain rate for maximum elongation, and higher flow stresses. These features are presented briefly and the significance of these observations on advancing the understanding of superplastic behavior in SePD processed nanocrystalline materials is discussed.

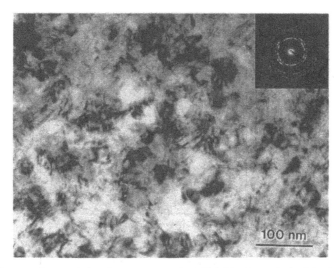

Figure 1. A transmission electron micrograph of TS-SePD processed Ni₃Al-Cr alloy.

TABLE 1. A summary of observation of superplasticity in nanocrystalline materials

Material	Grain Size* (nm)	Strain Rate (s⁻¹)	Temp. (°C)	Stress (MPa) ε=0.1	Max	Elongation (%)
Ti-6 Al-3.2 Mo [5,6]	60	5x10⁻⁴	600	150	-	600
Zn-Al [7]	80	5x10⁻⁴	120	18	-	230
Ni₃Al alloy [8]	50	1x10⁻³	650	400	1530	380
Ni₃Al alloy [8]			725	270	790	560
Al-Mg-Li-Zr alloy (1420) [9]	100	1x10⁻¹	250	106	154	330
Al-Mg-Li-Zr alloy (1420) [9]			300	33	146	850
Al-Cu alloy (2124) [10]	100	1x10⁻³	350	6	50	405
Ti-6 Al-4 V [11]	70	1x10⁻³	575	165	-	215

* Denotes the starting average grain size.

The stress-elongation behavior of nanocrystalline Ni₃Al-Cr alloy is shown in Figure 2(a) [8]. It shows very high flow stresses and strain hardening. Usually the flow stresses during superplasticity in microcrystalline materials vary from 1 to 50 MPa. In fact, the low flow stress is one of the features that has made the superplastic gas forming process possible. The low flow stresses also helps keep the extent of cavitation low in many aluminum and titanium based alloys. So, it is amazing to observe superplasticity with flow stresses exceeding 1 GPa in Ni₃Al-Cr alloys. The level of cavitation is important for the post-superplastic forming properties of material.

The high strain hardening was observed in many different alloys. The variation of stress-strain behavior as a function of strain rate is illustrated in Figure 2(b) [9]. Both the alloys, Ni₃Al-Cr and 1420 Al, exhibit superplasticity at significantly lower temperatures than the microcrystalline counterparts. The observation of optimum superplasticity at high strain rates is particularly noteworthy. The strain hardening coefficient, $d\sigma/d\varepsilon$, increases with increasing strain rate. An usual explanation for strain hardening during superplasticity is concurrent grain growth. While grain growth does occur in these nanocrystalline materials at superplastic temperatures, the extent of

234

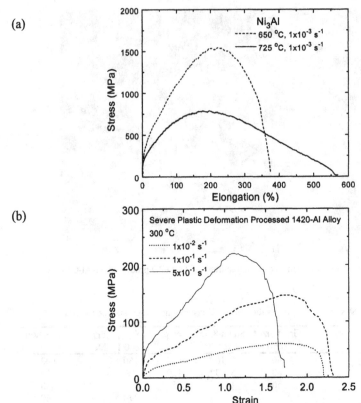

Figure 2. Stress-strain behavior of (a) a Ni₃Al-Cr alloy as a function of temperature [8] and (b) a 1420Al alloy as a function of strain rate [9]. Note the high strain rate for superplasticity in 1420Al alloy.

strain hardening observed cannot be explained by grain growth alone. Mishra et al. [9,12] have used the equation (1) and experimental TEM results after test to examine the extent of strain hardening by concurrent grain growth. For a constant strain rate-isothermal test, equation (1) gives a relationship between flow stress and instantaneous grain size as $\sigma \propto d^{p/2}$. For materials exhibiting a grain size exponent of 2, the flow stress is directly proportional to the grain size. It can be noted from Figure 2(a) that the flow stress at 650 °C increased by almost 5 times. However, the TEM investigation after testing showed that the grain size only increased by a factor of ~2 [12]. Clearly, there is an additional component to the strain hardening that needs to be resolved. Similar arguments can be made about the strain hardening trend in 1420Al alloy [9].

The ability of nanocrystalline materials to exhibit high ductility in spite of higher flow stresses can be gauged from a comparison of flow behavior with microcrystalline materials. An example of the influence of microstructural state on flow behavior is shown in Figure 3. The nanocrystalline Ni₃Al-Cr alloy exhibits equivalent ductility in spite of significantly higher flow stress. Similar trend can be noted from Table 2 for 1420 Al alloy. This aspect has important implication for cavitation. It suggests that at equivalent flow stress, the cavitation during superplastic deformation in nanocrystalline material is lower. Experimental investigation of cavitation during superplasticity in nanocrystalline materials is needed to further the understanding of influence of grain size on the relationship between high flow stresses and cavitation.

To gain a fundamental understanding of the grain size dependence of superplasticity in nanocrystalline range, it is important to consider the microstructural development during testing. Figure 4 shows dark field TEM micrographs of a 1420Al alloy deformed at 250 °C and 300 °C at a strain rate of 1×10^{-1} s^{-1} [9]. At both testing temperatures there was significant grain growth during test. In addition, there was considerable grain growth during the heat-up to 300 °C. A comparison of micrographs illustrates the difference in dislocation structure that evolved during superplastic deformation of 1420Al alloy. In Figure 4(a), the dislocations are straight, going across the grain, and the density is quite low. On the other hand, Figure 4(b) shows a lot of intragranular dislocations. There are a number of dislocation-dislocation and dislocation-particle

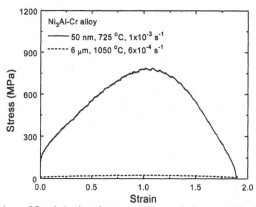

Figure 3. A comparison of flow behavior of nanocrystalline and microcrystalline Ni₃Al-Cr alloy. The nanocrystalline alloy exhibits equivalent ductility at significantly lower temperatures in spite of very high flow stresses.

TABLE 2. A comparison of some superplastic features in nanocrystalline and microcrystalline commercial Al-1420 alloy. Note the high flow stress in nanocrystalline state at 250 °C.

	100 nm grain size (Mishra et al. [9])		6 μm grain size (Kaibyshev [13])
% Elongation	330	850	700
Temperature (°C)	250	300	450
Strain rate (s^{-1})	1×10^{-1}	1×10^{-1}	4×10^{-4}
Flow stress (MPa) at 50% strain	188	50	4

interactions evident in this micrograph. Is the difference in dislocation configuration with grain size an indication of a transition? This question can be further addressed by comparing the experimental data with theoretical predictions.

The constitutive relationship for superplasticity in fine grained aluminium alloys can be expressed as [14],

$$\dot{\varepsilon}=40\frac{D_o E b}{kT}\exp\left(-\frac{84000}{RT}\right)\left(\frac{b}{d}\right)^2\left(\frac{\sigma-\sigma_o}{E}\right)^2 \qquad (2)$$

where, D_o is the pre-exponential constant for diffusivity, R is the gas constant and σ_o is the threshold stress. The important features of this constitutive equation are: a stress exponent of 2, an inverse grain size exponent of 2 and a temperature dependence close to the activation energy for grain boundary self diffusion. These features agree very well with the grain boundary sliding models of Ball and Hutchinson [15] and

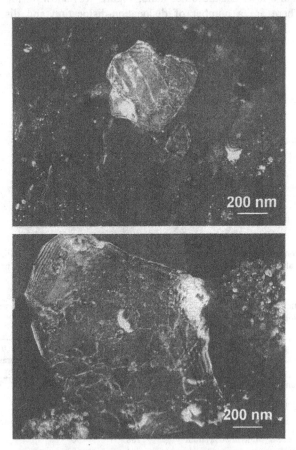

Figure 4. Dark field TEM micrographs of specimens deformed at (a) 250 °C and (b) 300°C and a strain rate of 1×10^{-1} s^{-1}. Note the intragranular dislocations and the difference in configuration with grain size. There is considerable dislocation-dislocation and dislocation-particle interactions at 300°C [9].

Mukherjee [16]. The mechanistic interpretation is that the superplastic mechanism in aluminium alloys is slip-accommodated grain boundary diffusion-controlled grain boundary sliding.

The experimental data for microcrystalline and sub-microcrystalline 1420Al alloy is compared with the empirical equation (2) and the theoretical predictions of Ball and Hutchison [15] and Mukherjee [16] grain boundary sliding models (Figure 5). It is interesting to note that the data for microcrystalline 1420Al agrees well with the theoretical predictions and equation (2). However, the lower temperature data of nanocrystalline 1420Al deviates from the overall trend in microcrystalline material. It is interesting to note that the data obtained at 300°C, where the grain growth during heating increased the starting grain size to 0.5 μm, matches quite well with the data for the microcrystalline alloy. This indicates slower kinetics in the nanocrystalline state on a normalized basis.

The observation of high flow stresses is valid regardless of whether the superplastic flow is grain boundary diffusion controlled or lattice diffusion controlled.

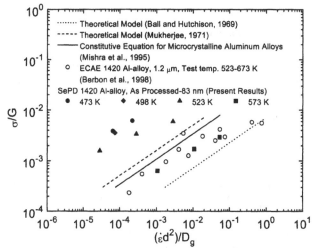

Figure 5. A comparison of the experimental stress-strain rate data with the empirical correlation and theoretical models. Note that while the data for microcrystalline and sub-microcrystalline starting grain size merge, the low temperature data with nanocrystalline state shows higher flow stresses [9].

Mukhopadhyay *et al.* [17] have observed lattice diffusion controlled superplastic flow in a microcrystalline Ni₃Al-Cr alloy. For slip accommodated grain boundary sliding, Sherby and Wadsworth [18] have classified the stress-strain rate data for microcrystalline superplastic alloys in two categories; (a) lattice diffusion controlled, and (b) grain boundary diffusion controlled. They expressed the empirical correlation for lattice diffusion controlled superplasticity as,

$$\dot{\varepsilon} = 5 \times 10^9 \left(\frac{D_L}{d^2} \right) \left(\frac{\sigma}{E} \right)^2, \tag{4}$$

where D_L is the lattice diffusion coefficient. The above relationship, which is a specific form of equation (1), can be used for comparison of data for Ni₃Al-Cr alloy. Figure 6 shows the variation of flow stress with temperature and grain size compensated strain rate for the nanocrystalline and microcrystalline Ni₃Al alloys. For comparison, the expected trend for equation 4 is also plotted. Equation 4 has been shown to apply to a number of disordered alloys. It can be noted that the data for nanocrystalline and microcrystalline alloys are not co-incident together after the grain size and temperature compensation have been applied (Figure 6). We note that the use of d^3 compensation would also not lead to a merging of the data. Clearly the flow stresses for superplasticity in nanocrystalline Ni₃Al is higher even on a normalized basis than that of the microcrystalline Ni₃Al. This is consistent with the findings of Mishra and Mukherjee [19,20] for Zn-22Al and Ti-6Al-4V alloys.

A reason for slower kinetics in nanocrystalline materials could be the dislocation configuration in ultrafine grains. As the grain size decreases, the density of intragranular dislocations decrease. The dislocations after being generated during the slip accommodation of grain boundary sliding, from the grain boundary triple junction

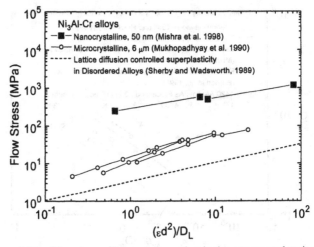

Figure 6. The variation of flow stress with temperature and grain size compensated strain rate. The flow stresses are higher than microcrystalline Ni₃Al alloy of same composition.

or ledges, would traverse the grain. Mishra and Mukherjee [19,20] have suggested that the difficulty associated with generation of dislocations in nanocrystalline materials can explain the high flow stresses. We suggest an additional concept that can result in slower kinetics of deformation in nanocrystalline materials at elevated temperatures. The velocity of straight dislocations (such as Figure 4(a)) would depend on the applied force on the slip plane and the drag force it feels from the end-nodes in the grain boundaries. If we assume that the drag forces are governed by the details of dislocation type in, the grain boundary, then the drag forces for a given configuration will be constant. The forward force on the dislocation can be expressed as,

$$F_f = \tau_a b d,$$
(3a)

where τ_a is the applied stress. The net force responsible for the forward movement of dislocation can then be written as,

$$F_n = \tau_a\, bd - 2F_d, \qquad (3b)$$

where F_d is the drag force acting on the dislocation from the nodes in the grain boundaries. From equation 3(b), it is easy to see that as the grain size becomes smaller the net force on the dislocation will reduce, which will directly influence the kinetics of dislocation motion and the overall deformation rate.

3. Conclusions

1. Some grain growth is inevitable during superplastic deformation of nanocrystalline materials as the temperature for onset of grain boundary sliding and grain boundary migration is similar.
2. The superplastic deformation kinetics are slower in nanocrystalline materials after grain size and temperature compensation.
3. The micromechanisms and constitutive relationships developed for superplasticity in microcrystalline materials are not simply scaleable to the nanocrystalline range.

4. Acknowledgments

The authors gratefully acknowledge the support from the National Science Foundation under grants NSF-DMR-9903321. We are thankful to Professor Ruslan Z. Valiev for providing the SePD materials.

5. References

1. Valiev, R.Z., Musalimov, R.Sh., and Tsenev, N.K. (1989) The non-equilibrium state of grain boundaries and the grain boundary precipitations in aluminium alloys, *Physica Status Solidi A* **115**, 451-7.
2. Valiev, R.Z., Korznikov, A.V., and Mulyukov, R.R. (1993) Structure and properties of ultrafine-grained materials produced by severe plastic deformation, *Mat. Sci. Eng.* **A168**,. 141-8.
3. Mukherjee, A.K., Bird, J.E., and Dorn, J.E. (1969) Experimental correlations for high temperature creep, *Trans. ASM.* **62**, 155-179.
4. Mukherjee, A. K. (1993) Superplasticity in metals, ceramics and intermetallics, in H. Mughrabi (ed.) *Plastic Deformation and Fracture of Materials*, VCH Verlagsgesellschaft mbH, Weinheim, pp.407-459.
5. Salischev, G.A., Valiakhmatov, O.R., Valitov, V.A., and Mukhtorov, S.K. (1994) *Materials Science Forum* **170-172**, 121-130.
6. Salishchev, G.A., Galeyev, R.M., Malisheva, S.P., and Valiakhmetov, O.R. (1997) Low temperature superplasticity of submicrocrystalline titanium alloys, *Materials Science Forum* **243-245**, 585-90.
7. Mishra, R.S, Valiev, R.Z, and Mukherjee, A.K. (1997) The observation of tensile superplasticity in nanocrystalline materials, Nanostructured Materials 9, 473-476.
8. Mishra, R.S., Valiev, R.Z., McFadden, S.X., and Mukherjee, A.K. (1998) Tensile superplasticity in a nanocrystalline nickel aluminide, *Mat. Sci. Eng.* **A252**, 174-8.
9. Mishra, R.S., Valiev, R.Z., McFadden, S.X., Islamgaliev, R.K., and Mukherjee, A.K. (1999) Observation of high strain rate superplasticity in a severe plastic deformation processed ultrafine grained 1420-Al alloy at low temperatures, to be published.
10. Mishra, R.S. (1998) unpublished research.
11. Mishra, R.S., Stolyarov, V.V., Echer, C., Valiev, R.Z., and Mukherjee, A.K., (1999) Mechanical behavior and superplasticity of a severe plastic deformation processed nanocrystalline Ti-6Al-4V Alloy, to be published.
12. Mishra, R.S., McFadden, S.X., Valiev, R.Z., and Mukherjee, A.K. (1999) Deformation mechanisms and tensile superplasticity in nanocrystalline materials, *JOM* **51(1)**, 37-40.

240

13. Kaibyshev, O.A. (1992), *Superplasticity of Alloys, Intermetallics and Ceramics*, Springer-Verlag, Berlin, pp.149.
14. Mishra, R.S., Bieler, T.R., and Mukherjee, A.K. (1995) Superplasticity in powder metallurgy aluminum alloys and composites, *Acta Metall. Mater.* **43**, 877-91.
15. Ball, A. and Hutchinson, M.M. (1969) Superplasticity in the aluminum-zinc eutectoid, *Met. Sci. J.* **3**, 1-7.
16. Mukherjee, A.K. (1971) The rate controlling mechanism in superplasticity, *Mat. Sci. Eng.* **8**, 83-9.
17. Mukhopadhyay, J., Kaschner, G., and Mukherjee, A.K. (1990) Superplasticity in boron doped Ni$_3$Al alloy, *Scripta Metall. Mater.* **24**, 857-62.
18. Sherby, O.D. and Wadsworth, J. (1989) Superplasticity-recent advanced and future directions, *Progress in Materials Science* **33**, 169-221.
19. Mishra, R.S., and Mukherjee, A.K. (1998) Superplasticity in nanomaterials, in A.K. Ghosh and T.R. Bieler (eds.) *Superplasticity and Superplastic Forming 1998*, TMS, Warrendale, pp.109-16.
20. Mishra, R.S., McFadden, S.X., and Mukherjee, A.K. (1999) Analysis of tensile superplasticity in nanomaterials, *Materials Science Forum* **304-306**, 31-38.

ON THE GRAIN-SIZE DEPENDENCE OF METAL FATIGUE:
OUTLOOK ON THE FATIGUE OF ULTRAFINE-GRAINED METALS

H. MUGHRABI
Institut für Werkstoffwissenschaften, Universität Erlangen-Nürnberg,
Martensstr. 5, D-91058 Erlangen, Federal Republic of Germany

1. Introduction

With the development of new materials processing techniques such as severe plastic deformation (SPD) with the equal-channel angular (ECA) extrusion technique [1], bulk structural materials of very fine grain size (UFG: ultrafine grain size) and extraordinary strength are becoming a reality. Among the mechanical properties of interest, the fatigue strength is considered to be of special importance. It is therefore timely to consider critically to what extent a drastically reduced grain size, compared to conventional grain size, will be expected to affect the fatigue behavior. Because of the lack of detailed studies so far, such considerations must of necessity be based not only on the limited data on fatigue of SPD-material but also largely on some basic considerations and "extrapolations" from work on materials of conventional grain size.

While the fatigue life of SPD-material, compared to that of conventional materials, certainly is the property of prime interest, an assessment of the cyclic deformation and fatigue properties of SPD-material must also cover other relevant aspects such as:

- cyclic slip mode
- cyclic hardening/softening
- cyclic saturation: cyclic stress-strain (CSS) behavior
- cyclic strain localization (persistent slip bands, shear bands)
- dislocation distributions and dislocation mechanisms
- fatigue damage: trans-/intergranular fatigue crack initiation, fatigue crack propagation
- fatigue life
- corrosion fatigue.

In the following, an attempt will be made to address all topics but the last one by summarizing some earlier work and some ideas on grain-size effects in fatigue with the aim to formulate the expectations with regard to the fatigue behavior of SPD-material. Then, these expectations will be compared with experimental data available so far. It will become obvious that there is a remarkable gap between expectation and hope, on the one side, and reality, on the other side. We shall confine ourselves largely to face-centred cubic (fcc) materials, with the emphasis on SPD-copper which is the SPD-material that has been studied in most detail to date, although it seems that other materials may have a bigger potential (see Sect. 2).

T.C. Lowe and R.Z. Valiev (eds.), Investigations and Applications of Severe Plastic Deformation, 241–253.
© 2000 *Kluwer Academic Publishers. Printed in the Netherlands.*

2. Grain-Size Effects in Fatigue: Previous Work and Some Basic Considerations

In order to put the fatigue behavior of very fine-grained material into a more general perspective, systematic studies on the grain-size dependence of the fatigue properties of different materials would be desirable, for example, in the manner in which monotonic Hall-Petch type strengthening has been studied in the past, compare, for example, Armstrong [2]. While there is a lack of more recent systematic fatigue studies of that kind, the refinement of grain size has been recognized long ago as one major microstructural measure in order to improve the fatigue properties. This aspect has been investigated some twenty or thirty years ago on a number of different materials such as, for example, low-carbon steels [3,4], α-brass [5,6], copper [6,7], aluminum [6] and aluminum alloys [8]. While grain refinement was beneficial in all cases mentioned, the type and degree of improvement varied from case to case. Thus, Thompson et al. [6] noted that "there is, in fact, good evidence for Petch-type strengthening not always carrying over into longer-life fatigue", see also Sect. 6. On the other hand, in aluminum alloys containing shearable precipitates and/or precipitate-free zones (PFZ) at the grain boundaries, damaging localized trans- and/or intergranular slip can be suppressed or reduced efficiently by decreasing the grain size and, hence, the slip path, thus leading to a significant retardation of fatigue crack initiation [8]. On the other hand, studies on single-phase fcc materials such as aluminum, copper and α-brass demonstrated that the cyclic slip mode can be important or more important than the Hall-Petch strengthening effect. In this sense, an interesting study was undertaken by Thompson and Backofen [6] almost 30 years ago. These authors performed stress-controlled fatigue tests on the three different metallic materials aluminium, copper and α-brass in which the cyclic slip mode varies (in the sense of decreasing cross slip) from (very) wavy to (very) planar.

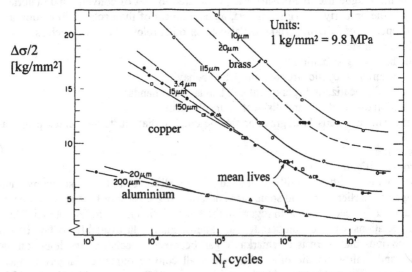

Figure 1. S-N plots, fatigued aluminum, copper and brass, different grain sizes.
After [6]. Courtesy of the authors.

The main results of Thompson and Backofen are displayed in Fig. 1, which shows the S-N curves for the three materials for different grain sizes. In this plot, an enhanced fatigue life is observed in the low-cycle fatigue (LCF) regime (number of cycles till failure: $N_f \lesssim 10^4$ cycles) in all cases, whereas in the high-cycle fatigue (HCF) range ($N_f \gtrsim 10^4$ cycles) only α-brass exhibits an improvement. It is evident that the fatigue performance is by far most improved in the case of the planar-slip material α-brass and only very moderate in the case of copper and aluminum. It was argued that the formation of dislocation cell structures (acting as obstacles against dislocation glide) masked the grain size effect in wavy-slip materials, whereas in a planar-slip material the grain boundaries are the only effective barriers to dislocation glide [6]. Low-carbon steels [3,4] which exhibit a positive grain size effect in spite of their wavy cyclic slip mode [6] seem to be the exception to this rule. In some cases, a complex situation exists: while grain size refinement seems to retard fatigue crack initiation, it can enhance or lower crack propagation rates in the near-threshold regime [9].

Based on the above studies, one could conclude that in order to exploit the improvement of fatigue properties by using SPD-processing, it would be more rewarding to focus on planar-slip alloys like α-brass, precipitation-hardened aluminium alloys or low-carbon steels. Nonetheless, we shall, in the following, concentrate the discussion on wavy-slip fcc materials such as in particular copper. The main reason is that almost all existing fatigue work on SPD-material refers to copper. Another reason is that the effects of grain size are sometimes more significant, when the fatigue life curves are not expressed in terms of the stress amplitude (S-N curve) but in terms of the plastic strain amplitude, i.e. in a Coffin-Manson plot [7,10,11]. A relevant discussion can be found in the paper by Lukáš and Kunz [7]. Here, we wish to first consider some general features of strain-life presentations before discussing the fatigue-life grain-size effects observed on copper for different "conventional" grain sizes [7,10,11] and on SPD-copper [12,13].

Figure 2 shows schematically a double-logarithmic plot of total strain amplitude $\Delta \varepsilon_t$ /2 ($\Delta \varepsilon_t$: total strain range) versus 2 N_f (number of load reversals till failure). The total strain amplitude is composed of the elastic strain amplitude $\Delta \varepsilon_{el}$ /2 and the plastic strain amplitude $\Delta \varepsilon_{pl}$ /2. In the terminology of Morrow [14], compare also Landgraf [15], the dependence of fatigue life N_f on the total strain amplitude $\Delta \varepsilon_t$ /2 is represented by the summation of the elastic and plastic strain resistance as follows:

$$\Delta \varepsilon_t /2 = \Delta \varepsilon_{el} /2 + \Delta \varepsilon_{pl} /2 = \frac{\sigma_f'}{E}(2N_f)^c + \varepsilon_f'(2N_f)^b \tag{1}$$

Here, E is Young's modulus, σ_f' denotes the fatigue strength coefficient, ε_f' the fatigue ductility coefficient, b the fatigue strength exponent ($b \approx -0.6$) and c the fatigue ductility exponent ($c \approx -0.1$). The fatigue lives in the asymtotic limits of LCF and HCF are given by the Coffin-Manson law ($\Delta \varepsilon_{pl}$ /2 vs. N_f) and by the Basquin law ($\Delta \varepsilon_{el}$ /2 vs. N_f), respectively. For practical purposes, it is important to know the fatigue coefficients σ_f' and ε_f'. Frequently, it is assumed that σ_f' and ε_f' are approximately equal to the

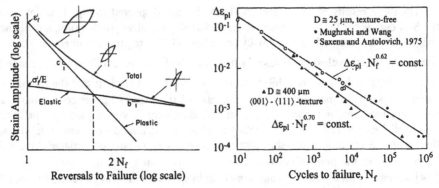

Figure 2 (on the left). Schematic strain-life diagram [15]. Courtesy of the author.
Figure 3. Coffin-Manson diagram, fatigued copper, two different grain sizes [10,11].

monotonic true fracture stress and true fracture strain (ductility), respectively [15]. Considering now the expected effect of grain refinement, it appears straightforward to assume that, in as-processed, heavily deformed SPD-material, σ_f' will be increased, and that this will lead to an improvement of HCF fatigue life. It is, however, more difficult to state how the fatigue ductility coefficient ε_f' will be affected by a reduced grain size and/or prior deformation. In the work of Agnew *et al.* [12,13] on SPD-Cu, a strongly reduced ductility was observed (true fracture strain of ca. 0.2 after ¼ cycle, see Section 4 and Fig. 5), whereas for more moderate pre-strains of 20% and 40%, the fatigue ductility coefficient ε_f' was found to be about twice as high and slightly lower, respectively, than for annealed copper in the work of Lukáš and Klesnil [16].

3. Dependence of Fatigue Life of Copper on Grain Size

After this excursion to the representation of fatigue-life data and the preceding brief discussion of the governing material parameters, some actual relevant data on grain-size effects in fatigued copper, shown in Figs. 3, 4a and 4b, will be discussed. Fig. 3 shows in a Coffin-Manson diagram the fatigue lives of copper polycrystals for two grain sizes (25 μm, 400 μm), as obtained in plastic-strain controlled tests [10,11]. For the finer grain size, a significant increase of fatigue life is observed which becomes more pronounced at larger fatigue lives. Conversely, it can be concluded that the fatigue life is reduced for the larger grain size. In this particular case, several factors have to be considered. First, the material with the larger grain size had a hard <001> – <111> fiber texture and also, for a given plastic strain amplitude, a larger cyclic saturation stress. While this by itself could be the reason for the reduced fatigue life, there was also a significant grain-size effect with regard to the fatigue failure mode in the sense that early intergranular fatigue cracking occurred in the material with the larger grain size, caused by the severe action of long persistent slip bands (PSBs), extending over almost the whole grain, against the grain boundaries (GBs) (compare also Section 5.3). It is this effect which is held mainly responsible for the observed behavior [11]. In this context, it

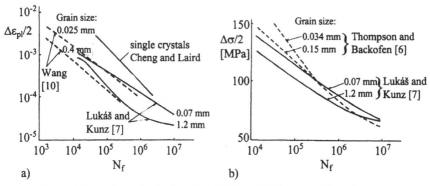

Figure 4. Comparison of grain-size effects in fatigue [7]. Courtesy of the authors. Coffin-Manson diagram. b) S-N curves

is interesting to note that Lukáš and Kunz had shown in an earlier study that the near-threshold macrocrack propagation rate was smaller for coarse-grained than for fine-grained material, as reported in [7]. This finding substantiates the previous argument, namely that the process of fatigue crack initiation (and early crack growth) is facilitated by PSB-GB cracking in the case of the larger grain size.

Lukáš and Kunz [7] have summarized the results of their own detailed plastic-strain controlled fatigue tests on copper of two different grain sizes (70 μm, 120 μm), extending up to numbers of cycles of $N = 10^7$, and the results of other authors' work [5,9,10] both in a Coffin-Manson plot (Fig. 4a) and in a Wöhler-type S-N plot (Fig. 4b). The latter plot is obtained by "translating" the plastic strain data to cyclic stress data via the cyclic stress-strain curve. It follows from a comparison of the two figures that while the S-N curve and the stress fatigue limit exhibit only a mild dependence on grain size, the Coffin-Manson plot and the plastic strain fatigue limit [16] show a strong dependence. With these results in mind, we now finally discuss the limited fatigue life data available on SPD-copper to date.

4. Fatigue Life of SPD-Material

While a number of studies on the cyclic deformation properties of fcc SPD-copper [12,13,17-19] and also SPD-nickel [20, 21] have now been performed, it seems that the only fatigue life data available so far are those of Agnew and Weertman [12] and of Agnew *et al.* [13] on SPD-copper. The data used for a Coffin-Manson plot were obtained in total-strain controlled tests, the plastic strain amplitudes were those recorded shortly before failure [12]. The authors describe the initial SPD-microstructure as one of subcells with a mean cell size of about 200 nm and at the same time, a lamellar-like fragmented structure (type B) in one case [12] and, in addition, in the other case, a structure of rather equiaxed subgrains with a mean size of 200–250 nm (type A).

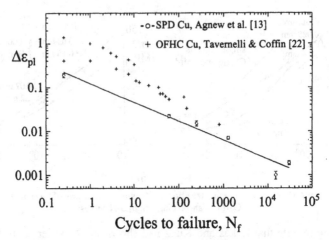

Figure 5. Coffin-Manson plot of fatigued SPD-copper [12]. Courtesy of the authors.

The earlier data [12] were presented in the form of a Coffin-Manson plot (for type B specimens), as shown in Fig. 5, and the later data (for type A material), obtained in stress–controlled tests, in the form of an S-N curve, shown in Fig. 6. In both cases, the authors also plotted data of other authors on fatigued copper of conventional grain size for comparison. It will be noted that, while the S-N data reveal a significant enhancement of fatigue life in the HCF regime, compared to the results of Thompson and Backofen [6] and Lukáš and Kunz [7] on copper of conventional grain size (Fig. 6), the LCF

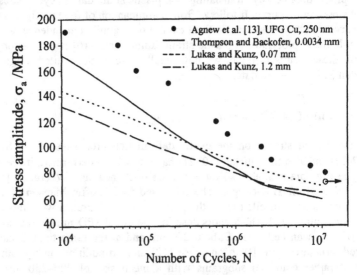

Figure 6. S-N curves of fatigued SPD-copper [13]. Courtesy of the authors.

fatigue life, as shown in the Coffin-Manson plot (Fig. 5), is reduced, compared to the data of Tavernelli and Coffin [22] and also to the data of refs. [7,10,11], compare Figs.

3 and 4a. The authors relate the positive effect in the HCF regime in the S-N plot (Fig. 6) to the more pronounced Bauschinger effect which had been studied earlier in some detail by Vinogradov et al. [17,18]. On the other hand, the authors attribute the disappointing LCF behavior (Fig. 5) to two effects, namely to pronounced softening during cyclic deformation [12,13] and to expressive "persistent" shear banding, extending over distances significantly larger than the nominal grain or "cell size", with extrusions and cracks. Here, we suggest that, in addition to the reasons given above, the unusually high cyclic flow stress level in SPD-material, at a given strain amplitude, will act to reduce fatigue life. The authors of refs. [17,18] also emphasize the heterogeneity of the microstructure of the fatigued specimens [12,13]. Furthermore, it is noted that while cyclic softening was very marked for type B specimens [12]. Vinogradov et al. [17,18] observed on type A specimens at lower plastic strain amplitudes that the substructure of the initial SPD-material was stable against cyclic softening and in fact exhibited slight cyclic hardening before approaching a state of saturation. Unfortunately, these authors did not state their (plastic) strain amplitudes properly. In another study on SPD-copper, Wu et al. [19] also noted the absence of cyclic softening at small plastic strain amplitudes.

Work on the fatigue of SPD-nickel at low plastic strain amplitudes ($\leq 10^{-3}$), performed by Bretschneider [20] and Thiele et al. [21], indicated a fairly stable cyclic deformation behavior with a saturation stress that was about thrice as large as in the case of nickel of conventional grain size, and the fatigue life at a given strain amplitude was strongly reduced. For the same material, Thiele et al. [21] found a decrease of internal stresses during cyclic deformation by X-ray diffraction. Finally, Bretschneider noted that systematic studies are hindered by scatter and poor reproducibility, emphasizing again the need for better control of the initial state of the as ECA-processed material [20].

5. Dislocation Arrangements and Glide Mechanisms: Some Simple Considerations with Respect to SPD-Material

In order to achieve a better understanding of the cyclic deformation and fatigue damage mechanisms, it is expedient to consider the development of the dislocation distribution by specific dislocation glide mechanisms, paying particular attention to the special case of ultrafine grain size in SPD-materials.

5.1. CONDITIONS FOR THE FORMATION OF DISLOCATION CELL STRUCTURES, UPPER LIMITS OF DISLOCATION DENSITY

There are some well established semi-empirical relationships between the shear stress τ, the dislocation density ρ and the dislocation cell size d, namely

$$\tau = \alpha G b \sqrt{\rho} \tag{2}$$

and

$$\tau = K G b / d^m. \tag{3}$$

Here, G is the shear modulus; α, K and m are characteristic constants. It is known that, depending on the structure and misorientation conveyed by a certain type of cell walls,

m can be as small as ½, corresponding to a Hall-Petch type relationship [23]. Normally however, m is close to one for typical cell structures and the constants α and K are approximately 0.5 and 10, respectively, compare the data of Staker and Holt for deformed copper [24]. If we assume for the sake of simplicity that τ corresponds to the shear stress on the plane of maximum shear stress, then the corresponding axial stress σ is equal to 2τ. Now, it has been shown that, for true strains of three or more, the axial flow stress σ of copper saturates at a value of approximately ≈ 450 MPa [25,26]. This value is in fair correspondence with the yield stress reported by Valiev [1] for ECA-processed SPD-copper. Since larger stresses can apparently not be achieved, the stress of $\sigma \approx 450$ MPa represents the stress level at which the smallest possible cell size can be obtained. Inserting numbers into Eq. (3), we obtain with $G = 45$ GPa, $m = 1$ and $b = 2.5 \times 10^{-10}$m a value $d_{min} \approx 500$ nm for the minimum cell size in copper. Comparing this value with the "grain" size D reported for SPD-copper, namely $D \approx 200\text{-}250$ nm [1,12,13,17,18], we find that $d_{min} \approx 500$ nm $\approx 2D$. In other words, since the smallest possible cell size in copper is about equal to or larger than the "grain" size in SPD-copper, the formation of cell walls within the grains of SPD-copper can hardly be expected. Rather, it seems that, in SPD-copper, the grains have taken the place of the cells. Eliminating τ between Eqs. (2) and (3), one obtains $\rho \approx K^2 / \alpha^2 d^2$. Replacing d by d_{min}, an upper limit of the dislocation density $\rho_{max} \approx 2 \times 10^{15}$m^{-2} is found. This value is in good accord with the upper limit of mean dislocation densities estimated in ref. [27].

5.2. CYCLIC SOFTENING OF PRE-DEFORMED MATERIALS

The ability of deformed wavy-slip materials to undergo cyclic softening is a unique property. Cyclic softening occurs when the yield strength exceeds the cyclic saturation stress that would normally have been attained at a given amplitude in an annealed material. Thus, materials which undergo cyclic softening readily have a largely history-independent cyclic stress-strain behavior, as first demonstrated by Feltner and Laird [28] for copper. Later, Lukáš and Klesnil [16] showed that this is only approximately true in the sense that a wavy-slip material like copper will only undergo cyclic softening ideally after pre-straining, if the pre-strains are not too large and if the plastic strain amplitude is not too small. There are a number of possible, rather different microstructural mechanisms that can contribute to cyclic softening:

    (i)   Mutual annihilation of dislocations of unlike sign.
    (ii)   Rearrangement of dislocations
    (iii)   Subgrain/cell structure coarsening
    (iv)   Cyclic strain localization in PSBs
    (v)   "Grain" boundary structure refinement/recovery
    (vi)   "Grain" size coarsening
    (vii)   "Large-scale" shear banding.

In metals of conventional grain size, the softening mechanisms (i) to (iv) commonly play a role. Of these, mechanisms (i) to (iii) are essential for a material in order to exhibit a well-defined state of cyclic saturation and thus a unique cyclic stress-strain behavior. By contrast, mechanism (iv) represents a damaging deformation mode. In material of ultrafine grain size, the additional mechanisms (v) to (vii) must be considered.

Mechanism (v) is believed to play a role in the recovery of the defects of the non-equilibrium grain boundary structure. Mechanism (vi) is related to mechanism (iii) but now applies to grain boundaries rather than to subgrain boundaries. Mechanism (vii) is a damaging localized deformation mode extending over a larger number of grains that has been observed only in fatigued SPD-material [12,13,18].

Quite generally, the described softening mechanisms can only become effective, if a certain plastic strain amplitude is exceeded so that the dislocations are displaced sufficiently to allow the necessary interactions. This is most easily shown for mutual dislocation annihilation (mechanism (i)). The plastic shear strain amplitude is given by

$$\gamma_{pl} = \pm \rho b L \,, \tag{4}$$

where L represents the average glide path of a dislocation in a half-cycle. In saturation, the dislocation density ρ is constant and limited to a saturation value [29]

$$\rho \le \frac{1}{2L \cdot y}. \tag{5}$$

Here, y is the so-called annihilation distance [27] which depends on the dislocation character and which has been established as $y_s \approx 50$ nm and $y_e \approx 1.6$ nm for screw and edge dislocations, respectively, in copper at room temperature [27]. Eliminating ρ between Eqs. (4) and (5), one can show easily in this simple model that dislocation annihilation will only occur, if γ_{pl} exceeds a critical value given by

$$\gamma_{pl}^{crit} = \pm \frac{b}{2y}. \tag{6}$$

If dislocations are arranged in groups of n dislocations in a planar-slip material, then the critical shear strain amplitude is larger by a factor n [29]. Thus, it is easily understood that while cyclic softening is typical of low strain amplitudes, it becomes less effective, when the strain amplitude becomes too small (for a given type of dislocations) and approaches the critical strain amplitude (Eq. (6)). This is in line with the experimental observations made on SPD-material by Agnew et al. [12,13], showing that, while cyclic softening first increases with increasing strain amplitude, it decreases subsequently.

5.3. CYCLIC STRAIN LOCALIZATION IN PERSISTENT SLIP BANDS (PSBS)

A characteristic feature of the cyclic deformation of fcc metals is the instability of the cyclically hardened dislocation microstructure against further cyclic straining, when a critical state of hardening (which depends on the strain amplitude in strain-controlled tests) is reached. Subsequently, the microstructure breaks down on a local scale in thin lamellae parallel to the active glide plane. Then, the cyclic strain becomes localized in these lamellae, called PSBs, as a new and softer microstructure (wall "ladder" structure in fcc metals) develops in these PSBs, compare the earlier references [10,11,29].

The questions to be addressed are whether PSBs are expected to develop in very fine-grained SPD-material and what consequences this would have. In earlier studies, it had been shown that, in grains of polycrystals, PSBs are rarer than in single crystals because of the constraints by neighboring grains and that the PSBs are less active, because the local shear strain amplitude in the PSBs is reduced [10,11]. Quite recently, these findings have received new support from some work by Kawazoe et al. [30] on copper with a grain size of about 8 μm. These authors concluded that the appearance of PSBs was extremely restricted. Thus, it appears improbable that PSBs of the same type as in materials of conventional grain size can develop in SPD-material such as copper. Moreover, the typical thickness and the typical wall spacings of PSBs in single-phase fcc metals and alloys are of the order of 1 μm, *i.e.* significantly larger than the grain size of some 100 nm in SPD-material. However, in precipitation-hardened alloys with shearable precipitates, very much thinner PSBs can develop, possibly also in very fine-grained material.

Even if PSBs develop, they should be much less efficient in producing either transgranular stage I shear cracks and/or intergranular cracks by PSB-GB cracking for the following reasons. The severeness of PSB activity can be expressed through the height e of the extrusions that emerging PSBs produce (at the surface). As was shown in a microstructural model by Essmann *et al.* [31], the extrusion height e is proportional to the specimen diameter in the case of single crystals and to the grain size D in the case of polycrystals. Typically, one finds [32]

$$e \approx 3 \times 10^{-4} D. \tag{7}$$

Thus, for a grain size of some 100 nm, the extrusion height would only be ca. 0.03 nm.

For PSB-GB cracking, it can be shown that the PSB can be considered in terms of a double dislocation pile-up acting against two opposite grain boundaries [33]. The local effective shear stress τ_E necessary to cause the grain boundary to crack is given by

$$\tau_E = const. \frac{G}{D}. \tag{8}$$

Thus the critical effective stress for PSB-GB cracking in SPD-material would be larger by orders of magnitude, compared to PSB-GB cracking in materials of conventional grain size. Hence, PSB-GB cracking should not be a problem in SPD-material.

5.4. "LARGE-SCALE" SHEAR BANDING

While PSBs in their classical form have not been observed in fatigued SPD-material, another type of microstructural instability, associated with (cyclic) strain localization has been observed in the form of "persistent" shear bands extending over distances much larger than the grain size [12,13,18,19]. These bands are reminiscent of micro-bands commonly found after severe monotonic (complex) deformation such as cold rolling [34] and are considered to be typical of a microstructural instability occurring during deformation in materials subjected to a "strain-path change" after severe deformation under constraint. The occurrence of this kind of strain localization suggests that even the small plastic strains imposed in a cyclic deformation test suffice to trigger the instability. It furthermore appears probable that type B material [13], compare Sect. 4, is more prone to shear banding, since the initial microstructure already contains "precursors" of shear bands in the form of lamellar cell structures.

6. Closing Remarks, Directions for Future Work

The limited results available today on the fatigue of SPD-materials are interesting. They have provided insight into new phenomena and, at the same time, raised a number of questions. This should stimulate further research. It is suggested that further work should concentrate on the following topics:

- The reproducibility of ECA-processing with regard to microstructural control should be improved.
- Some further clarification on the nature of the "grain" or "cell" boundaries is needed. Is SPD-material truly polycrystalline down to grain sizes of 100 nm?
- Suitable heat/annealing treatments, compare [12,13,17,18,20,21], must be explored further in order to remove some of the initial microstructural instability of as-processed SPD-material.
- The role of textures in fatigue needs to be considered.
- Fatigue tests should be performed in both strain and stress control and over a wider range of amplitudes than hitherto.
- Cyclic-stress strain data (saturation stress versus plastic strain amplitude) are necessary and are completely lacking at present.
- Materials other than copper (such as planar-slip alloys, (precipitation-hardened) light metal alloys, carbon steels) and titanium (and its alloys), as shown at this Workshop by Stolyarov et al. [35], would appear to be more promising and should be investigated (see Sect. 2).
- On the theoretical side, some modelling specific to ultrafine grain size is necessary, compare Sect. 5. In addition, Hall-Petch type modelling of fatigue strengthening by grain refinement is needed, taking into account that, in fatigue, in contrast to monotonic deformation, pile-up effects will be much smaller, because the plastic strain amplitude is always confined to the microstrain range, thus limiting the displacements of the glide dislocations.

7. Acknowledgments

Sincere thanks are due to many colleagues, in particular to Sean Agnew and Julia Weertman for providing unpublished data and to Waltraud Kränzlein and Heinz Werner Höppel for their help in the preparation of this manuscript.

8. References

1. Valiev, R. (1997) Structure and mechanical properties of ultrafine-grained metals, *Mater. Sci. Eng.* **A234**, 59-66.
2. Armstrong, R.W (1999) Dislocation mechanics description of polycrystal plastic flow and fracturing behaviors, in *Mechanics and Materials: Fundamentals and Linkages*, edited by M. A. Meyers, R. W. Armstrong and H. Kirchner, John Wiley & Sons, Inc. pp. 363-398.
3. Oates, G. and Wilson, D.V. (1964) The effects of dislocation locking and strain ageing on the fatigue limit of low-carbon steel, *Acta metall.* **12**, 21-33.

252

4. Taira, S., Tanaka, K. and Hoshina, M. (1979) Grain size effect on crack nucleation and growth in long-life fatigue of low-carbon steel, in *Fatigue Mechanisms*, edited by J.T. Fong, ASTM STP 675, American Soc. for Testing and Materials, Philadelphia, Pa., pp. 135-173.
5. Forrest, P.G. and Tate, A.E.L. (1964-65) The influence of grain size on the fatigue behaviour of 70/30 brass, *J. Inst. Metals* 93, 438-444.
6. Thompson, A.W. and Backofen, W.A. (1971) The effect of grain size on fatigue, *Acta metall.* 19, 597-606.
7. Lukáš, P. and Kunz, L. (1987) Effect of grain size on the high cycle fatigue behaviour of polycrystalline copper, *Mater. Sci. Eng.* 85, 67-75.
8. Starke, E.A. and Lütjering, G. (1978) Cyclic plastic deformation and microstructure, in *Fatigue and Microstructure*, American Society for Metals, Metals Park, Ohio, pp. 205-243.
9. Gerberich, W.W. and Moody, N.R. (1979) A review of fatigue fracture topology effects on threshold and growth mechanisms, in *Fatigue Mechanisms*, edited by J.T. Fong, ASTM STP 675, American Soc. for Testing and Materials, Philadelphia, Pa., pp. 292-341.
10. Wang, R. (1982) Untersuchungen der mikroskopischen Vorgänge bei der Wechselverformung von Kupferein- und vielkristallen, *Doctorate Thesis*, University of Stuttgart.
11. Mughrabi, H. and Wang, R. (1988) Cyclic stress-strain response and high-cycle fatigue behaviour of copper polycrystals, in *Basic Mechanisms in Fatigue of Metals*, edited by P. Lukáš and J. Polák, Elsevier, Amsterdam, pp. 1-13.
12. Agnew, S.R. and Weertman, J.R. (1998) Cyclic softening of ultrafine grain copper, *Mater. Sci. Eng.* A244, 145-153.
13. Agnew, S.R., Vinogradov, A. Yu., Hashimoto, S. and Weertman, J.R. (1999), Overview of fatigue performance of Cu processed by severe plastic deformation, submitted to *Mater. Sci. Eng. A*.
14. Morrow, JoDean (1965) Cyclic plastic strain energy and fatigue of metals, in *Internal Friction, Damping and Cyclic Plasticity*, ASTM STP 378, American Soc. for Testing and Materials, Philadelphia, Pa., pp. 45-87.
15. Landgraf, R.W. (1970) The resistance of metals to cyclic deformation, in *Achievement of High Fatigue Resistance in Metals and Alloys*, ASTM STP 467, American Soc. for Testing and Materials, Philadelphia, Pa., pp. 3-36.
16. Lukáš, P. and Klesnil, M. (1973) Cyclic stress-strain response and fatigue life of metals in low amplitude region, *Mater. Sci. Eng.* 11, 345-356.
17. Vinogradov, A., Kaneko, Y., Kitagawa, K., Hashimoto, S. and Valiev, R. (1997) Cyclic response of ultrafine-grained copper at constant plastic strain amplitude, *Scripta mater.* 36, 1345-1351.
18. Vinogradov, A., Kaneko, Y., Kitagawa, K., Hashimoto, S. and Valiev, R. (1998) On the cyclic response of ultrafine-grained copper, *Mater. Sci. Forum* 269-272, 987-992.
19. Wu, S.D., Wang, Z.G., Li, G.Y., Alexandrov, I.V. and Valiev, R.Z. (1999) Cyclic deformation characteristic of submicrometer copper processed by ECAE technique, in *FATIGUE'99, Proc. of the Seventh International Fatigue Congress*, Vol. 1, Higher Education Press, Beijing, P.R. China & EMAS Ltd., West Midlands, UK, pp. 247-252.
20. Bretschneider, J. (1999) private communication.
21. Thiele, E., Bretschneider, J., Hollang, L., Schnell, N. and Holste, C. (1999) Influence of thermal treatment and cyclic plastic deformation on the defect structure in ultrafine-grained nickel, in *the present proceedings*.
22. Tavernelli, J.F. and Coffin, L.F. (1959) A compilation and interpretation of cyclic strain fatigue tests on metals, *Trans. Am. Soc. Metals* 51, 438-453.
23. Abson, D.J. and Jonas, J.J. (1970) The Hall-Petch relation and high-temperature subgrains, *Metal. Sci. J.* 4, 24-28.
24. Staker, M.R. and Holt, D.L. (1972) Dislocation cell size in deformed copper, *Acta metall.* 20, 569-576.
25. Lloyd, D.J. and Kenny, D. (1978) The stress-strain behaviour of copper over a large strain range, *Scripta metall.* 12, 903-907.
26. Luyten, J., Delaey, L. and Aernoudt, E. (1978) Strength and substructure evolution in drawn copper-alumina wires, *Mater. Sci. Eng.* 32, 193-196.
27. Essmann, U. and Mughrabi, H. (1978) Annihilation of dislocations during tensile and cyclic deformation and limits of dislocation densities, *Phil. Mag. A*, 40, 731-756.
28. Feltner, C.E. and Laird, C. (1967) Cyclic stress-strain response of f.c.c. metals and alloys, I. Phenomenological experiments, II. Dislocation structures and mechanisms, *Acta metall.* 15, 1621-1632 and 1633-1653.

29. Mughrabi, H., Ackermann, F. and Herz, K. (1979) Persistent slipbands in fatigued face-centered and body-centered cubic metals, in *Fatigue Mechanisms*, edited by J.T. Fong, ASTM STP 675, American Soc. for Testing and Materials, Philadelphia, Pa., pp. 69-105.

30. Kawazoe, H., Yoshida, M., Basinski, Z.S. and Niewczas, M. (1999) Dislocation microstructures in fine-grained Cu polycrystals fatigued at low amplitude, *Scripta mater.* **40**, 639-644.

31. Essmann, U., Goesele, U. and Mughrabi, H. (1981) A model of extrusions and intrusions in fatigued metals I. Point-defect production and the growth of extrusions, *Phil. Mag. A* **44**, 405-426.

32. Mughrabi, H., Wang, R., Differt, K. and Essmann, U. (1983) Fatigue crack initiation by cyclic slip irreversibilities in high-cycle fatigue, in *Fatigue Mechanisms: Advances in Quantitative Measurement of Physical Damage*, edited by J. Lankford, D.L. Davidson, W.L. Morris and R.P. Wei, ASTM STP 811, American Soc. for Testing and Materials, Philadelphia, Pa., pp. 5-45.

33. Liu, W., Bayerlein, M., Mughrabi, H., Day, A. and Quested, P.N. (1992) Crystallographic features of intergranular crack initiation in fatigued copper polycrystals, *Acta. metall. mater.* **40**, 1763-1771.

34. Bay, B., Hansen, N., Hughes, D.A. and Kuhlmann-Wilsdorf, D. (1992) Evolution of fcc deformation structure in polyslip, *Acta metall. mater.* **40** 205-219.

35. Stolyarov, V.V., Latysh, V.V., Kolobov, Yu.R., Valiev, R.Z., Zhu, Y.D. and Lowe, T. (1999) The development of nanostructured SPD Ti for medical application, in *the present proceedings*.

PLASTICITY AND WORK-HARDENING AT 300–4.2 K OF NANO-STRUCTURED COPPER AND NICKEL PROCESSED BY SEVERE PLASTIC DEFORMATION

V.Z. BENGUS, E.D. TABACHNIKOVA, R.Z. VALIEV*, I.V. ALEXANDROV*, V. D. NATSIK
B. Verkin Institute of Low Temperature Physics and Engineering of Ukraine Academy of Sciences, Kharkov, 310164, Ukraine
*Ufa State Aviation Technical University, K. Marks st., Ufa, 450000 Russia

Abstract

Mechanical properties of bulk nanostructured Cu and Ni with 200 nm grains (manufactured by the equal channel angular pressing) are measured under uniaxial compression at 300, 77 and 4.2 K. Measured mechanical characteristics exceed several times those of coarse grained polycrystalline Cu and Ni.

1. Introduction

There are two well known methods to increase the yield stress and strength of pure metals. One is by work-hardening by preliminary plastic deformation that creates a rather uniformly fragmented high density dislocation structure that hinders plastic flow [1]. Another method is to prepare polycrystals with the smallest possible dimension of grains by thermomechanical treatment [1]. In this case the yield stress increases with decreasing average grain size (typically according to the Hall-Petch relation [2]). For example, the yield stress of pure polycrystalline Cu at the room temperature doubles from 44 to 88 MPa by the decreasing grain size from 90 to 12 µm [3].

The second method of strengthening has been further developed using the equal channel angular (ECA) pressing method. This deformation method induces uniform plastic deformation by high intensity simple shear during extrusion through two intersecting channels. It causes intensive fragmenting of the grains of polycrystals, saturating their volume with deformation defects. Subsequent thermal treatment by isothermal annealing can result in bulk samples of pure Cu and Ni with average grain sizes of 200 nm and having substantially increased hardness [4].

Bulk nanostructured pure metals are of great interest for technical applications at low temperatures, specifically, in cryogenics. However, there has been almost no study of low temperature mechanical properties and work hardening mechanisms in nanostructured materials [5]. Therefore, the aim of this work is to use uniaxial compression to study low temperature mechanical properties of bulk pure Cu and Ni that has been prepared by ECA pressing to have an average grain size of about 200 nm.

T.C. Lowe and R.Z. Valiev (eds.), Investigations and Applications of Severe Plastic Deformation, 255–260.

2. Materials and Methods

Samples of Cu (99.98% purity) and Ni (99.99% purity) were prepared by ECA pressing up to an equivalent true plastic shear strain of 3.5 to create rods 14 × 14 × 160 mm having an average grain size of 200 nm [4]. Specimens of about 2 × 2 × 7 mm were cut by electroerosion from these rods. Figures 1 (a, b) and 2 (a, b) show microstructures of these samples observed by TEM (JEM–2000EX electron microscope at the 200 kV voltage) and corresponding electron diffraction patterns taken from an 0.5 μm^2 area.

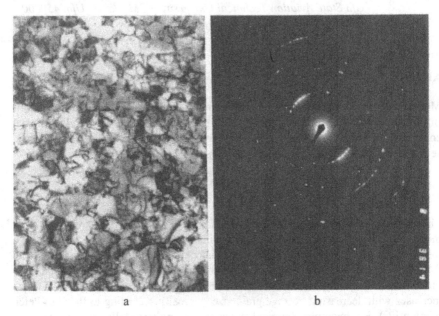

a   b

Figure 1. Typical transmission electron micrograph – the scale is 500 nm in 1 cm – (a) and corresponding diffraction pattern (b) of ECA pressed Cu.

These specimens were strained by uniaxial compression at 300, 77 (in liquid nitrogen) and 4.2 K (in liquid helium) at a speed of 0.28 mm/min (strain rate of about 6.7×10^{-4} s^{-1}) in a testing machine with a 10 kN/mm rigidity. Compression flow stresses σ were calculated as the ratio of load P to the initial specimen cross-sectional area. Magnitudes of plastic strain ε were taken from compression diagrams (P, t) by marking-out from the point on the diagram the initial deviation from a straight line. A proof stress corresponding to this point is considered as a conditional yield stress σ_χ. Values of this characteristic do not exceed the commonly accepted 0.2% offset yield stress $\sigma_{0.2}$. Metallographic observations of surfaces of deformed specimens were carried-out by optical microscope MIM–7.

3. Experimental Results

3.1. COPPER

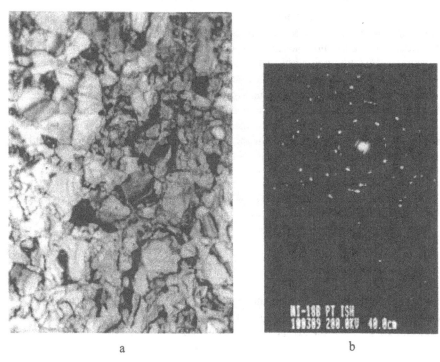

a   b

Figure 2. Typical transmission electron micrograph – the scale is ~ 450 nm in 1 cm – (a) and corresponding diffraction pattern (b) of ECA pressed Ni.

Initial parts of typical uniaxial compression "σ – ε" curves for the Cu deformed at 300, 77 and 4.2 K are shown in Figure 3. Analysis of these curves as well as

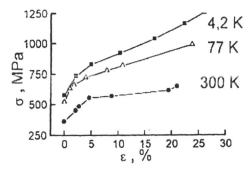

Figure 3. Initial parts of stress-strain curves "σ – ε" of the 200 nm mean grain size Cu compressed at 300, 77 and 4.2 K with the 6.7 × 10⁻⁴ s⁻¹ strain rate.

metallographic observations of deformed specimens surfaces points out typical features of low temperature plastic deformation of bulk nanostructured Cu. They are: retaining of high plasticity (more than 50% elongation) in nanostructured specimens with decreasing temperature from 300 to 4.2 K; increasing of the conditional yield stress from 365 up to 580 MPa with decreasing temperature from 300 to 4.2 K; appearance at 4.2 K of a jump-like plastic deformation along the deformation curve beginning from $\varepsilon = 5\%$. Note that amplitudes of first jumps of the load are from 10 to 25 N and increase with increasing ε. At $\varepsilon \approx 25\%$ the amplitude of the load jump reaches 60 N, correspondings to a 0.1% shortening of the specimen. Metallographic observations of the surface of deformed specimens show localized shear bands especially pronounced at 4.2 K. Values of measured conditional yield stresses σ_χ of our nanostructured Cu are nearly 5 to 10 times higher than values of yield stresses of conventional polycrystalline Cu known in the literature [3]. For comparison values of measured conditional yield stresses at 300, 77 and 4.2 K for the nanostructured Cu are given in the Table 1 together with yield stress data for conventional polycrystalline Cu from [3].

TABLE 1. Conditional yield stresses σ_c for a polycrystalline [3] and the nanostructured (this work) copper at low temperatures

Average grain size, µm	σ_y, MPa			
	300 K	77 K	4.2 K	Reference
90	44	45.6	45.6	[3]
30	61	67	70	[3]
12	88	106	108	[3]
0,2	365	520	580	This work

The observed increase of σ_χ in the nanostructured Cu is not described by the Hall-Petch relation, as can be seen in Figure 4 showing "$\sigma_\chi - d^{-0,5}$" dependences for 300, 77 and 4.2 K (d – is the average grain size).

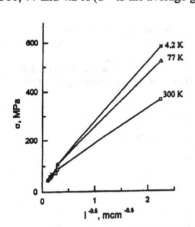

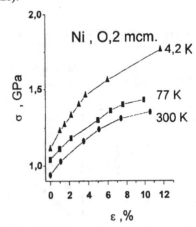

Figure 4. Hall-Petch plots of Cu for 300, 77 and 4.2 K.

Figure 5. Initial parts of stress-strain curves "$\sigma - \varepsilon$" of the 200 nm mean grain size Ni compressed at 300, 77 and 4.2 K with the $6.7 \times 10^{-4}\,s^{-1}$ strain-rate.

3.2. NICKEL

Low temperature plastic deformation of the nanostructured Ni has the same features as Cu. Values of conditional yield stresses of Ni are shown in the Table 2. As in the

TABLE 2. Conditional yield stresses σ_χ for the nano-crystalline Ni at low temperatures

T, K	σ_y, MPa
300	940
77	1040
4.2	1110

case of the nanostructured Cu they exceed by many times the yield stresses of polycrystalline Ni [6]. Typical low temperature uniaxial compression "$\sigma - \varepsilon$" curves for the nanostructured Ni specimens are shown in Figure 5.

4. Discussion of Results

The low temperature mechanical properties of the bulk nanostructured specimens of Cu and Ni are superior to those of conventional polycrystals of these metals. This indicates the possibility of employing nanostructured states in metals at low temperatures.

Consideration of the shapes of the "$\sigma - \varepsilon$" curves shows two forms: "parabolic" and 3 - stage ("s-like" as in FCC single crystals of the easy glide orientation) [7]. The forms of these curves usually depend on the relationship between the height and transverse dimensions of the specimens [7]. "Parabolic" curves are observed for the short specimens and 3-stage deformation curves may be observed when the ratio of a specimen height to transverse dimensions is nearly 3:1.

Such distinctions can be explained in terms of a source model of work-hardening [7], recognizing the different behavior of glide bands in each case. In short specimens, free development of glide bands is impeded at the butt-ends of a specimen. Therefore, the average contribution of glide bands to the macroscopic deformation of a specimen increases during deformation, giving rise to the "parabolic" form of the "$\sigma - \varepsilon$" curve [7]. The three-stage curve corresponds to a constant average contribution of glide bands to deformation, which is typical when they freely exit the specimen lateral sides.

It was known that by decreasing the average grain size of polycrystals, their "stress-strain" curves may become 3-staged at small grain sizes, even in BCC metals [8]. In the framework of the model [7], this may be an indication that at small grain sizes glide bands may contribute to the macroscopic strain in a way that does not change during subsequent straining. This supposition needs experimental verification in ECA processed nanostructured materials.

Three-stage deformation curves may provide evidence that plastic deformation of ECA-processed nanostructured materials is essentially limited by activity of dislocation sources at non-equilibrium grain boundaries [9, 10]. This is likely a result of the unique spatial distribution of internal stresses in such materials in comparison with conventional polycrystals. This possibility also needs experimental validation and theoretical consideration.

5. References

1. Hertzberg, R.W. (1983) *Deformation and Fracture Mechanics of Engineering Materials*, John Wiley & Sons, N.Y.
2. *Physical Metallurgy* (1983) R.W. Cahn and P. Haasen (eds.), North-Holland Physics Publish., Amsterdam.
3. Conrad, H. (1961) Role of grain boundaries in processes of creep and long-term failure, in John E. Dorn (eds.), *Mechanical Behavior of Materials at Elevated Temperatures*, McGraw-Hill Book Company, N. Y., pp. 96–149.
4. Akhmadeev, N.A., Valiev, R.Z., Kopylov, V.I., Mulyukov, R.R (1992) Formation of submicrograin structure in copper and nickel with a use of severe shear deformation, *Metally* 15, 96–101 (in Russian).
5. Gray III, G.T., Lowe, T.C., Cady, C.M., Valiev, R.Z., Aleksandrov, I.V. (1997) Influence of strain rate & temperature on the mechanical response of ultrafine-grained Cu, Ni, and Al-4Cu-0.5Zr, *Nanostructured Materials* 9, 477–480.
6. Thompson, A.W. (1977) Effect of grain size on work hardening in nickel, *Acta Metall.* 25, 83–86.
7. Bengus, V.Z. (1981) A source model for work hardening of single crystals, *Czech. J. Phys.* B 31, 125–129.
8. Honeycombe, R.W.K. (1968) The Plastic Deformation of Metals, EDW and ARNOLD Publishers Ltd.
9. Kaibyshev, A., Valiev, R.Z. (1987) Grain Boundaries and Properties of Metals, Moscow, *Metallurgy*, pp. 147–151 (in Russian).
10. Valiev, R.Z. (1995) Approach to nanostructured solids through the studies of submicron grained polycrystals, *Nanostructured Materials* 6, 73–82.

COPPER GRAIN BOUNDARY DIFFUSION AND DIFFUSION INDUCED CREEP IN NANOSTRUCTURED NICKEL

YU.R. KOLOBOV, G.P. GRABOVETSKAYA, M.B. IVANOV, R.Z. VALIEV[1], T.C. LOWE[2]
Institute of Strength Physics and Material Science, Russian Academy of Science; Pr. Academichesky 2/1, 634021 Tomsk, Russia.
[1]*Institute of Physics of Advanced Materials, Russian Academy of Science; K. Marx 12, 450000 Ufa, Russia.*
[2]*Los Alamos National Laboratory, Los Alamos, New Mexico USA.*

1. Introduction

Nanostructured (NS) materials (grain size d ≤ 100 nm) are currently being intensively investigated. This special attention is due to the distinct physical, mechanical and other properties compared to when they are coarse-grained (CG) [1]. In particular, it has been found that diffusion coefficients (D) in NS materials exceed by several orders of magnitude the respective values in CG materials [2–4]. However, the physical reasons for the anomalously high values of D in these materials are debatable. The authors of [4] suggest that the self-diffusion in nanostructured Ni prepared by the inert gas condensation method occurs mainly along the boundaries of clusters and on the surface of pores located there. Clusters (having size 1–10 μm) consist of the complexes of nanoparticles having grain size d ≤ 100 nm. Diffusion along nanoparticles boundaries inside a cluster does not differ from grain boundary diffusion in CG materials, while diffusion along cluster boundaries closely resembles surface diffusion.

The author of [5] suggests that the difference between diffusion-dependent properties of coarse-grained materials and nanostructured materials is connected only with the high relative volume fraction of grain boundaries in the latter.

There is no conventional point of view on the character of the diffusion processes in NS materials. It remains difficult to interpret the experimental data because of the small grain size, the possibility of the overlap of diffusion fluxes from adjacent boundaries, and relaxation and the migration of GBs during diffusion annealing. So it is interesting to carry out diffusion experiments on nanostructured (NS) materials (grain size about 0.1–0.3 μm) produced by severe plastic deformation, in particular by the method of equal-channel angular pressing (ECAP) [6]. Samples of materials produced by ECAP do not contain pores and are relatively large compared to NS produced by other methods. At the same time the state of GB in these materials are characterized as existing in a non-equilibrium state, as indicated by the significant distortion of the lattice near the GBs and corresponding elastic stresses [7].

Lately an essential change of the GB state in metals was found to be caused by diffusion fluxes of impurity atoms from an external source (*e.g.* a coating) [8]. Such

T.C. Lowe and R.Z. Valiev (eds.), Investigations and Applications of Severe Plastic Deformation, 261–265.
© *2000 Kluwer Academic Publishers. Printed in the Netherlands.*

treatment during plastic deformation under certain conditions results lead to the development of diffusion induced grain boundary sliding (DIGBS) and, consequently, to the essential decrease of creep resistance and to the increase of polycrystal plasticity [9]. The effect of DIGBS is shown [8-9] to only take place in the non-stable diffusion regime $\Re_1$ following Harrison's classification, updated in [10]. This regime is transient from diffusion only in GB (regime ∇) to quasi-steady grain boundary diffusion with essential leakage of diffusant into the bulk of grain (regime B_2) and represents non-equilibrium impurity distribution in near grain boundary region. In the $\Re_1$ regime there are large internal stresses in the GB, relaxation of which results in formation of lattice and grain boundary dislocations in near grain boundary areas (dislocation walls) [8]. This and the non-equilibrium state of GB, are the most probable physical origin of the apparent decrease in grain boundary shear resistance.

However, as noted above, some doubt about the suitability of the conventional approach (Harrison classification) for the treatment of the experimental data on diffusion in NS materials was stated in [5]. This doubt arises in part from concern about the possible overlap of the diffusion impurity fluxes from adjacent GBs from the very beginning of the diffusion process. Furthermore, in NS materials produced by ECAP, grain boundary impurity diffusion occurs in regime B_2' (according to the classification [11]), as opposed to regime B_2 as observed for coarse-grained materials in which diffusion takes place along many GB and the penetration depth of an impurity from a surface into the sample considerably exceeds the grain size. Besides by the analogy to [10] and [11] it is also possible to assume behavior in regime B_1' in NS materials and, therefore, speculate that such materials undergo grain boundary impurity in a manner similar to coarse-grained materials, enabling the creep activation effect to be observed.

In this association the comparative study of the impurity diffusivity and creep acceleration effect under the action of grain boundary diffusion fluxes of impurities from an external source in NS and coarse-grained state is of great interest.

2. Experimental

Nickel of commercial purity both in coarse-grained (grain size ~ 20 μm) and nanostructured states (grain size ~ 0.3 μm) was studied. The nanostructured state was formed by equal channel angular pressing [6].

Creep testing of NS nickel was carried out in temperature interval of 398–573 K under constant loading in vacuum of 10^{-2} Pa using a PV–3012 machine.

The diffusion annealing of nickel under a copper coating was performed in vacuum of 10^{-2} Pa for 3–5 hours at 423, 573, 773, 823 and 873 K for coarse-grained specimens and at 398, 423, 443 and 573 K for NS specimens.

Copper concentration in nickel at different depths was measured by secondary-ion mass spectroscopy (SIMS). The total absolute error in the measurement of copper concentration in a layer did not exceed 20%.

The grain boundary migration rate was estimated by light microscopy from changes of average grain size during annealing. Averages were calculated over 1000 grains.

3. Results and Discussion

After annealing at T = 773, 823 and 873 K it is possible to obtain the concentration profiles (Figure 1, curve 1–3) and to calculate copper grain boundary diffusion coefficients in coarse-grained nickel (D_b). It is taken into account, that during nickel annealing at 773, 823 and 873 K experimentally measured grain boundary migration rates are $1.5 \cdot 10^{-11}$, $0.6 \cdot 10^{-10}$ and $1.9 \cdot 10^{-10}$ m/s, respectively. With this in mind, diffusion coefficients may be calculated by [11]:

$$C = C \cdot \exp(-x\sqrt{V / D_b \delta}), \qquad (1)$$

where C_0 and C are the copper concentration at the source surface and at the depth x, respectively; V is grain boundary migration rate. The x value is assumed to be 30 μm, this approximately corresponds to the distance below the surface at which measurements of grain boundary migration rate were performed. D_b values calculated from (1) are given in Table 1. These data were used to calculate the activation energy for the copper grain boundary diffusion in coarse-grained nickel Q_b = 124.7 kJ/mol.

Assuming that the parameters of the Arrhenius dependence are constant in the entire temperature range the $D_b \delta$ values were calculated for lower temperatures using the experimental data (Table 1).

TABLE 1. Coefficients of grain boundary diffusion of copper in NS and coarse-grained nickel

Material	Ni	Ni (NS)
Temperature, K	$D_b \delta^*$, m^2/s	D_b, m^2/s
398	$4.64 \cdot 10^{-29}$ **	$5.06 \cdot 10^{-15}$
423	$4.31 \cdot 10^{-28}$ **	$9.6 \cdot 10^{-15}$
443	$2.14 \cdot 10^{-27}$ **	$2.2 \cdot 10^{-14}$
573	$4.65 \cdot 10^{-24}$ **	$D_b \delta = 1.4 \cdot 10^{-20}$
773	$4.0 \cdot 10^{-21}$	
823	$1.26 \cdot 10^{-20}$	
873	$3.9 \cdot 10^{-20}$	

* usually is considered to be δ=(0,5-1,0) nm, ** the calculated values.

The copper concentration profiles for NS nickel after annealing at 398–573 K for 3 h are shown on Figure 1 (curves 4–7). The D_b values for copper diffusivity in NS nickel were calculated using these data with the assumption that grain boundaries do not migrate in the temperature interval 393–443 K (according to transmission electron microscopy data). During annealing at 573 K the grain boundaries migrate at the rate of $7 \cdot 10^{-11}$ m/s. Along immobile grain boundaries, in the case where no leakage of the diffusant occurs from grain boundaries into the bulk of grains, the diffusion coefficients may be calculated by the formula [11]:

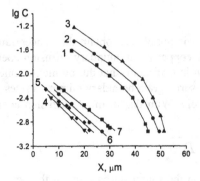

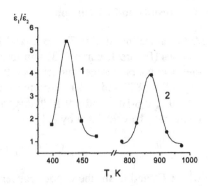

Figure 1. Copper concentration profiles in coarse-grained (curves 1-3) and NS (curves 4-7) nickel after diffusion annealings at temperatures 773(1), 823(2), 873(3) and 398(4), 423(5), 443(6), 573(7) K (3 hours).

Figure 2. Temperature dependence curves of the creep acceleration effect value in NS (curve 1) and coarse-grained (curve 2) nickel under the action of copper grain boundary diffusion fluxes from the surface.

$$C = C_0 \cdot \mathrm{erfc} \frac{x}{2\sqrt{D_b t}}, \qquad (2)$$

where t is time of diffusion annealing, x is a distance from the sample's surface to the layer having a copper content of about $C = 0.1$ at.%, corresponds to the SIMS method resolution.

The $D_b\delta$ values for the migrating grain boundaries were calculated from (1), as well as in coarse-grained nickel. As it is seen from the table, if the GB diffusion width is considered to be $\delta \sim 1$ nm and to be temperature independent, the D_b values of copper in NS nickel are higher by 4–6 orders of magnitude than those in coarse-grained nickel. The activation energy of the grain boundary diffusion determined using the temperature dependence of D_b for copper in NS nickel curve slope is 60.2 kJ/mol. This value is approximately half as much as that in coarse-grained nickel.

The activation energy of grain boundary diffusion is close to activation energy of free surface diffusion. This result has been discussed in [2–4], where diffusion in NS materials containing pores was investigated. However there are no pores in NS nickel produced by ECAP in the present experiment. For this reason it is possible to assume, that the significant increase of grain boundary diffusion permeability in NS nickel is connected with the special non-equilibrium (high-energy) state of the grain boundaries.

The influence of grain boundary diffusion fluxes, *e.g.* the effect creep acceleration of polycrystals [9], may have their own unique features in NS materials. We have found the steady creep rate of NS nickel increases under the action of grain boundary diffusion fluxes of copper from an external source. The effect takes place in temperature interval of 398–473 K, in which volume diffusion is assumed to be "frozen" (Figure 2, curve 1). The magnitude of the effect was estimated by the ratio $\dot{\varepsilon}_1 / \dot{\varepsilon}_2$ ($\dot{\varepsilon}_1$ - steady creep rate in NS (Figure 2, curve 1) or coarse-grained (Figure 2, curve 2) [8] states under the action of grain boundary diffusion fluxes of copper; $\dot{\varepsilon}_2$ – steady creep rate under the same loading in vacuum). The effect maximum is seen to take place at a much lower temperature in comparison to the coarse-grained state (Figure 2).

4. Conclusions

The experimental study of copper diffusivity in nickel indicates that GB diffusion coefficients in nanostructured materials are 4–6 orders of magnitude higher than in coarse-grained materials. The measured activation energy of grain boundary diffusion is close to that for surface diffusion, despite the lack of the residual porosity in the materials investigated.

The creep acceleration effect under the action of grain boundary diffusion fluxes of copper in nanostructured nickel takes place at much lower temperatures compared to coarse-grained nickel.

5. Acknowledgements

The work is financially supported by Russian Fond of Fundamental Research (grant #98–02–16517) and Europe commission (grant #97–1243).

6. References

1. Birringer, R. and Gleiter, H. (1988) Nanocrystalline Materials, In: *Encyclopedia of Materials.-Sci. and Eng. Suppl.*, ed. R.W.Cahn, Pergamon Press, pp. 339–349.
2. Gleiter, H. (1995) Nanostructured materials: State of the Art and Perspectives, *Nanostructured materials*. Proceedings of the second international conference on nanostructured materials, Stuttgart university, Germany 6, pp. 3–14.
3. Larikov, L.N. (1995) Diffusion processes in nanocrystalline materials, *Metallofiz. Nov. Technol.* 17, 3–29.
4. Bokstein, B.S., Brose, H.D., Trusov, L.I., and Khvostantseva, T.P. (1995) Diffusion in nanocrystalline nickel, *Nanostructured materials*. Stuttgart university, Germany 6, pp. 873–876.
5. Klotsman, S.M. (1993) Diffusion in nanocrystalline materials, *Fiz. Met. Metalloved.* 76, 5–18.
6. Valiev, R.Z., Korznikov, A.V., Mulyukov, R.R. (1992) Structure and properties of metallic materials with submicrocrystalline structure, *Fiz. Met. Metalloved.* 73, 373–384.
7. Horita, Z., Smith, D.J., Furukawa, M., Nemoto, M., Valiev, R.Z., and Langdon, T.G. (1996) Evolution of grain boundary structure in submicrometer-grained Al-Mg alloy, *Materials Characterization* 37, 285–294.
8. Kolobov, Yu.R. (1998) Diffusion-controlled processes on grain boundaries and metal polycrystal plasticity, Novosibirsk: Nauka.
9. Kolobov, Yu.R. (1994) Grain-boundary diffusion and polycrystals plasticity, *Trans. Mat. Res. Soc. Jpn.* 16B, 1397–1401.
10. Mishin, Yu.M., Razumovskii, I.M. (1988) Mathematical models and methods of identification of individual grain boundary diffusion parameters, in Bokstein B.S (ed.) Structure and properties of inner surfaces in metals, M. Nauka, pp.96–131.
11. Kaur, I., Mishin, Yu., Gust, W. (1995) Fundamentals of Grain and Interphase Boundary Diffusion, John Wiley & Sons Ltd.

STRUCTURE AND DEFORMATION BEHAVIOR OF SPD CU-BASED NANOCOMPOSITE

W. BUCHGRABER[1], R.K. ISLAMGALIEV[2], YU.R. KOLOBOV[3],
N.M. AMIRKHANOV[2]
[1]*Austrian Research Center Seibersdorf, Austria*
[2]*Institute of Physics of Advanced Materials,*
Ufa State Aviation Technical University, Ufa, Russia
[3]*Institute of Strength Physics and Materials Science RAN, Tomsk, Russia*

Abstract

Severe plastic torsion straining was used in the present work for processing of the metal matrix $Cu + 0.5\%Al_2O_3$ nanocomposite. The mean grain size, particle size distribution and elastic strain were studied by TEM and XRD. It is shown that high ultimate strength (680 MPa) and microhardness (2300 MPa) as well as high thermal stability and electrical conductivity are the features of the nanocomposite samples. The decrease of the creep rate by an order magnitude and the increase of the time to failure by a factor of 4-5 is revealed in nanocomposite as compared the extruded sample.

1. Introduction

Metal matrix nanocomposites have attracted the growing interest of experts in material science due to expectation of attractive combination the enhanced strength properties and thermal stability [1]. The development of severe plastic deformation methods enables these expectations to be examined experimentally. Results of investigations of the structure, thermal stability and deformation behavior of the $Cu + 0.5\%Al_2O_3$ nanocomposite processed by internal oxidation and severe plastic torsion straining are considered in the present work.

2. Experimental

Extruded samples of the $Cu+0.5\%Al_2O_3$ composite (GlidCop Al-15) were processed by torsion straining under high pressure [2] in two regimes. In the first, the initial materials were deformed under the pressure 6 GPa resulting in samples, 10 mm in diameter and 0.2 mm thick. In the second regime, the initial samples were subjected to torsion straining under the pressure 2.8 GPa. In this case their diameter was 15 mm and thickness was 0.3 mm.

TEM studies were conducted on electron microscopes JEM-100B and EM-125K at an accelerating voltage of 100 and 125 kV, respectively. The investigation of Al_2O_3

T.C. Lowe and R.Z. Valiev (eds.), Investigations and Applications of Severe Plastic Deformation, 267–272.
© *2000 Kluwer Academic Publishers. Printed in the Netherlands.*

particles size and their distribution in a matrix was made by using the method of extraction replicas. The foils were prepared by jet electropolishing. Electron diffraction patterns were taken from an area of 2 μm^2. The grain size was determined from the dark field image as an average value of the maximum size and its cross section. The mean grain size in the sample was obtained by averaging over more than 100 grains.

The investigations of thermal stability were conducted by studying the dependence of microhardness, elastic strain and electrical resistivity on annealing temperature as well as heat release. The measurements of microhardness were made by Vickers' method using a PMT-3 device under a load of 100 g. The DSC investigations were conducted on a differential scanning calorimeter Perkin Elmer DSC7, the rate of heating being 20°C/min. The study of elastic strain was performed by means of the X-ray diffraction on a device DRON-4 [3]. The electrical resistivity was determined by the potentiometric four-probe method at the constant current at room temperature. Tensile tests were conducted on a modified testing machine PV-3012 M. Creep tests during tension were conducted on samples having gauge dimensions 0.35x2x6 mm^3 at the temperature 423 K in vacuum under constant load 285 MPa.

3. Results and discussion

TEM investigations of the Cu+0.5%Al$_2$O$_3$ composite subjected to torsion straining under the pressure 6 GPa revealed a highly dispersed structure with a mean grain size of about 80 nm (Fig.1). This is smaller than the grain size of 150 nm in pure copper after similar treatment [4]. The value of elastic strains in the Cu+0.5%Al$_2$O$_3$ nanocomposite determined by means of X-ray diffraction was 1×10^{-3}. This agrees with values of elastic strain observed in pure copper after torsion straining [5]. At the same time, after torsion straining under pressure 2.8 GPa the Cu+0.5%Al$_2$O$_3$ composite had a highly dispersed structure as well, but a grain size of 200 nm (Fig.1b). After annealing at 400°C for 1 hour the first recrystallized areas appeared in this sample. Distinct boundaries were observed between these areas and the non-recrystallized ones with high density of dislocations. By means of the method of replicas it was established that the area of such portions was not more than 5-6%.

After annealing at 600°C for 1 hour the area of non-recrystallized portions was 10%. A large number of annealing twins were observed inside of recrystallized grains. The grain size was about 2-3 μm. After annealing at 800°C for 1 hour there occurs complete recrystallization of the copper matrix of the Cu+0.5%Al$_2$O$_3$ composite. The mean grain size was 2.5 μm. Annealing at 1000°C for 1 hour increases the mean grain size of the copper matrix up to 3.5 μm.

The study of replicas shows the Al$_2$O$_3$ particles, less than 20 nm in size, uniformly distributed within the material volume (Fig.2a). At the same time, in the matrix there are areas where coarser Al$_2$O$_3$ particles are observed. Size distribution of such particles is shown in Fig.2b. Annealing at the temperature 800°C for 1 hour does not change the size of particles and their distribution.

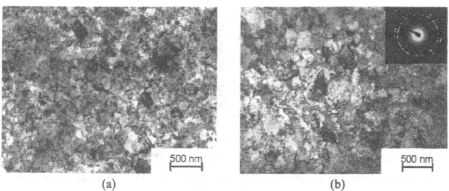

(a)   (b)

Figure 1. Microstructure of the Cu + 0.5% Al$_2$O$_3$ composite after torsion straining under pressure: (a) 6 GPa; (b) 2.8 GPa.

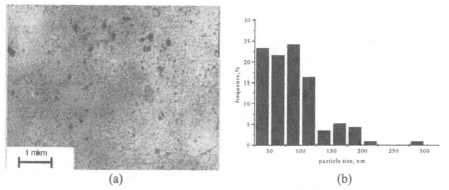

(a)   (b)

Figure 2. Structure of the Cu+0.5%Al2O3 composite after torsion straining: (a) micrograph of replicas; (b) histogram of size distribution for coarse Al$_2$O$_3$ particles.

Very important feature of the Cu+0.5% Al$_2$O$_3$ nanocomposite is high thermal stability of microhardness up to 500^{0}C (Fig.3). After annealing at higher temperatures (650-800^{0}C) the H$_v$ decreases slightly but its value remains higher than in the pure coarse grained copper by a factor of 3.5 approximately. The comparison of H$_v$ and the structure of annealed samples shows that microhardness of nanocomposite is determined mainly by the Al$_2$O$_3$ particles and weakly depends on the size of copper matrix grains and the state of their boundaries. Though after annealing at 800^{0}C the mean grain size of copper in the Cu + 0.5%Al$_2$O$_3$ nanocomposite is increased up to 2.5 μm, the sizes of the Al$_2$O$_3$ particles and their distribution are not changed and the H$_v$ value of the nanocomposite after the mentioned annealing remains high - 1.4 GPa. The results of DCS investigations showing the absence of heat release at heating up to 700°C (Fig.3) also confirm high thermal stability of the given composite. However relaxation of elastic strain and recovery of electrical resistance in the Cu+0.5%Al$_2$O$_3$ nanocomposite (Fig.4) precede grain growth observed at 400°C.

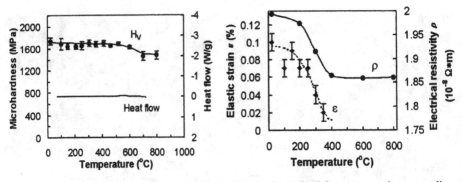

Figure 3. Dependence of microhardness and DSC signal for the Cu+0.5%Al$_2$O$_3$ nanocomposite on annealing temperature.

Figure 4. Dependence of elastic strain (ε) and electrical resistivity (ρ) on annealing temperature for the Cu+0.5%Al$_2$O$_3$ nanocomposite (T$_{test}$ = 20°C).

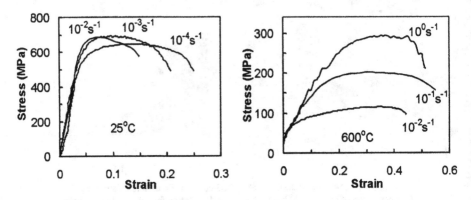

Figure 5. Tensile stress-strain curves for the Cu + 0.5%Al$_2$O$_3$ nanocomposite with grain size of 80 nm tested at room temperature at strain rates of 10^{-2}, 10^{-3}, 10^{-4} s^{-1}.

Figure 6. Tensile stress-strain curves for the Cu + 0.5%Al$_2$O$_3$ nanocomposite tested at temperature of 600°C at strain rates of 10^0, 10^{-1}, 10^{-2} s^{-1}.

Tensile tests at ambient temperature of the Cu+0.5%Al$_2$O$_3$ composite with a mean grain size of 80 nm reveal an attractive combination of high strength (680 GPa) and good ductility (25%) (Fig.5). This value of strength is significantly higher than 520 MPa in the samples with a grain size of 200 nm. It should be noted that the given nanocomposite demonstrates high ultimate strength up to 300 MPa even at 600°C with ductility of 40% (Fig.6).

The results of the creep investigations of the Cu+0.5%Al$_2$O$_3$ composite contained high dispersed ceramic particles is of special interest.

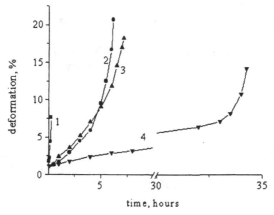

Figure 7. Creep curves of Cu (1) and composite Cu+0.5%Al₂O₃ (2-4) at 423 K under the load 285 MPa: (2) – after extrusion; (3) after extrusion + rolling 70%; (4) – after extrusion + torsion straining.

From Fig. 7 it is seen that under load 285 MPa only the third accelerated stage of creep is observed in commercially pure copper in the state after torsion straining. There are three stages on the creep curves of the Cu-0.5%Al₂O₃ composite: non-established, stationary and accelerated. The creep rate at the established stage of the composite depends significantly on preliminary treatment.

The Cu+0.5%Al₂O₃ composite has the highest creep rate (4.4×10^{-6} s^{-1}) in the state after rolling deformation by 70% (curve 3). The creep rate at the established stage of the extruded sample is a bit less and equals to 3.8×10^{-6} s^{-1}. However, the failure of the extruded sample occurs in less time (curve 2). The lowest creep rate at the established state (4×10^{-7} s^{-1}) and the largest period to failure are observed in the Cu-0.5%Al₂O₃ composite after torsion straining (curve 4).

Thus, additional torsion straining decreases the creep rate at the established stage of the Cu-0.5%Al₂O₃ composite by an order of magnitude and increases the time to failure by a factor of 4-5. Additional cold rolling deformation by 70% almost does not influence the creep character of the extruded Cu-0.5%Al₂O₃ composite. One can assume that the increase of low temperature creep resistance of the Cu-0.5%Al₂O₃ composite in the state after torsion straining is caused by the decrease in the grain size to 200 nm. The latter leads to the increase in the grain boundaries areas which become additional barriers to dislocation movement during low temperature creep.

4. Concluding remarks

1) Structure of the Cu+0.5%Al₂O₃ nanocomposite processed by torsion straining is characterized by small grain size of 80 nm, highly dispersed Al₂O₃ particles of 20 nm and high elastic strain of 1×10^{-3}.

2) Relaxation of elastic strain and recovery of electrical resistance during heating of the Cu+0.5%Al₂O₃ nanocomposite precede the grain growth observed at 400°C.

3) The $Cu+0.5\%Al_2O_3$ nanocomposite demonstrates an attractive combination of high strength (680 MPa), microhardness (2300 MPa) as well as good ductility and high electrical conductivity.

4) The creep rate is less by an order of magnitude and the time to failure is higher by a factor of 4-5 in the $Cu+0.5\%Al_2O_3$ nanocomposite as compared the extruded sample.

5. Acknowledgements

The authors gratefully acknowledge the support from INTAS under the grant N97-1243. We would like also to express our thanks to Dr. A.V. Sergeeva and A.R. Kilmametov for their help in tensile tests and XRD, respectively.

6. References

1. Alexandrov, I.V., Zhu, Y., Lowe, T., Islamgaliev R.K., and Valiev, R.Z. (1998) Microstructures and properties of nanocomposites obtained through SPD-consolidation of powders, *Metallurgica and Materials Transactions A* **29**, 2253-2260.

2. Valiev, R.Z., Alexandrov, I.V., and Islamgaliev, R.K. (1998) Processing and mechanical properties of nanocrystalline materials prepared by severe plastic deformation, in G.M.Chow and N.I.Noskova (eds.), *Nanocrystallyne Materials: Science and Technology. NATO ASI*, Kluwer Academic Publishers, Dordrecht, pp.121-142.

3. Alexandrov, I.V., Zhang, K., Kilmametov, A.R., Lu, K., and Valiev, R.Z. (1997) The X-ray characterization of the ultrafine-grained Cu processed by different methods of severe plastic deformation, *Mater.Sci.Eng.* **A234-236**, 331-334.

4. Lian, J., Valiev, R.Z., and Baudelet, B. (1995) On the enhanced grain growth in ultrafine grained metals, *Acta metall. mater.* **43**, 4165-4170.

5. Islamgaliev R.K., Chmelik, F., and Kuzel, R. (1997) Thermal stability of submicron grained copper and nickel, *Mat.Sci.Eng.A* **A237**, 43-51.

MICROSTRUCTURAL REFINEMENT AND MECHANICAL PROPERTY IMPROVEMENT OF COPPER AND COPPER-AL$_2$O$_3$ SPECIMENS PROCESSED BY EQUAL CHANNEL ANGULAR EXTRUSION (ECAE)

SHANKAR M.L. SASTRY
Washington University
Campus Box 1185, One Brookings Drive, St. Louis, MO. 63130, U.S.A.

Abstract

Pure copper and copper alloys containing different volume fractions of nano size Al$_2$O$_3$ dispersoids were processed by ECAE at 25-250°C, annealed at 25-500°C, and microstructural refinement and mechanical property improvement were studied. In samples processed by ECAE to strains of 5, hardness values increased by 100% in copper specimens and 40-50% in Cu-Al$_2$O$_3$ specimens. A drastic increase in hardness was observed after a strain of 1, and hardness values leveled off beyond a strain of 2. Samples were annealed at 25-500°C, to determine the extent of recovery of hardness values. In ECAE processed copper samples, significant reduction in hardness occurred between 100-200°C, and higher temperature annealing resulted in hardness values close to the values before ECAE processing. However, in Cu-Al$_2$O$_3$ specimens, the increased hardness was maintained in samples annealed at as high as 500°C. Strength and ductility values determined from 3 point bend tests showed strength values of 150-200 MPa for copper and 650-800 MPa for Cu-Al$_2$O$_3$ samples with excellent ductility in copper as well as Cu-Al$_2$O$_3$ samples. The observed results were analyzed in terms of the magnitudes of strengthening contributions from dislocation substructure, fine grains, and incoherent dispersoids.

1. Introduction

Extending microstructural refinement down to the nanoscale by different methods has received significant attention in recent years because of the potential benefits of fine grain size on physical and mechanical properties of materials [1,2]. Until recently, Nanoparticle synthesis and processing methods (such as physical vapor deposition (PVD), solution phase synthesis (SPS), mechanical milling (MM)), and thermomechanical processing methods such as cold/hot working followed by annealing were the primary methods used to produce fine grained materials. Severe Plastic Deformation methods developed recently [3,4] offer yet another route for the production of ultra fine grained materials. Fine grained materials produced by particulate processing routes such as PVD, SPS, and MM contain absorbed and adsorbed impurities, which generally result in poor inter-particle and inter-agglomerate bonding and significant consolidation related defects. Evaluation of properties of such

273

T.C. Lowe and R.Z. Valiev (eds.), Investigations and Applications of Severe Plastic Deformation, 273–279.
© 2000 *Kluwer Academic Publishers. Printed in the Netherlands.*

materials masks the 'intrinsic' effects of ultra fine grain structure and leads to erroneous conclusions regarding material behavior.

Using SPD methods, ultra fine grained materials have successfully been produced. In pure metals produced by SPD methods, in the as deformed state, room temperature strengthening is determined by contributions from dislocation substructure as well as fine grains. In order to separate the effects due to dislocations and fine grains, samples have to be annealed to reduce the dislocation density to the levels encountered in fully annealed materials. Exposure of SPD processed metals to such annealing treatments invariably result in significant grain growth also; consequently it has not been possible to separate the dislocation substructure and fine grain effects on deformation behavior. The present study is an attempt to isolate the above effects on deformation behavior. Copper samples containing controlled amounts of stable incoherent second phase particles were deformed by SPD and annealed with the expectation that the second phase particles would stabilize the fine grain structure resulting from recrystallization occurring during deformation or post deformation annealing.

2. Experimental Procedure

99.99% pure copper and powder metallurgy processed and internally oxidized Cu-0.15 wt.%Al, Cu-0.25 wt.% Al, and Cu-0.60 wt.% Al alloys were used as the starting material for SPD. 0.25 in. diameter extruded rods of pure copper annealed at 400°C were the starting specimens for the study. The Cu-Al alloys were produced as powders and were internally oxidized to produce uniform dispersion of fine incoherent Al_2O_3 particles in copper matrix by a proprietary process. The dispersoids are stable at temperatures close to the melting point of the copper matrix and are effective in retarding grain growth. The powders were hot isostatically pressed and/or extruded. 0.25 in diameter rods machined from the HIPed and extruded billets were used for SPD experiments. Equal channel angular extrusion experiments were carried at 25-250°C in a 90° equal channel die an extrusion rate of 0.1 in./min. 0.25 in dia. and 1.0 in. long specimens were produced by multiple extrusions. ECAE processed samples were annealed at 200-500°C to determine the stability of grain structure and retention of strength and hardness upon high temperature annealing.

Mechanical properties of the ECAE processed and annealed samples were determined by microhardness measurements and three-point bend testing. Microstructures at each stage of ECAE processing and upon high temperature annealing were characterized.

3. Results and Discussion

The microstructures of Cu and Cu-0.25 Al_2O_3 specimens prior to extrusion are shown in Figures 1 a,b, and c. Copper specimens have a fully recrystallized microstructure with a grain size of about 40 mm. The hot isostatically pressed (HIPed) specimens have a bimodal distribution of large and small grains. The conventionally extruded specimens have a grain size of 20-40 μm. After 3 ECAE passes at room temperature, there is significant grain distortion and refinement in all the three alloys. The grain sizes in the

Cu   HIPed Cu-0.25 Al₂O₃   Extruded Cu-0.25 Al₂O₃

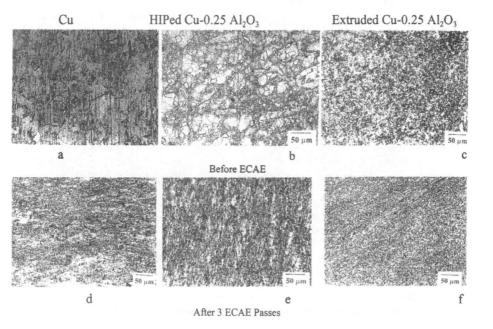

Before ECAE

After 3 ECAE Passes

Cu-0.25Al₂O₃ . Hardness increases by more than 200% in copper after 2 ECAE passes and then levels off. The hardness increase after ECAE in HIPed and extruded Cu-0.25Al₂O₃ is significantly less - only 10-25%. These results indicate that in ECAE Cu-0.25Al₂O₃, the individual strengthening contributions are not linearly additive, a characteristic feature of alloys in which multiple strengthening contributions operate.

HIPed Cu-0.25Al₂O₃ specimens were ECAE processed at 180°C to determine the effect of processing temperature on microstructure development. The microstructures shown in Figure 3 indicate that grain distortion and refinement are significantly less severe at 180°C than at 25°C due to dynamic recovery at the higher temperature.

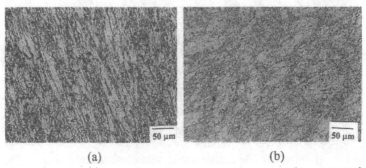

(a)   (b)

Figure 3. Microstructures of Cu-0.25 Al₂O₃ after 1 ECAE pass at (a) 25°C, and (b) 180°C

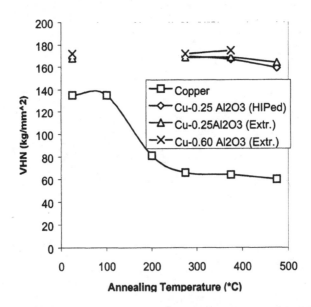

Figure 4. Effect of post-ECAE annealing on hardness of Cu and Cu-0.25 Al₂O₃

The effect of post-ECAE annealing on hardness of Cu and Cu-0.25 Al$_2$O$_3$ specimens is shown in Figure 4. In ECAE processed copper samples, significant reduction in hardness occurs between 100-200°C, and higher temperature annealing results in hardness values close to the values before ECAE processing. However, in Cu-Al$_2$O$_3$ specimens, the increased hardness resulting from ECAE is maintained in samples annealed at as high as 500°C.

The microstructures of post-ECAE annealed specimens are shown in Figure 5. Copper specimens annealed at 380°C show a fully recrystallized microstructure with bimodal grain size distribution. Upon annealing at 400°C, significant grain growth occurs and microstructure reverts back to the initial microstructure. The Cu-0.25Al$_2$O$_3$ specimens annealed at 400°C show partially recrystallized microstructure with 1-2 μm recrystallized fine grains and elongated 10-20 μm elongated unrecrystallized grains.

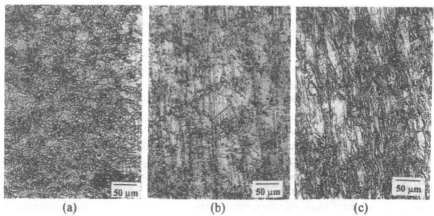

(a)   (b)   (c)

Figure 5. Microstructures of (a)) ECAE processed Cu annealed at 300°C/15 min, (b) ECAE processed Cu annealed at 400°C/30 min, and (c) ECAE processed Cu-0.25Al$_2$O$_3$ annealed at 400°C/30 min.

Figures 6a and 6b are the load-displacement curves of fine grained copper specimens produced by ECAE processing. Load-displacement curve of fine grained copper specimen produced by nanoparticle processing is shown in Figure 6c for comparison. The curve shown in Figure 6c is representative of the behavior exhibited by specimens prepared by IGC, SPS, and POLYOL processes; the specimens failed before the onset of any plastic deformation. In contrast, the specimens processed by ECAE, exhibited significant plastic deformation before fracture (Figures 6a and 6b). Although the stress state in the three-point loaded specimens is a mixed tension and compression through the thickness of the specimen, it serves as a screening test to determine the approximate values of yield stress and the extent of plastic deformation under tensile loading conditions. The absence of any plastic elongation in particulate-processed specimens is due to the deleterious effects of impurities, poor inter-particle bonding and other process-induced defects. Nanoparticle ensembles have large surface areas making them susceptible to impurity intake relative to coarse powder. The surface and bulk impurities could have deleterious effect on densification.

278

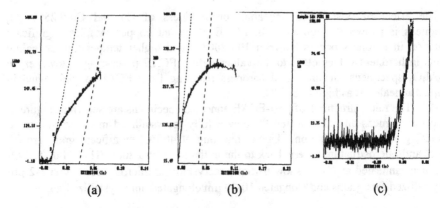

(a)   (b)   (c)

Figure 6. Load-displacement curves from 3-point bend tests: (a) ECAE processed copper, (b) ECAE processed Cu- $0.25Al_2O_3$, and (c) Nanoparticle processed copper.

The fracture surfaces of particulate processed and ECAE processed nanocrystalline copper specimens were examined by scanning electron microscopy. Whereas the particulate processed specimens fracture by inter particle and interagglomerate separation, the ECAE processed specimens exhibit ductile-dimple fracture typical of specimens exhibiting significant plastic deformation.

The microhardness values, and strength and ductility values determined from 3 point bend tests for differently processed copper specimens are shown in Figure 7. The particulate processed specimens have low yield strength values, and fracture without plastic elongation. For most metals, yield strength is approximately equal to one third of the Vickers hardness value. Measured compressive yield strengths agreed well with those determined from microhardness tests. However, the yield strength values determined from tensile and bend tests are significantly lower than those predicted from microhardness values in particulate-processed copper. Excellent agreement between the measured and predicted values is observed in ECAE processed copper and Cu-Al_2O_3 specimens.

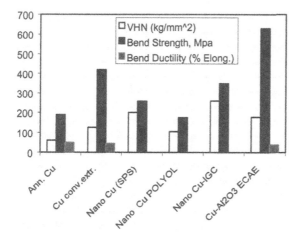

Figure 7. Microhardness, bend strength and bend ductility of differently processed copper Cu-Al$_2$O$_3$.

4. Summary and Conclusions

Equal channel angular extrusion results in significant grain refinement in copper and copper alloys containing stable incoherent alumina particles. Upon high temperature annealing of ECAE processed alloys, complete recrystallization and significant grain growth occur in pure copper. In copper alloys containing second phase particles, recrystallization and grain growth are significantly reduced. A good combination of high strength and ductility is achieved in ECAE processed alloys.

5. References

1. R & D Status and Trends in Nanoparticles, Nanostructured Materials, and Nanodevices in the United States, Proceedings of the May 8-9, 1997 Workshop, Arlington, VA, Richard W. Siegel, Evelyn Hu, and M.C. Roco, Eds.
2. Workshop on Global Assessment of R & D Status and Trends in Nanoparticles, Nanostructured Materials, and Nanodevices, February 10, 1998, Arlington, VA.
3. Segal, V.M. (1995) Materials processing by simple shear, *Materials Science and Engineering* **A197**, 157.
4. Valiev, R.Z. (1997) Structure and mechanical properties of ultrafine-grained metals, *Materials Science and Engineering* **A234-236**, 59.

STRUCTURE AND MECHANICAL PROPERTIES OF ULTRAFINE-GRAINED CHROMIUM PRODUCED BY SEVERE PLASTIC DEFORMATION PROCESSING

V. PROVENZANO[1], N.A. KRASILNIKOV[2], D.V. PAVLENKO[2], D.G. RICKERBY[3], A.P. ZHILYAEV[4]

[1] *Naval Research laboratory, Code 6323, Washington, DC 20375-5343 USA*
[2] *Ulyanovsk State University, Ulyanovsk, 432700 Russia*
[3] *Joint Research Center, I-21020 Ispra (VA), Italy*
[4] *Institute for Physics of Advanced Materials, USATU, Ufa, 450000 Russia*

1. Introduction

Due to their elevated temperature properties, such as high strength, melting temperature and oxidation resistance, the chromium-based alloys are potentially very attractive for variety of structural applications, including their use as low activation materials for fusion reactor components [1]. To date, however, the potential of this class of alloys has not been translated into practice, because, though they are strong and hard, they are inherently brittle. In fact, they exhibit little or no ductility at low temperatures together with an unacceptably high ductile-to-brittle transition (DTBT) temperature. The sixties and early seventies, marks the period when mechanical properties of these alloys were actively studied and their inherently brittle behavior was amply demonstrated. However, the mechanical results from some of the studies, published during this period, appear to suggest that the low temperature ductility and fracture toughness properties of these alloys could be markedly improved by grain refinement. Among the results reported during this period, are those of Wilcox [2], Gilbert [3], and Wain [4] and their respective co-workers, all of which point to the critical role of grain size on the resulting mechanical properties. For example, Wilcox and co-workers showed that decreasing the grain size of chromium from 37 to 2.4 μm by the addition of a few percent of thoria particles resulted in a significant decrease in DTBT temperature (from 140 to 15°C) together with significant improvements both in the ductility and strength [2]. However, in spite of these promising results, reported in the above referenced studies, the overall picture concerning the chromium-based alloys was far from clear. In fact, data reported by other researchers showed contradictory and confusing trends regarding the beneficial effect of grain refinement on the mechanical properties. Adding to this uncertainty, was the lack of systematic studies. In fact, in a number of studies, the effect of grain size was either masked or not clearly separated from other effects, such as solid solution and dispersion strengthening, or prestraining. Finally, no clear consensus was reached

281

T.C. Lowe and R.Z. Valiev (eds.), Investigations and Applications of Severe Plastic Deformation, 281–287.
© *2000 Kluwer Academic Publishers. Printed in the Netherlands.*

among the various researchers concerning the different effects and proposed mechanisms of ductility.

Although preceded by about a decade of active research, the middle seventies is roughly the time when the interest in the chromium-based alloys for structural applications suddenly declined. This rapid and greatly diminished interest was most likely caused by the realization that, up that point, no effective and reliable way was found to significantly improve the ductility and fracture toughness properties of these alloys. The past few years, however, have seen a renewed interest in high temperature materials that are intrinsically brittle at low and at moderate temperatures, such as the intermetallics and the refractory metals and their alloys. For example, it is known that phase modification [5–7] or the addition of small amounts of VII and VIII groups elements increases the plasticity of chromium alloys. Also, this renewed interest has been partly motivated by the research and development efforts in the area of nanostructured materials, especially some novel and unique mechanical properties exhibited by these fine-grained materials.

Nanostructured materials, characterized by small grain sizes (100 nm or less) and large surface to volume ratios, often exhibit unique and novel properties relative to those of the coarse-grained counterparts. Among the novel properties are the mechanical properties of nanocrystalline metals and alloys. For example, Valiev and his co-workers have observed low temperature superplastic behavior in some aluminum alloys and substantial ductility in some intermetallics when the grain size is reduced to nanoscale dimensions [8–10]. Besides these results, other investigators have reported observing Hall-Petch softening, increased ductility, and the absence of strain hardening in nanocrystalline metals when the grain size is reduced below some critical dimension (in the range of 5–15 nm, depending on the metal) [11–13]. The enhanced low temperature ductility observed in nanocrystalline metals has been largely rationalized by the lack dislocation activity at these small grain dimensions and to the onset of creep deformation mechanisms, such as grain boundary sliding [13]. The diminishing role of dislocations and the increasing role of grain boundaries in the deformation processes, as the grain size decreases, are consistent with theoretical results from modeling studies by Van Swygenhoven and Schiotz and their respective co-workers [14, 15]. Taken together, these experimental results and model predictions suggest that mechanical properties of intrinsically brittle materials can be significantly improved by refining their microstructure down to nanoscale dimensions.

In this paper we will present mechanical property data obtained on pure chromium, which microstructure was refined by severe plastic deformation (SPD) processing. Besides the mechanical property results, we also present results from the microstructural analysis obtained on metal samples after the grain refinement process and subsequent annealing. Finally, possible mechanisms to explain the experimental results will be discussed.

2. Materials and Experimental Procedure

In this study, pure chromium (99.7%) with a coarse-grain microstructure and average grain size of 80 μm, was used as the starting materials. Using Bridgman anvils, the two metals were subjected to torsion straining under high quasi-hydrostatic pressure in order to obtain ultrafine-grained (UFG) microstructure. Three different processing temperatures were used to produce the UFG microstructure: room temperature, 470°C

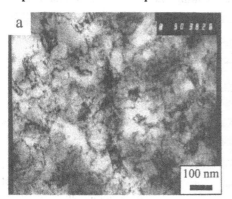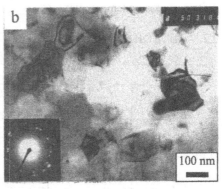

Figure 1. Microstructure of pure chromium subjected to torsion straining at (a) room temperature and (b) at temperature of 540°C.

and 540°C. The samples that were processed at room temperature, under a pressure of 6 GPa, were disk-shaped samples with dimensions of 10 mm in diameter and 0.2 mm in thickness, whereas the samples processed at the higher temperatures were also disk-shaped with a diameter of 20 mm and the same thickness. The microstructure of the deformed samples was examined by transmission electron microscopy (TEM). The strength property data were obtained by performing bend tests by means of cutting edge loading with traverse velocity of 1 mm/min in temperature range of 20–200°C. The samples used for bend tests were parallelepipeds with dimensions of $2 \times 0.2 \times 14$ mm. The output data, consisting of "load-traverse displacement" curves, was PC controlled and recorded. Subsequently, the data was recalculated and expressed in terms of "stress-bending displacement" curves.

3. Results and Discussion

3.1. TRANSMISSION ELECTRON MICROSCOPY (TEM)

TEM analysis showed that the microstructure of Cr samples processed at room temperature consists of a highly dispersed fragmented structure, whose fragments have a mean size of 100 –150nm (Figure 1a). The same TEM micrograph shows also that there is high density of lattice dislocations up to 10^{15} m^{-1}. The microstructure was stable during annealing treatments up to 400°C temperature. However, for the samples deformed at 540°C, some structure recovery was observed. As demonstrated by the TEM micrograph shown in Figure 1b. In this latter case, a well-defined granular-type structure was observed with a mean grain size of 0.3 μm. In addition, numerous spots, arranged in circles, showing azimuth spreading, were observed in the SAED patterns. This is characteristic of high angle misorientation between adjacent fragments (grains). The grain boundaries in such microstructure have a high density of dislocations, whereas the grain interiors are almost free of dislocations. Such type of microstructure is indicative of recovery processes taking place during intense plastic deformation at the higher processing temperature.

3.2. MICROHARDNESS

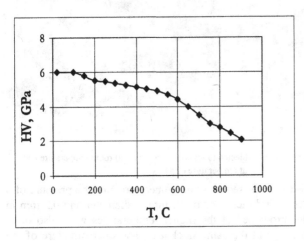

Figure 2. Microhardness of UFG chromium as a function of annealing temperature.

As stated earlier, in its initial state, pure chromium had an average grain size of about 80 μm and a microhardness of 1.75 GPa. It was found that the microhardness UFG chromium processed at room temperature is more than 3 times higher (6 GPa) than the coarse-grained counterpart. It is believed that this increase in the microhardness is a direct result of the much finer microstructure of the deformed samples. Further, within the experimental error, the microhardness value across the sample was fairly constant. This is consistent with the uniform microstructure observed in processed material. Figure 2 shows the microhardness of UFG chromium processed at room temperature as a function of annealing temperature. It should be noted that a relatively high microhardness value was retained up to 600°C temperature. Only for annealing temperatures above 900°C, the microhardness approaches the value of starting coarse-grained chromium. Also, at this point it useful to note that UFG chromium samples processed at higher temperature had lower microhardness values than those processed at room temperature. In fact, the microhardness was about 4 GPa for samples deformed at 470°C and 2.6 GPa for those deformed at 540°C. This latter value is still about 1.5 times higher than that of the starting material.

TABLE 1. Results of bending tests for UFG chromium.

Processing Temperature	Number of turns	Temperature of bending test	σ_Y	σ_U	σ_R	$K=\sigma_U-\sigma_R$
Non-deformed materials			430	550	530	20
540° C	1	20° C	724	1010	1010	0
540° C	5		840	1340	1340	0
Non-deformed materials			356	517	410	107
540° C	1	100° C	560	696	332	364
540° C	5		610	984	508	476
Non-deformed materials			300	507	386	111
540° C	1	200° C	480	760	456	304
540° C	5		600	1100	610	490

Here σ_Y is yield strength; σ_U is ultimate strength; σ_R is fracture stress or stress in unbroken samples bent on the angle of 90°.

3.3. MECHANICAL RESULTS AND FRACTOGRAPHY

The most reliable method for determining the mechanical properties of brittle materials, such as chromium and chromium-iron alloy, is bending testing [15]. Bending tests under the same conditions were also performed on chromium samples processed at 540°C for 1 or 5 turns, whose UFG microstructure had a grain size is in the range of 0.3–0.8 μm. All the mechanical test results are summarized in the above table of values.
Figure 3 shows "Stress-bending displacement" curves for different test temperatures. The basic difference in mechanical behavior of UFG chromium, compared to the starting material, is the absence of significant strengthening in UFG chromium for all testing temperatures. However, the flow stress is much lower for case of coarse-grained chromium. Increasing the testing temperature to 200°C leads to an increase in the plasticity coefficient $K = \sigma_U - \sigma_R$ [15] (see the Table 1 and Figure 3b). Furthermore, similar behavior has been observed in other UFG and nanocrystalline materials [16]. It has been suggested that such a behavior is due to enhanced grain boundary diffusivity in the fine-grained material. In fact, it has been shown that in UFG materials grain

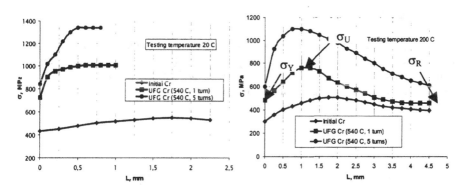

Figure 3. "Stress- bending displacement" curves for UFG chromium tested at the room temperature and at 200°C.

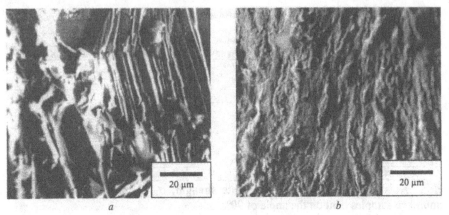

Figure 4. Fracture surface of chromium samples broken at room temperature: a – in initial state; b – with ultrafine-grained microstructure.

boundary diffusion coefficient is higher by 2–3 orders of magnitude when compared to the corresponding coarse-grained materials [17]. It has been proposed that the increased diffusivity results from the fact that the grain boundaries in materials subjected to severe plastic deformation, are in highly strained and in a non-equilibrium state This fact would explain the relaxation processes starting at relatively low homologous temperatures. Additionally, enhanced diffusivity through the triple junctions, which is normally 2-3 orders of magnitude higher than GB diffusivity, may also be contributing to the increased diffusivity [18].

Figure 4a shows the fracture surface of a coarse-grained chromium sample, whereas Figure 4b shows the fracture surface of a UFG chromium sample processed at 540 °C. From Figure 4, one can observe the essential difference in two fracture surfaces, corresponding to the two different samples. For coarse-grained chromium, fracture features are "river-like" patterns that is typical for brittle fracture By contrast, the fracture features of the UFG sample are more "cup-like" (Figure 4b), that is indicative of a ductile fracture [19]

3.4. TEXTURE MEASUREMENTS

Texture measurements were carried out on UFG pure chromium and on UFG Cr-5Fe alloy processed at 540° C for 5 turns. The pole figures (PF) of pure chromium are fairly symmetrical with axial texture pattern. As mentioned earlier, this texture is typical of UFG metals after processing by torsion straining.

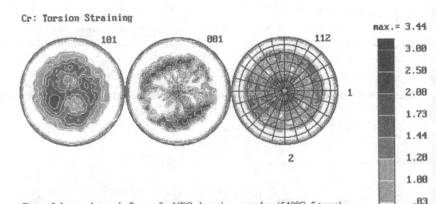

Figure 5. Incomplete pole figures for UFG chromium samples (540°C, 5 turns)

4. Conclusions

UFG chromium samples have been obtained by using severe plastic torsion straining. The mean grain size was about 100 nm for the samples processed at room temperature. Increasing the processing temperature to 540°C results in chromium samples with an average grain size of 0.3 μm. Compared to the starting coarse-grained material, UFG chromium has higher microhardness at room temperature and higher bend strength for test temperatures up to 200°C. Also, the UFG microstructure exhibited a high degree of thermal stability. The study of mechanical behavior during bending tests and the corresponding fracture features observed by SEM have revealed a visible tendency for increasing ductility of the processed UFG chromium, when compared to the coarse-grained counterpart. X-ray study revealed an axial texture that forms during torsion straining, that is typical of many metals with cubic crystal structure.

5. References

1. Stamm, H., et al. (1998) Thermomechanical characteristics of low activation chromium and chromium alloys, *J. Nucl. Mater.* Part B **263**, 1756–1761.
2. Wilcox, B.A., Veigel, N.D., and Clauer, A.H. (1972) Ductile-brittle transition of chromium, *Met. Trans.* **3**, 273–283.
3. Gilbert, A., Reid, C.N., and Hahn, G.T. (1963) Observation on the fracture of chromium, *J. Inst. Metals* **92**, 351-356.
4. Henderson, F., Bullen, F.P., and Wain, H.L. (1970) The Preparation and properties of chromium wires, *J. Inst. Metals.* **98**, 65–70.
5. Kovsh, V., et al. (1973) Influence of cyclic deformation on dislocation structure and mechanical behavior of molybdenum, chromium and tungsten, *Problems of Strength* **11**, 15–20.
6. Igolkina, L.S., et al. (1986) Quenching and aging in chromium alloy, *Physics of Metals* **8**, 32–37.
7. Trefilov, V.I., et al. (1986) Deformation hardening and dislocation structure evolution in polycrystalline BCC metals, *Physics of Metals* **8**, No.2, 89–97.
8. Valiev, R.Z. (Ed.) (1996) Ultrafine grained materials produced by severe plastic deformation, *Special issue of Annales de Chimie–Science des Materiaux* **21**.
9. Valiev, R.Z.(1997) Structure and mechanical properties of ultrafine grained metals, *Mater. Sci. Eng.* **A234–236**, 59–66.
10. Lian, J.S., Valiev, R.Z., and Baudelet, B. (1995) On the enhanced grain-growth in ultrafine grained metals, *Acta Metall. Mater.* **43**, 4165–4170.
11. Sanders, P.G., Eastman, J.R., and Weertman, J.R. (1997) Elastic and tensile properties of nanocrystalline copper and palladium, *Acta Materialia* **45**, 4019–4025.
12. Siegel, R.W. and Fourger, G.E. (1995) Mechanical properties nanophase metals, *NanoStruct. Mater.* **6**, 205–216.
13. Hahn, H. and Padmanabhan, K.A. (1997) A model for deformation of nanocrystalline materials, *Phil. Mag.* **A76**, 125–130.
14. Van Swygennhoven, H. and Caro, A. (1997) Plastic behavior of nanophase Ni: a molecular dynamics computer simulation, *Appl. Phys. Lett.* **71**, 1652–1654.
15. Schiotz, J., Tolla, F.D., and Jacobsen, K.W. (1998) Softening of nanocrystalline metals at very small grain size, *Nature* **391**, 561–563.
16. Zolotorewski, V.S. (1983) *Mechanical Properties of Metals*, Metallurgy Publ., Moscow, (in Russian).
17. Mishra R.S., McFadden S.X., and Mukherjee A.K (1998) Analysis of tensile superplasticity in nanocrystalline, In: *Proceeding of Towards Innovation in Superplasticity II*, Eds. T. Sacuma, T. Azawa, K. Higashi, Kobe, Japan, 31–38.
18. Kolobov, Yu.R. (1998) *Diffusion-controlled Processes on Grain Boundaries and Plasticity of Metals*, Nauka Publ., Novosibirsk. (in Russian).
19. Gordeeva, T.A., Zhegina, I.P. (1978) *Fracture analysis in materials evaluation*, Mashinostroenie Publ., Moscow (in Russian).

CYCLIC STRESS-STRAIN RESPONSE OF PB-SN AND ZN-AL EUTECTIC ALLOYS FINE-GRAINED BY EQUAL CHANNEL ANGULAR PRESSING

Y. KANEKO[1]*, A. VINOGRADOV[1]*, K. KITAGAWA[1], S. HASHIMOTO[2]*
[1]Dept. of Mechanical Systems Engineering, Kanazawa University
Kanazawa 920-8667, Japan
[2]Dept. of Engineering Physics and Mechanics, Kyoto University, Kyoto
606-8501, Japan
*Now at the Dept. of Intelligent Materials Engineering, Osaka City University, Osaka 558-8585, Japan

1. Introduction

Superplastic deformation usually requires testing temperature above $0.5 T_m$ when an average grain size is less than 10 μm (T_m is the melting point) [1]. This is because the superplastic deformation is connected closely with thermally-activated processes such as dislocation climb and grain-boundary diffusion. It has long been recognized that the grain size affects the temperature at which the superplastic deformation occurs. To achieve low temperature superplasticity, several Al-alloys have been fine-grained by the equal channel angular pressing (ECAP) technique [2]. The properties of the superplastic materials have primarily been studied in monotonic straining while the limited results have been reported on their cyclic behavior [3]. Hence, it seems reasonable to utilize the superplastic alloys fine-grained by ECAP to perform a precise fatigue experiment at room temperature. In the present work, we carried out cyclic tests at room temperature on the ECAP fabricated Pb-62%Sn and Zn-22%Al eutectic alloys. The attention is paid particularly to the strain-rate dependence of the stress amplitude.

2. Experimental

Average grain sizes of the ECAP produced Pb-Sn and Zn-Al bulk materials were of 5 μm and 1μm, respectively. To illustrate the typical behavior of coarse-grain metals we used the annealed copper polycrystals with an average grain size of 75 μm. The samples with the gage part of $2 \times 2.5 \times 5$ mm^3 were shaped by spark erosion and then they were polished mechanically and electrolytically. Tensile tests were done under the initial strain rate of $4 \times 10^{-3} s^{-1}$. The strain-rate-change tests were carried out to clarify the strain-rate dependence of the flow stress and to determine the region of possible superplasticity. The cyclic experiments were carried out at room temperature under a constant plastic strain amplitude, ε_{pl}, controlled at $\varepsilon_{pl} = 5 \times 10^{-4}$. To explore the strain-rate dependence of the cyclic stress-strain response, the frequencies of cycling were set at 0.05, 0.15, 0.5 or 1.5 Hz. From the stress-strain data, we calculated the stress amplitude

T.C. Lowe and R.Z. Valiev (eds.), Investigations and Applications of Severe Plastic Deformation, 289–295.
© 2000 Kluwer Academic Publishers. Printed in the Netherlands.

and the Bauschinger energy parameter [4] β_E, which is particularly sensitive to the structural changes during fatigue: $\beta_E = (4\sigma_a\varepsilon_{pl} - S)/S$, where σ_a is the stress amplitude and ε_{pl} is the plastic strain amplitude and S is the area of hysteresis loop. Hence, the β_E value means a ratio of the elastically recoverable energy to the energy loss due to plastic deformation.

3. Results

3.1 MONOTONIC TENSILE DEFORMATION

The ECAP Pb-Sn and Zn-Al alloys showed a fairly large ductility in monotonic straining: the nominal strains at fracture were greater than 1 and 2 in Pb-Sn and in the Zn-Al alloy, respectively. Annealed copper was fractured at the strain of 0.5. The strain hardening in the Pb-Sn and the Zn-Al alloys was low compared to annealed copper.

Creep properties are usually characterized by a relationship between the flow stress and the strain rate. It is possible that the strain rate also affects the cyclic deformation bahaviour of the materials showing the high ductility at room temperature. The strain rate dependence of the flow stress under monotonic deformation is shown in Fig.1 for both the Pb-Sn and Zn-Al alloys. The flow stress increased with increasing strain rate. The m-value of the Zn-Al alloy was almost constant ($m \approx 0.25$) in the chosen range of strain rates, where the m-value is given by the empirical relation $\sigma_f = A\dot{\varepsilon}^m$. On the other hand, the m-value in Pb-Sn was about 0.5 at $\dot{\varepsilon} = 10^{-4}$ s^{-1}. This value corresponds typically to superplastic deformation.

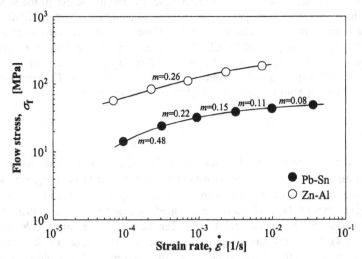

Figure 1. Relationship between flow stress and strain rate under monotonic tensile deformation.

3.2. CYCLIC DEFORMATION TESTS

Figure 2 shows the cyclic hardening curves at 0.15Hz. The cyclic hardening curve of copper could be divided naturally into rapid hardening and saturation stages as that for single crystal specimens. On the other hand, the stress amplitudes of the Pb-Sn and Zn-Al alloys were almost unchanged during cycling. These features of the stress amplitude were common for all the frequencies tested. The constancy of the stress amplitude, which was observed in the Pb-Sn and Zn-Al alloys, was consistent with the result obtained previously on ECAP copper [5]. However, it should be noted that some ECAP copper samples showed cyclic softening [6,7] which became pronounced particularly under high strain amplitudes. The reason of softening or non-softening behavior has been in the focus of the recent study [8] where it was shown that the rate of softening is the issue of initial, severely pre-deformed, structure rather than an effect of strain amplitude solely.

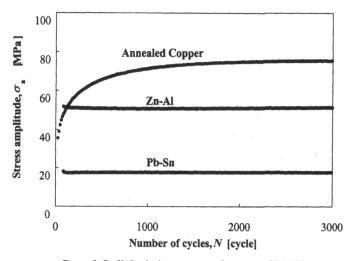

Figure 2. Cyclic hardening curves at a frequency of 0.15 Hz.

Figure 3 presents the saturated stress amplitude plotted against the strain rate. In the Zn-Al alloy, the stress amplitude increased apparently with increasing strain rate: the stress amplitude at $\dot{\varepsilon} = 8.8 \times 10^{-3} \mathrm{s}^{-1}$ was about two times higher than that at $\dot{\varepsilon} = 1.9 \times 10^{-4} \mathrm{s}^{-1}$. The stress amplitudes of the Pb-Sn alloy and copper were almost independent of the strain rate. This result of the cyclically deformed Pb-Sn alloy appears inconsistent with the strain-rate dependence of the flow stress under monotonic deformation (Fig.1). It should be noticed that the cyclic stress amplitudes were always lower than the flow stresses under monotonic deformation at the same strain rate.

In the Pb-Sn and Zn-Al alloy, the β_E values seem to increase slightly during cyclic deformation. On the contrary, the β_E value tends to decrease in copper after an initial rapid increase. (This characteristic change in the β_E value should be associated with

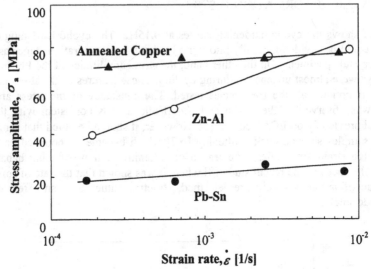

Figure 3. Saturation stress amplitude plotted against strain rate in the cyclic deformation tests under a plastic strain amplitude of 5×10^{-4}.

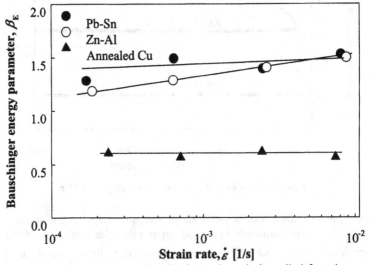

Figure 4. Bauschinger energy parameter plotted against strain rate in the cyclic deformation tests under a plastic strain amplitude of 5×10^{-4}.

formation of the persistent slip band [4].) To investigate the strain-rate dependence of the Bauschinger effect, the saturated β_E values are plotted against the strain rate, Fig.4. The β_E value of the Zn-Al alloy increases with the increasing strain rate while no clear strain rate dependencies are recognized in the Pb-Sn alloy and copper. At any strain rates tested, the Pb-Sn and Zn-Al alloys exhibited significantly higher Bauschinger effect than that in copper. Similarly, a high β_E value has also been observed in UFG copper produced by ECAP [5]. Hence, in this sense, the ECAP grain refinement generally results in the high Bauschinger effect. It is also probable for the present eutectic alloys that the residual stresses [10] between grains can play a significant role in the Bauschinger effect. This possibility is supported by the fact that the β_E values are almost the same in the Pb-Sn and Zn-Al alloys, irrespective of the difference in their grain size.

3.3. FATIGUE CRACK PROPAGATION

The fatigue crack behavior is of key importance for assessment of fatigue properties. Although it is not a main subject of the present work, it is worth referring briefly to the crack appearance in order to take a broader view on the discussion which has been opened in [8]. The crack tends to propagate either along the interphase boundary (Fig. 5a, Pb-Sn) or along the shear bands which are clearly visible in the Zn-Al alloy (Fig. 5b). Similarly to those in UFG Cu these bands are oriented at about 45° to the loading axis and extend over a large number of grains.

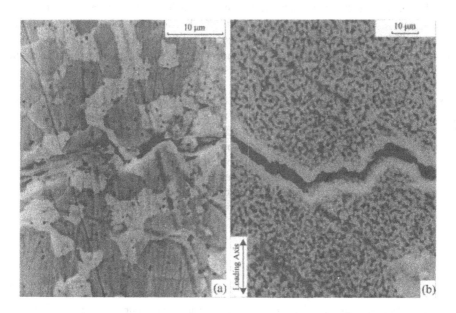

Figure 5. A view of the fatigue crack in the ECAP (a) Pb-Sn and (b) Zn-Al alloys.

4. Discussion

Studies on the superplastic deformation deal usually with the steady state creep deformation under monotonic straining. It has been proposed that superplastic deformation involves dislocations piling-up at the grain boundary and subsequent climb along the boundary plane (see reviews [1,10]). However, a direct application of this approach to cyclic deformation is difficult because the present plastic strain amplitude ($\varepsilon_{pl}=5\times10^{-4}$) is far too small compared to superplastic deformation under monotonic straining. Since the edges of the hysteresis loops in both alloys were sharply pointed, the applied plastic strain could be accommodated during the initial strain hardening stage: it can be said that the steady-state creep deformation does not occur in our cyclic tests. For this reason, the stress amplitude under cyclic deformation is lower than the flow stress under tensile testing. Hence, the initial strain hardening stage — which comes after the strain reversal — should play an important role in the cyclic stress-strain response.

The cyclically-deformed Pb-Sn alloy did not show any significant strain-rate dependence of the stress unlike the monotonic creep deformation. One can explain this as follows. It is possible that the substantial part of the applied strain is accommodated by to-and-fro motion of dislocations because the strain amplitude is very small in comparison with monotonic deformation. This assumption is also supported by the fact that the hysteresis loop did not exhibit the features of the steady-state creep. Hence, the contribution of the thermally-activated processes such as dislocation climb may be insignificant in the present cyclic tests, resulting in the low strain-rate dependence of the stress amplitude. On the other hand, it is recognized that the Zn-Al alloy showed the strain-rate dependence of the stress amplitude and the β_E value. Although the detail mechanism producing such a discrepancy with the Pb-Sn alloy is still ambiguous, we can qualitatively understand the strain-rate dependence in the Zn-Al alloy from the standpoint of the difference in grain sizes between the two materials. A mean free path of mobile dislocations would be reduced in the Zn-Al alloy having smaller grain size. Hence, the accommodation of the applied plastic strain requires some thermally-activated process such as dislocation climb and/or the increase in the dislocation density. At low strain rate, the Zn-Al alloy could deform partly with an aid of thermally-activated mechanisms. However, since a relative contribution of latter processes should be reduced at a higher strain rate, the dislocation density must increase, that, in turn, leads to higher elastic interaction between dislocations, to higher back stresses and to the β_E increase.

5. Summary

1. The stress amplitudes, σ_a, of the fine-grain ECAP Pb-Sn and Zn-Al alloys were almost constant during cycling under a constant plastic strain amplitude of 5×10^{-4}.
2. No significant effect of strain rate was recognized in the Pb-Sn alloy, although the flow stress under monotonic straining depended strongly on the strain rate. In contrast, Zn-Al revealed that σ_a and the Bauschinger energy parameter increased with strain rate.
3. The strain-rate dependence of the cyclic stress amplitude and the β_E value may be interpreted by assuming the different role of thermal activation in the materials under investigation: in Pb-Sn having a greater grain size the thermally activated mechanisms are inisignificant in contrast to Zn-Al where these processes are supposed to be active.

6. References

1. Langdon, T.G., (1991) *Mater.Sci.Eng.,* **A137** 1.
2. Furukawa, M., Ma, Y., Horita, Z., Nemoto, M., Valiev, R.Z., and Langdon, T.G. (1998) *Mater.Sci.Eng.,* **A241**, 122.
3. Bowden, J.W. and Ramaswami, B. (1990) *Metall.Trans.* **21A,** 2497.
4. Abel, A. (1978) *Mater.Sci.Eng.* **36,** 117.
5. Vinogradov, A., Kaneko, Y., Kitagawa, K., Hashimoto, S., Stolyarov, V., and Valiev, R. (1997) *Scripta Mater* **36,** 1345.
6. Vinogradov, A., Kaneko, Y., Kitagawa, K., Hashimoto, S., and Valiev, R. (1998) *Mater.Sci.Forum* **269-272,** 987.
7. Agnew, S.R.and Weertman, J.R. (1998) *Mater.Sci.Eng.,* **A244,** 145.
8. Agnew, S.R., Vinogradov, A., Hashimoto, S., and Weertman, J.R. (in press) *J.Electronic Materials.*
9. Kaneko, Y. , N. Ishikawa, A. Vinogradov and K. Kitagawa, *Scripta Mater.* **38** (1998) 1609.
10. Arieli, A. and Mukherjee, A.K. (1981) *Metall. Trans. A*, **13A** 717.

GRAIN GROWTH IN ULTRAFINE-GRAINED COPPER PROCESSED BY SEVERE PLASTIC DEFORMATION

R.K.ISLAMGALIEV[1], N.M.AMIRKHANOV[1], K.J.KURZYDLOWSKI[2], J.J.BUCKI[2]

[1]Institute of Physics of Advanced Materials, Ufa State Aviation Technical University, Ufa, Russia
[2]Department of Materials Science and Engineering, Warsaw University of Technology, Warsaw, Poland

Abstract

Grain growth in ultrafine-grained (UFG) copper during isothermal annealing and heating at a constant rate was studied by TEM, DSC and microhardness measurements. Three stages of thermal structure evolution in UFG copper were identified. These stages are connected with partial relaxation of the defects, migration of non-equilibrium grain boundaries and grain growth. The values of the activation energy in Stage II and Stage III were estimated.

1. Introduction

It is known that the evolution of microstructure during heating of UFG materials fabricated by severe plastic deformation (SPD) [1] is accompanied by different relaxation processes due to both the grain growth and the relaxation of a non-equilibrium defect structure of samples [2, 3]. At the same time, a character of grain growth in UFG materials depends significantly on a type of heat treatment [4]. In this context, the aim of the present work is to study kinetics of recovery and grain growth in UFG copper, occurring during isothermal annealing and linear heating.

2. Experimental

UFG copper samples, (99.99%), were processed by torsion straining under high pressure of 6 GPa at room temperature [1]. The structure of samples was studied by means of electron microscope JEM–100B. Thin foils were made using a method of jet electropolishing. Electron diffraction patterns were taken from an area of 2 μm^2. A mean grain size was determined by means of an intercept method.

Calorimetric investigations were conducted on differential scanning calorimeters Perkin Elmer DSC7 in pure Ar atmosphere. In scanning regime a rate of heating was 20° C/min. Isothermal experiments were conducted at the annealing temperature of 250° C.

T.C. Lowe and R.Z. Valiev (eds.), Investigations and Applications of Severe Plastic Deformation, 297–302.
© 2000 Kluwer Academic Publishers. Printed in the Netherlands.

Linear heating experiments, the investigations of microstructure and measurements of microhardness were conducted on UFG samples subjected to heating in argon atmosphere (at the rate 20° C/min) to one of the following temperature values: 100, 130, 150, 185, 220, 270, 300° C. The samples were subsequently air cooled. Isothermal annealing was conducted in a laboratory SNOL type furnace in a liquid bath (potassium-sodium saltpeter), and in an oil bath at 200° C. Specimens after isothermal annealing were subjected to quenching in water.

Measurements of Vicker's microhardness, H_V, were conducted on a PMT–3M device under a load of 100 g. X-ray diffraction studies were performed on a DRON–4 device using CuKα radiation. A full width at half maximum (FWHM) of peaks was calculated using a standard technique.

3. Experimental results

TEM studies of copper samples subjected to SPD reveal the formation of an equiaxed grain structure with a grain size of about 0.18 μm (Figure 1a). Moreover, a great number of spots uniformly arranged in circles on an electron diffraction pattern, indicate the formation of grain boundaries (GBs) of a high angle type. It should be also pointed out, that the significant azimuthal elongation of spots observed on electron diffraction patterns indicates a high level of internal stresses. Despite that, the majority of grains are free of lattice dislocations.

3.1. HEATING AT A CONSTANT RATE

After heating to 185° C at a constant rate 20° C/min no noticeable grain growth was revealed in UFG copper (Figure 1b). After heating to 220° C (Figure 1c) an abnormal grain growth, leading to a bimodal grain size distribution, starts. Grains of about 1.5 μm size are formed within ultrafine grained "matrix" (20–30% of observed areas). During further heating to 270 and 300° C the increase in both the mean grain size of coarse grains (up to 2–2.5 μm) and their total volume fraction (up to 30–40%) was observed. At the same time, a growth of fine grains was not revealed.

The DSC studies of the copper sample subjected to SPD revealed an isothermal peak in a calorimetric signal over the temperature interval of 130–270° C, with a maximum at 200° C (Figure 1f). This is accompanied by a change in the enthalpy by 0.56 J/g.

The plot of changes in microhardness of UFG copper during heating at a constant rate of 20° C/min is shown in Figure 1f. At room temperature the microhardness of samples subjected to SPD is 1770 ± 90 MPa. After heating to 150° C no changes in H_V were observed. During heating to 185° C and 270° C a microhardness value decreased by approximately 200 MPa and 600 MPa, respectively.

3.2. ISOTHERMAL EXPERIMENTS

On the basis of experiments with heating at the constant rate, the following four temperatures were chosen to study kinetics of microhardness at isothermal annealings:

200, 250, 300, 350° C. It is seen that three stages can be distinguished on the $H_\varsigma(t)$ plots for temperatures 200° C and 250° C (Figure 2). In Stages I and III a value of H_ς changes slightly, in Stage II a sharp drop in its value is observed.

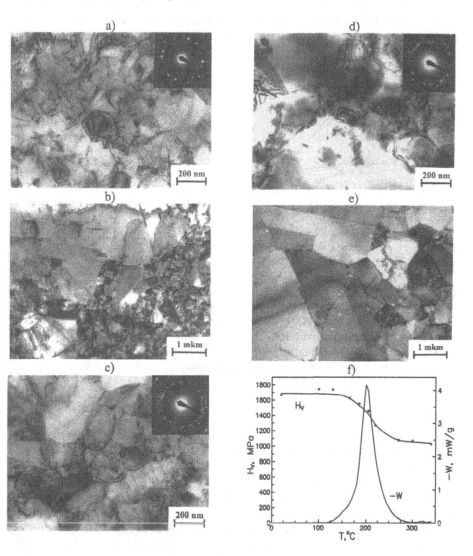

Figure 1. Bright field image and SAED for UFG copper: (a) as-deformed; (b) after heating to 185 C with rate of 20 C/min; (c) after heating to 220 C, (d) after annealing at 250 C, 4 sec; (e) after annealing at 250 C, 50 sec; (f) the DSC signal with subtracted base line and microhardness values for UFG copper (heating rate 20 C/min)..

In specimens typical for Stage I, after annealing at 250° C for 4 sec (Figure 1d), no noticeable grain growth occurs and the mean grain size is about 0.18 µm. At the same time, the reduction in azimuthal elongation of spots on the diffraction pattern as well as the decrease of FWHM in X-ray peaks of the most intense reflections (111), (200), (222) by approximately 26% indicates a partial relaxation of internal stresses in the material [4]. In Stage II one can observe the onset of the abnormal grain growth and the appearance of thickness fringes on some GBs. In particular, after annealing at 250° C for 50 s, relatively large grains of a mean size about 0.8 μm were revealed (~15–20% of

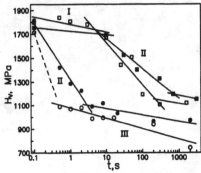

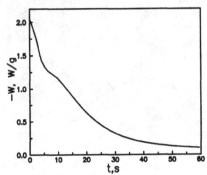

Figure 2. Kinetics of changes in microhardness as a function of annealing time for UFG copper annealed at: ■–200 °C, □–250 °C, ●–300 °C, ○–350 °C.

Figure 3. The DSC isothermal signal of UFG copper at 250 °C.

the investigated material) (Figure 1e). In Stage III, after heat treatment at 250° C for 30 min, the grains with a mean size of about 0.8 μm occupy about 90% and the grains with a size of about 5–10 μm occupy 10%.

The calorimetric isothermal annealing of UFG copper at 250° C (Figure 3) reveals an additional peak of heat release which starts at time of annealing of 5 sec.

4. Discussion

The present investigations allow us to identify three stages of recovery during isothermal annealing of UFG copper (Figure 2) which are related to different relaxation processes.

In particular, in Stage I the grain size remains practically constant, despite the fact that the level of internal stresses decreases significantly. This indicates the relaxation of the non-equilibrium defect structure. In Stage II, a significant drop in a value of microhardness is observed. At the same time, an additional peak of heat release revealed on the isothermal DSC signal can be associated with abnormal grain growth [5]. A similar abnormal grain growth during migration of non-equilibrium grain boundaries was observed in TEM studies of UFG structure typical of Stage II (Figure 1e). Stage II starts in a period of ~4 – 5 sec after the onset of isothermal annealing and coincides with the additional peak in the DSC signal during isothermal annealing (Figure 3). This coincidence allows us to interpret the latter as the indication of the abnormal grain growth. According to TEM studies, in Stage III a growth of grains within the whole volume occurs.

Let us analyze in detail the activation energy of recovery in Stages II and III. The $H_V(t)$ data at each stage were approximated by the least squares method by the function $H_V = f(t) = a * \ln (t) + b$. Let us assume that in each recovery stage a value of microhardness was changed according to the following expression [4]

$$\frac{dH_V}{dt} = c \cdot F(H_V) \cdot \exp(-\frac{Q}{kT}) \quad (1)$$

where Q is the activation energy, K is a Boltsman constant, and $F(H_V)$ is a function of microhardness. Equation (1) was used to analyze changes in the microhardness data ΔH_{Vi} at the annealing temperatures T_i (i = 1, 2, 3, 4). From the condition $\Delta H_{V1} = \Delta H_{V2}$ for the same stage of recovery, using a_i parameters of approximation function $f(t)$ Q values can be estimated. These values were 55 kJ/mol (0.6 eV) and 98 kJ/mol (1 eV), respectively. The error in determination of the activation energy was about 20%.

As mentioned above, only partial relaxation of internal stresses by 26% occurs at the end of Stage I before migration of GBs. It means that the GBs migration starts in non-equilibrium state of sample structure. At the same time internal stresses in UFG pure metals are caused in general by the existence of non-equilibrium GBs [1, 6]. Consequently, Stage II can be associated with migration of non-equilibrium GBs. This assumption is supported by comparison of obtained values of the activation energy in Stages II and III with the activation energy of migration of non-equilibrium GBs in UFG copper, 71 kJ/mol [7], and the one of migration of GBs in coarse-grained copper, 107 kJ/mol [8]. These values suggest that in Stage II mainly migration of non-equilibrium GBs occurs, accompanied by their transformation to a more equilibrium state, whereas the activation energy in Stage III is close to the activation energy of normal grain growth in copper.

Now let us analyze the character of microstructure recovery during linear heating of UFG copper at the constant rate 20° C/min. Up to the temperature of heating 180° C relaxation of the structure takes place on a limited scale (Figure 1b). The intense abnormal grain growth starts above 185° C (Figure 1c). At the same time, heat release starts at the temperature about 130° C. The essential portion of the peak is observed over the temperature range 130–185° C, and attains maximum at 200° C. These findings directly confirm that during constant heating the relaxation processes connected with transformation of the defect structure precede grain growth. Another specific feature of microstructure changes during heating is that at the end of the heat release range, namely at temperature of 270° C, the growth of fine grains is not observed, while in the structure of the UFG copper in Stage III after isothermal annealing coarse grains are observed. Consequently, the peak on the DSC curves over the temperature range 130–270° C can correspond only to Stage I and Stage II of recovery.

Let us note that some other works devoted to the analysis of results of calorimetric investigations also describe a step like character of structure recovery. In particular, it was shown in [2] that during heat treatment of nanocrystalline Pt the relaxation processes of internal stresses and/or non-equilibrium GBs occur first, while grain growth occurs at higher temperatures. Different stages of recovery were also reported in [9] for heavily deformed Fe.

5. Conclusions

1. The experimental investigations carried out reveal three stages of microstructure recovery in UFG copper connected with: partial relaxation of the defect structure, migration of non-equilibrium grain boundaries and grain growth, respectively. The values of the activation energy in Stage II was estimated to be equal to 55 kJ/mol and 98 kJ/mol was obtained for Stage III.

2. The DSC investigations in scanning regime demonstrate a broadening exopeak with a value of heat release of 0.56 J/g observed in the temperature range of 130° to 270° C. It is related to partial relaxation of the defects of grain boundaries and abnormal grain growth attributed to migration of non-equilibrium GBs. Moreover, isothermal experiments show an additional peak indicated the onset of abnormal grain growth as well.

6. Acknowledgments

This work was supported by the INTAS project # 97–1243.

7. References

1. Valiev, R.Z., Alexandrov, I.V., and Islamgaliev, R.K. (1998) Processing and mechanical properties of nanocrystalline materials prepared by severe plastic deformation, in G.M.Chow and N.I.Noskova (eds.), *Nanocrystallyne Materials: Science and Technology. NATO ASI,* Kluwer Academic Publishers, Dordrecht, pp. 121–142.
2. Tschope, A., Birringer, R., and Gleiter, H. (1992) Calorimetric measurements of the thermal relaxation in nanocrystalline platinum, *J.Appl.Phys.* **71,** 5391–5394.
3. Islamgaliev, R.K., Chmelik, F., and Kuzel, R. (1997) Thermal stability of submicron grained copper and nickel, *Mat.Sci.Eng.A* **A 237,** 43–51.
4. Amirkhanov, N.M., Islamgaliev, R.K., and Valiev, R.Z. (1998) Thermal relaxation and grain growth upon isothermal annealing of ultrafine-grained copper produced by severe plastic deformation, *Fiz. Metall. Metalloved.* **86,** 99–105 (in Russian).
5. Chen, L.C., and Spaepen, F. (1991) Analysis of calorimetric measurements of grain growth, *J.Appl.Phys.* **69,** 679–688.
6. Islamgaliev, R.K., and Valiev, R.Z. (1999) TEM investigation of elastic strain near grain boundaries in ultrafine-grained copper, *Fiz. Metall. Metalloved.* **87,** 46–52 (in Russian).
7. Lian, J., Valiev, R.Z., and Baudelet, B. (1995) On the enhanced grain growth in ultrafine grained metals, *Acta mater.* **43,** 4165–4170.
8. Hirth, J.P., and Lothe, J. (1982) *Theory of Dislocations,* McGraw-Hill, New York.
9. Kurzydlowski, K.J., Ralph, B., Chojnacka, A., and Bucki, J.J. (1996) A quantitative description of recrystallization and grain growth in single phase bcc iron, *Acta mater.* **44,** 3005–3013.

INVESTIGATION OF DUCTILITY AND DAMAGE ACCUMULATION BY TWO STAGE DEFORMATION USING ECAE/ECAD AND THE TENSILE TEST

R. LAPOVOK, R. COTTAM, G. STECHER, R. DEAM, E. SUMMERVILLE
CSIRO Manufacturing Science and Technology, Melbourne, Victoria 3072, Australia

Abstract

Lower-Bound Ductility (LBD) is the critical strain at the moment of fracture as a function of stress state and temperature. Measurements of LBD are normally performed by testing in a hyperbaric chamber. There is an alternative approach for determination of the LBD based on the cumulative nature of damage. The basic idea of this new approach is to apply two successive loadings with different stress ratios that lead to fracture, and then to calculate the LBD function.

The processes of equal channel angular extrusion and drawing are used to introduce some damage in the specimen for a specified stress - temperature history. The amount of plastic deformation is determined by the angle between the two intersecting channels, while the stress ratio is varied by applying different back-pressures. The rest of the damage required to cause fracture is introduced by a subsequent tensile test. The LBD diagram for continuously cast Al 6061 has been deduced using this method.

1. Introduction

In the last three decades the local strain approach has turned out to be a useful instrument for fracture prediction of components under uniaxial loading. According to this concept, fracture starts when the local plastic strain reaches some critical value. The critical local strain, as a function of stress state, temperature and strain rate, called the lower-bound ductility (LBD), has to be obtained experimentally [1]. The systematic investigation of critical strains over the wide range of compressive stresses was started by Bridgman [2]. Lately a few experimental rigs have been designed to perform mechanical testing in chambers with high hydrostatic pressure at elevated temperatures, [3]. Although it is very convenient to change the stress path using the hydrostatic pressure of a working liquid in a chamber, few such testing machines exist around the world. Problems are especially severe if high pressure is accompanied by high temperature.

Since LBD has wide application, an alternative to the method of testing in hyperbaric chambers with high hydrostatic pressure is presented here. The method is

T.C. Lowe and R.Z. Valiev (eds.), Investigations and Applications of Severe Plastic Deformation, 303–312.
© 2000 *Kluwer Academic Publishers. Printed in the Netherlands.*

based on the concept of the accumulative nature of damage and on the dependence of damage accumulation on deformation processing history.

2. The Damage Mechanics Approach

The phenomenological description of damage accumulation in isotropic materials can be characterized by a scalar function $\omega(\tilde{\sigma}, \tilde{\varepsilon}, T)$, introduced by Kachanov, [4], where $\tilde{\sigma}$ and $\tilde{\varepsilon}$ are the stress and plastic strain tensors and T is the temperature. The damage, ω, is equal to zero for undamaged material and increases to the value 1 at the point of fracture.

The basic concept of damage mechanics is that damage is proportional to the extent of accumulated plastic shear strain, ε. It has been shown by many experiments, [5], that the damage / shear strain dependence for monotonic loading can be described by the power law, where coefficients are functions of stresses and temperature. The simple equation for calculating damage by monotonic loading with a constant stress state can be written in the following form, [1]:

$$\omega = \left(\varepsilon / \varepsilon_{cr}\right)^a \tag{1}$$

where a is the intensity of damage accumulation.

In the case of sequences of N different monotonic loadings with a constant stress state within each loading, damage can be calculated by summation of damage at every monotonic stage of loading, [1].

A further simplification can be made assuming linear damage mechanics. The intensity of damage accumulation a in the equation (1) will be taken to be equal to 1.

The strain criterion of damage is based on the comparison of plastic strain accumulated by any particle of material along its loading path with the critical intensity of plastic shear strain obtained experimentally. The lower-bound ductility $\varepsilon_{cr}(\tilde{\sigma}, T, ...)$ is the basic physical property in the damage criterion.

It has been shown, [1], that LBD depends on two main parameters: the stress index and temperature. The stress index is the ratio of hydrostatic pressure to equivalent shear stress:

$$\overline{\sigma}/\sigma_e = \frac{1}{3}\sigma_{ii} \left/ \sqrt{\frac{1}{2}\left(\sigma_{ij} - \overline{\sigma}\delta_{ij}\right) \cdot \left(\sigma_{ij} - \overline{\sigma}\delta_{ij}\right)} \right. \tag{2}$$

were δ_{ij} is the Kronecker delta

Thus the basic physical property represented by lower-bound ductility will be considered as a function of the stress index and the temperature, $\varepsilon_{cr}(\overline{\sigma}/\sigma_e, T)$.

The dependency of LBD on strain rate is also considered in this study.

3. A New Method for Lower-Bound Ductility Determination

The basic idea of this new approach is to apply two successive loadings which lead to fracture ($\omega = 1$) and then to calculate the lower-bound ductility function. The stress index is kept constant along each loading path and failure occurs at a particular value of the critical strain (strain at fracture). This value of critical strain at a known stress ratio gives a point on the lower-bound ductility diagram.

Equal Channel Angular Extrusion (ECAE) and Equal Channel Angular Drawing (ECAD) with and without back pressure have been used to introduce some damage, ω_1, in the specimen at a specified stress - temperature history. The rest of the damage, ω_2, required to cause fracture has been introduced by a subsequent tensile test. Based on the concept of the accumulative nature of damage ($\omega = \omega_1 + \omega_2$), the damage, ϖ_1, introduced at the first stage of deformation is calculated as $1 - \omega_2$.

Using equation (1) under the assumption of the linear model of damage accumulation, the equation for LBD definition can be written:

$$\varepsilon_2 / \varepsilon_{cr}\left((\overline{\sigma}/\sigma_e)_2\right) = 1 - \varepsilon_1 / \varepsilon_{cr}\left((\overline{\sigma}/\sigma_e)_1\right) \tag{3}$$

where ε_2 is the equivalent plastic strain introduced in the material by the ECAE or ECAD process; ε_1 is the equivalent plastic strain introduced in the material by the following tensile test; $(\overline{\sigma}/\sigma_e)_2$ is the stress index in the ECAE or ECAD process; $(\overline{\sigma}/\sigma_e)_1$ is the stress index in tension; and $\varepsilon_{cr}\left((\overline{\sigma}/\sigma_e)_1\right)$ is the critical strain in tension for heat treated material. Therefore, the only unknown variable is the critical strain, $\varepsilon_{cr}\left((\overline{\sigma}/\sigma_e)_2\right)$, which is the value of LBD function at the stress index defined by ECAE/ECAD processes. This value of critical strain gives a point on Lower-Bound Ductility Diagram.

3.1. EXPERIMENTAL PROCEDURE.

The original billet was continuously cast Al 6061 and has an equiaxed grain structure with the grain size changing from 80 μm in the center of the billet to 70 μm near the surface and the corresponding hardness changing from 50 HV to 53 HV respectively. T4 pre-heat treatment was given to all specimens to ensure that the material had maximum ductility and a homogeneous microstructure particularly in terms of the precipitate state, and to eliminate the segregation effects resulting from the casting process. The microstructure of specimens after heat treatment was analysed for the distribution of the second phase, grain size and grain orientation. The consistency and homogeneity of the microstructure has been verified by optical microscopy and also by hardness testing. The hardness after the heat treatment became relatively constant (32.7 – 33.5 HV) across the section of the billet in both directions compared to the as received material.

Three dies, with angles 90°, 120° and 150°, were designed and constructed to conduct the laboratory tests using an INSTRON machine to apply the load. A frame containing a hydrocylinder for applying back pressure and back tension to the samples during processing was designed, made and fitted to the INSTRON.

Square bars of 10x10 mm cross section were prepared and extruded/drawn through the dies at different speeds of the ram equal to 4, 40 and 400 mm/min with and without back pressure/tension. Molybdenum di-sulphide grease was used as lubricant to minimize friction during extrusion and drawing. Tensile specimens with 20 mm gauge length and 4 mm diameter were prepared from the extruded or drawn material. The specimens were polished to remove any surface damage imparted onto the material during machining. Tensile testing was conducted after polishing using the same equivalent strain rate as in the ECAE/ECAD processes.

3.2. ANALYSIS OF THE STRESS STATE IN THE TENSILE SPECIMEN TEST

Initially, the tensile test has been done for heat treated material to obtain the critical strain in tension at the stress index close to 0.58. The velocity of the ram has been kept at three different levels of 4, 40 and 400 mm/min to produce strain rates equal to 0.002, 0.02 and 0.2 respectively. The critical strain at fracture has been calculated as the ration of initial and necking cross-sectional areas, $\varepsilon_{cr} = \ln S_0 / S$.

The stress index at the uniform stage is a constant equal to 0.58. The stress index at the necking stage can be expressed using the Davidenkov-Spiridonova's formulas for stress components definition, [6]:

$$\overline{\sigma}/\sigma_e = \sqrt{3}\left(1/3 + \left(r_f^2 - r^2\right)/2r_f R\right) \qquad (4)$$

where r_f, is the radius of the neck at the moment of fracture and R is the radius of curvature of the contour of the neck.

Dividing the deformation pass in two parts such as uniform deformation before necking and deformation during necking the average stress index during tensile testing has been calculated. It has the value close to 0.64. Therefore, the only point on the LBD diagram that has been obtained from tensile testing alone is represented by the following data: $\varepsilon_{cr} = 1.28$ for $\dot{\varepsilon} = 0.002$ and $(\overline{\sigma}/\sigma_e) = 0.633$; $\varepsilon_{cr} = 1.35$ for $\dot{\varepsilon} = 0.02$ and $(\overline{\sigma}/\sigma_e) = 0.637$; $\varepsilon_{cr} = 1.38$ for $\dot{\varepsilon} = 0.2$ and $(\overline{\sigma}/\sigma_e) = 0.638$.

3.3. EQUAL CHANNEL ANGULAR EXTRUSION (ECAE)

ECAE is the process developed and patented 1977 in Russia by Segal, [7]. The general principle of this process is shown in Fig. 1a. The tool consists of a die with two channels of the same cross-section intersecting at an angle ϕ. A well lubricated billet of almost the same cross section is pushed by the punch, through the intersection of channels. In both channels, the billet moves as a rigid body while all deformation is localised in the small area around the channels' meeting line, [8]. As a result of this process a large uniform plastic strain is imposed in the material without reduction of the initial cross-section. The metal is subjected to a simple shear strain under relatively low pressure.

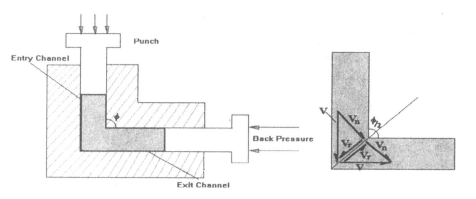

Figure 1. A sketch of a cross section of ECAE.
a – ECAE with an angle 90°; b- material velocities during extrusion.

As the properties of a deformed material are dependent on the stress history, the extrusion of material in the equal channel die can be done against a punch producing some prescribed back-pressure, as in Fig.1a. In this case, the shear deformation is accompanied by some level of compressive stresses. The use of square section channels for the ECAE used in these experiments, results only in plane strain deformation. The extruded material can be considered as a rigid-perfectly plastic isotropic medium. Following Segal, [8], the deformed state is defined, Fig. 1b:

$$\varepsilon_e = \Delta V_r / V_n = 2\cot\phi/2 \tag{5}$$

Using a constitutive equation for rigid-perfectly plastic isotropic material the components of a stress tensor can be defined, [8]. Substitution of stress components in formula (3) gives the stress index during ECAE, $(\overline{\sigma}/\sigma_e)_2$, with back pressure, σ_2, by the following expression:

$$(\overline{\sigma}/\sigma_e)_2 = (\sigma_2 / \tau_s + \cot\phi/2)/\sqrt{3} \tag{6}$$

where τ_s is the yielding shear stress.

The negative values of stress index are realized during ECAE.

3.3.1. Influence of the Die Angle on ECAE

Three grid specimens have been prepared to investigate the strain distribution experimentally. These specimens have been subjected to interrupted extrusion through each die. The distortion of the grid is homogeneous even in the surface layers, Fig. 2b. All cells change their shape from rectangular to rhomboidal as soon as they pass the channel connection line. One side of the rhomboid is parallel to the bisection of the angle ϕ.

Visible texture is produced after the first extrusion through the dies with 90° and 120°, Fig. 2a. Initially equiaxed grains become elongated in the direction parallel to the channel connection line. The amount of deformation imparted into the material by

308

extrusion through the 150° die is not large enough to produce a visible texture after the first extrusion.

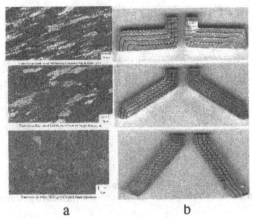

a   b

Figure 2. The ECAE of grid specimens through the dies with angle 90°, 120° and 150°.
a - microstructure of the extruded specimens; b - distortion of the grid observed by the interrupted extrusion.

3.4. EQUAL CHANNEL ANGULAR DRAWING (ECAD)

Suriady & Thomson, [9], were first to deform material by drawing rather than by extrusion through equal channel angular dies. They studied the change in microstructure of a specimen after multi - drawing it through a die with 135° angle according to different schedules. The general principle of this process is shown in Fig. 3a.

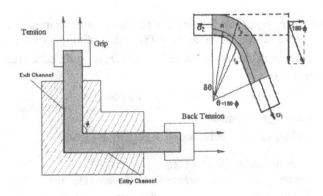

a   b
Figure 3. A sketch of the cross-section of ECAD.
a – ECAD die with angle 90°;
b - presentation of ECAD process as a combination of bending with tension.

In the equal channel angular drawing, Fig. 3a, material undergoes a plastic deformation which can be presented as a bending under tension process, Fig. 3b. The specimen has to be initially pre-bent to fit the channels. Therefore, the metal flow in the width direction during the following drawing is restricted by a significant large bulk of rigid material. Taking this into account, the strain in the width direction is neglected and the problem is considered as a plain-strain problem. The stress and strain distribution will be obtained by using the assumption that transverse cross sections that were originally plane and normal to the centre line remain so after deformation.

The problem of pure bending under plane strain conditions combined with tension is considered by Hill, [10]. If the bending is carried out while the bar is stressed by tension σ_2 applied to the ends, it results in uniform pressure $p = \sigma_2 h / r_a \phi$ applied over the inner surface of the bar. The stresses components in this case are defined, [10], and the stress index can be calculated:

$$\bar{\sigma}/\sigma_e = \left(1 - 2 \cdot \ln \frac{r_b}{r}\right)\Big/\sqrt{3} \quad \text{for} \quad (R \leq r \leq r_b) \tag{7}$$

$$\bar{\sigma}/\sigma_e = \left(-\frac{p}{\tau_s} - 1 - 2 \cdot \ln \frac{r}{r_a}\right)\Big/\sqrt{3} \quad \text{for} \quad (r_a \leq r \leq R) \tag{8}$$

where R satisfies:

$$R^2 = r_a r_b \exp(-p / 2\tau_s) \tag{9}$$

The positive values of stress index are realised during ECAD.

Integration along the bending angle of equivalent plastic strain increment gives the value of the accumulated equivalent plastic strain:

$$\varepsilon_e = 2\left|\left(1 - R^2/r^2\right)\theta\right|, \quad \theta = \pi - \phi \tag{10}$$

The value of the accumulated equivalent plastic strain in the tensile specimen made from the drawn material can be estimated by averaging of ε_e between radii defining the surfaces of the tensile specimen

3.4.1. Influence of the Die Angle on the ECAD

Three grid specimens have been prepared to investigate the strain distribution experimentally. These specimens have been subjected to interrupted drawing through each die. The distortion of the grid during drawing, Fig. 4b, proves the hypothesis about the bending deformation mode for ECAD. There is a clearly visible neutral line where cells do not change their shape and dimensions. Cells located above this line are stretched along the drawing directions proportional to the distance from the neutral line. All other cells located below the neutral line are shrunk, becoming smaller the further they are from the neutral line. The level of strain introduced by bending increases as the angle of channel intersection decreases. The difference in texture produced after the first drawing in outer and inner layers of the specimen is shown in Fig. 4a.

3.5. RESULTS OF TESTING IN ECAE/ECAD AND TENSION

The clearly visible difference in the necking of tensile specimens after preliminary extrusion and drawing through different dies provides additional proof of the different

310

level of damage introduced in the material prior tensile testing. The level of damage introduced into the material during the tensile test was calculated, (1), using measured values of the necking area of fractured specimens. The dependence of damage introduced into the material in tension for three different angles of the extrusion or drawing dies is shown in Fig. 5.

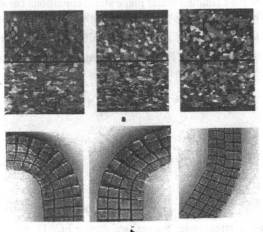

Figure 4. The ECAD of grid specimens through the dies with angle 90°, 120° and 150°.
a - microstructure of the drawn specimens; b - distortion of grid by interrupted drawing.

The results presented in Fig. 5 show that at small strain rates a higher level of damage can be introduced into the material. It is also clear, that the smaller the die angle, the higher the level of damage that is introduced into the material during extrusion (Fig.5a) or drawing (Fig. 5b). The more damage introduced into the material by ECAE or ECAD, the less damage is subsequently required by the tensile test to fracture the specimen.

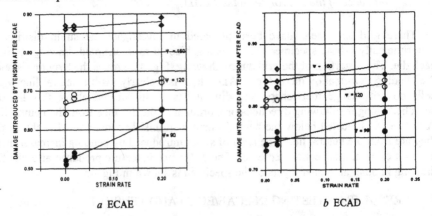

a ECAE   *b* ECAD

Figure 5. Damage introduced into the material during tensile testing after preliminary (a)ECAE and (b) ECAD as a function of strain rate for three different angles of drawing dies

3.6. DEFINITION OF LOWER-BOUND DUCTILITY DIAGRAM

To obtain points on Lower-Bound Ductility diagram, the following variables were calculated:
- damage introduced into the material by ECAE/ECAD processes as the difference between value at the fracture, 1, and damage introduced by the following tension;
- plastic strain introduced into the material by ECAE/ECAD processes, (5), (10);
- stress index in the plastic zone during either the ECAE or ECAD process, (6), (7), (8);
- critical strain at the stress index experienced by material during ECAE/ECAD processes as the ratio of plastic strain to the damaged introduced by transforming (1).

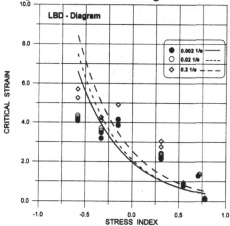

Figure 6. Room Temperature Lower - Bound Ductility calculated from two – step deformation.

The graphical presentation of the critical strain versus the corresponding stress index shown in Fig. 6 gives the LBD diagram. These data are shown together with expected exponential fits to the data, [1], for the different strain rates (0.2, 0.02, 0.002 s^{-1})

As expected, the LBD increases for negative values of stress index and decreases for positive values of stress index. The difference between LBD diagrams for different strain rates does not seem significant, which reinforces the hypothesis that LBD at room temperature is independent of strain rate.

4. Conclusion

This work has developed a new experimental technique and supporting theoretical framework for the determination of Lower-Bound Ductility. This approach is based on the cumulative nature of damage. Two successive loadings with different stress ratios, namely ECAE/ECAD and tensile test, are applied till fracture of the specimen and the LBD function is calculated.

The LBD diagram at room temperature has been deduced using this method for continuously cast Al 6061. These results have shown the same trend as those obtained in hyperbaric chambers.

312

5. References

1. Lapovok R., Smirnov S. and Shveykin V. (1996) Ductility defined as critical local strain, *Proceedings of First Australasian Congress on Applied Mechanics*, Melbourne, Australia, 1996, pp 181-185.
2. Bridgman, P.W. (1970) *The Physics of High Pressure*, Dover Publications, New York.
3. Bogatov, A.A., Smirnov, S.V., Bykov, V.I. and Nesterenko, A.V. Author's Certificate No.1422090 (USSR), GO1N 3/10 A device for testing specimens under three axis loading.
4. Kachanov, L.M. (1986) *Introduction to Continuum Damage Mechanics*, Kluwer Academic Publishers, Dordrecht.
5. Lemaitre, J. and Chaboche, J. (1975) A non-linear model of creep-fatigue damage accumulation and interaction, in: J. Hulte (ed.), *Mechanics of Visco-Plastic Media and Bodies*, Springerverlag, Berlin, pp. 291-300.
6. Davidenkov, N.N. and Spiridonova, N.I. (1945) Analysis of stress state in the neck of the tensile specimen, *Zavodskaya Laboratoriya*, 6, pp. 9-37, (in Russian).
7. Segal V.M. (1977) The Method of Material Preparation for Subsequent Working, Patent of the USSR, No 575892.
8. Segal V.M., Reznikov V.I., Drobyshevskiy A.E. and Kopylov V.I. (1981) Plastic working of metals by simple shear, *Translation. Russian Metallurgy*, 1, pp. 99-105.
9. Suriadi, A.B. and Thomson, P.F. (1997) Control of deformation history for homogenizing and optimizing mechanical properties of metals, *Proceeding of Australiasia-Pacific Forum on Intelligent Processing & Manufacturing of Materials*, IPMM, pp. 920-926.
10. Hill, R. (1986) *The Mathematical Theory of Plasticity*, Oxford University Press, Oxford.

SEVERE PLASTIC DEFORMATION OF FE-NI INVAR ALLOY AND FE-NI MARAGING STEELS

A.M. GLEZER, V.V. RUSANENKO, V.I. ISOTOV, V.I. KOPYLOV,
A.F. EDNERAL
G.V. Kurdyumov Institute of Metal Physics and Functional Materials
State Science Center of Iron and Steel Industry

Abstract

The equal channel angular pressing method was used to strengthening the Fe-36%Ni invar alloy and of some industrial maraging steels by severe plastic deformation. The features of the dislocation structure formed on this process, as well as mechanical and thermophysical properties of deformed alloys have been studied.

1. Introduction

The classical invar Fe-Ni alloys have fcc lattice and exhibit low strength in annealed state. The improvement of strength can be achieved either by alloying, providing precipitation or solid-solution hardening, or by work hardening. The alloying is generally attended with decreasing of invar properties [1]. In contrast work hardening can provide not only strength increasing but improvement of thermophysical (invar) properties as well. Cold rolling, drawing or extrusion realized however a large degree of plastic deformation required for this improvement. Unfortunately these methods of plastic flow result in unwanted cross-section thinning or strong extension of invar alloy. This problem can be overcome by application of equal channel angular (ECA) pressing as a method of severe plastic deformation of materials [2].

The most important results for an understanding of processes taking place under large degree of plastic deformation were obtained by V. Rybin *et al.* [3]. The conception of limited deformation structure, which relates to fragmented dislocation structure, and of significant role of rotational mode of plastic deformation at the final stages of flow process have been developed on a basis of the experimental data for bcc and fcc metals. In so doing, it was calculated that the main features of the limited fragmented structure did not depend on the type of crystal lattice and did not depend on the method of plastic deformation.

In this work the ECA pressing method was used to strengthening the Fe-36% Ni invar alloy and some industrial maraging steels to remain during severe plastic deformation the initial cross-section of worked material. The features of the dislocation structure formed at this process, as well as mechanical and thermophysical properties of alloys have been studied.

T.C. Lowe and R.Z. Valiev (eds.), Investigations and Applications of Severe Plastic Deformation, 313–318.
© 2000 Kluwer Academic Publishers. Printed in the Netherlands.

2. Experimental

Ingots of 14x14 mm square section were subjected to ECA pressing at room temperature. Fig.1 shows the scheme of ECA pressing at the angle between canals to be 90° used at present work. The degree of plastic deformation in a single pass was about 70%. A variation of deformation shear in opposite directions at alternation of passes was carried out. This allowed to obtain a homogeneously deformed material without pronounced crystallographic texture. The number of passes was 1, 2, 4, 8 and 12. In order to compare the dislocation structure of the alloy after ECA pressing and after some another method of large plastic deformation the rods of 12x14 mm section were cold rolled in 12 passes into 3.5 mm thick bands at room temperature.

The structure after different stages of deformation process was characterized by transmission electron microscopy (TEM), optical microscopy and X-ray diffraction. Methodical aspects of misorientation measurements was published in [4]. The specimens for structural investigations were cut from the rods deformed along three basic sections: along A containing the shear deformation direction, along B and along C (see Fig.1). The measurement of mechanical properties (yield stress, fracture stress, plasticity, Vickers hardness) and of thermal expansion coefficient in the range 20°-300°C was made.

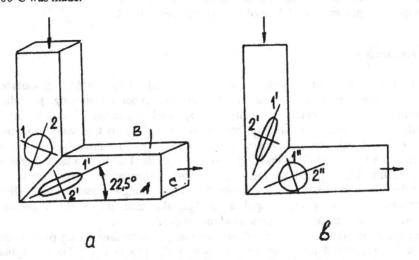

Figure 1. The scheme of ECA pressing used in this work.

3. Results and discussion

The initial structure of Fe-36% Ni alloy consisted of uniaxial polyhedral grains of mean size 100 μm whereas after the first pass of ECA pressing the grains transform into ellipsoids with long axis arranged at the angle 22.5° to the longitudinal axis of the deformed specimen.

The predominant type of substructure after first pass is the ribbon-like dislocation structure (Fig.2). It consists of "ribbons" of width 0.1-0.5 μm with grain boundaries providing the low-angle (up to 2-3°) with mainly twist component of misorientation. The high density (10^{10}-10^{11} cm^{-2}) of random distributed dislocation was observed within the ribbons. As was shown by TEM trace analysis [5], the ribbons were mainly arranged along the {111} slip planes of fcc lattice. It is typical that in the ribbon packet the misorientation was not summarized but compensated each other.

The next revealed structural feature after the first pass was the presence of kink bands. They represented the regions 1-3 μm wide with ribbon structure. The extra shear as large as several tens of degrees (up to 40°) was recorded at high angles to the habit plane of the ribbons (fig.3). As a rule kink bands were grouped into packets inside of which the regions with the opposite direction of kink were alternated. Apparently, the kink bands were connected with relaxation of high level of deformation stresses.

Let us consider in detail the structure of the shear bands. The main feature of the subboundaries involved into the shear bands was they were arranged parallel each other, but sometimes were broken inside the bands. Such "broken" subboundaries responsible for the rotation misorientation were essentially the partial disclinations. In some cases the misorientation of the "broken" subboundaries attained to 10-13° that indicated a reasonably

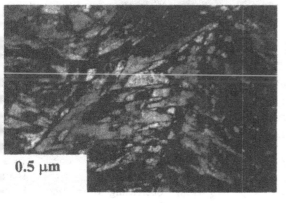

0.5 μm

Figure 2. The typical ribbon dislocation structure of invar alloy after ECA pressing in one pass.

0.5 μm

Figure 3. The kink bands in Fe-36%Ni alloy after ECA pressing (one pass).

high power of the rotation mode of deformation.

After cold rolling with the same degree of deformation the character of the dislocation structure was in principle identical. By this is meant that in highly deformed fcc crystals the regularities of dislocation structure formation did not depend principally on deformation method.

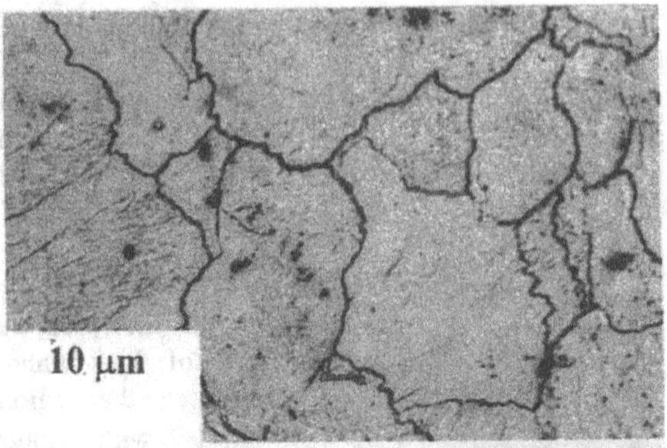

Figure 4. The grain structure of invar alloy after ECA pressing in two passes (with opposite shear direction).

0.5 μm

Figure 5. The typical fragmented structure in Fe-36% Ni alloy after severe deformation by ECA forming in 12 passes.

The second pass of ECA pressing was shown to restore the polyhedral shape of grains. However the grain boundaries became curved with the sharp local bends and steps (fig.4). The "ribbon" structure occupied the major volume of the grains. The main difference from the ribbon structure originated after the first pass was in the appearance of the cross sub-boundaries inside the ribbons. The number of kink bands in the grains after two passes was more than that after one pass.

An increase of number of passes to 4, 8 and 12 resulted in the disappearance of the initial grain boundaries. The substructure became more homogeneous and fragmented during which the fragment shape approached to an equilibrium one with increasing of number of passes. The dislocation density inside the fragments decreased to 10^9-10^{10} cm^{-2} but the misorientation between the neighboring fragments increased to 10° after 12 passes. The size of the equilibrium fragments was about 0.2 μm (fig.5). The process of the transition from the ribbon structure to the fragmented one was accompanied by sharp decreasing of number of the kink bands. It is unlikely that the structural state of Fe-36%Ni alloy obtained in this investigation after the maximum number of passes (12) by ECA pressing may be called nanocrystalline one. First, the fragment size was about 200 nm and, second, the misorientation angle between individual fragments (about 10°) made impossible the consideration of them as normal grains.

Fig. 6 shows the variation of mechanical and thermophysical properties as well as the level of inner stresses characterized by X-ray diffraction line width as number of passes increases in invar alloy. The strength characteristics increased monotonically with increasing of number of passes but the maximum increase occurred in an early stage of deformation. So we had a chance of yield stress changing of invar alloy from 650 to 800 MPa by varying of the pass number. In this case the plasticity of the deformed alloy was retained at the rather high level. The thermal expansion coefficient did not change monotonically with the pass number. The pressing in one pass resulted in decreasing of the positive value $+0.4 \times 10^{-6}$ K^{-1} to negative value -0.4×10^{-6} K^{-1}. After deformation in two passes the coefficient increased to zero but after four passes it was close to the initial one.

Thus the combination of the strength and thermophysical characteristics obtained in this work correspond to the parameters of high-strength invar alloys which can be turned to practical use. Besides, the experiments on ECA pressing of the industrial maraging steels Fe-18% Ni-2% Mo-2% Ti, Fe-18% Ni-9% Cr-5% Mo-0.7% Ti, Fe-12% Cr-4% Ni-14% Co- 5% Mo-1% Ti were carried out in this work. It was established that after ECA pressing the

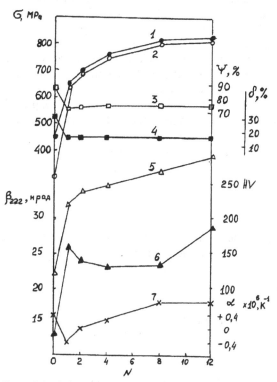

Figure 6. Variation of fracture stress (1), yield stress (2), plasticity (3,4), Vickers hardness (5), X-ray diffraction line β_{222} width (6) and thermal expansion coefficient α (7) with the number of passes N after ECA pressing of Fe-36%Ni alloy.

deformed structure was nearly the same (especially the first composition of maraging steel) as the structure of the invar Fe-Ni alloy. The dislocation fragment width coincided with the width of initial lath martensite crystals but the length of the fragments was in several times less. The yield stress of maraging steels was found to be increased to value of 2500 MPa after ECA pressing.

4. Conclusions

1. The effect of ECA pressing at room temperature with the channel angle 90° on the structure, mechanical and thermophysical properties of the invar Fe-36% Ni alloy and of the commercial Fe-Ni-Cr-Co-Mo-Ti maraging steels was analyzed.
2. It was shown that the strength of the materials being studied can be essentially increased (in two times for invar alloy and up to 2500 MPa for maraging steels) depending on the number of passes on ECA pressing. Moreover, the negative, positive or zero values of thermal expansion coefficient can be influenced by varying of the number of passes
3. It was established that the main features after 70% first pass are the extended shape of grains and the ribbon substructure. The ribbon dislocation structure is characterized by high density of free dislocation and by broken subgrain boundaries with misorientation 2-3° (up to 13° in some cases). The plane of the subgrain boundaries coincided with the {111} dislocation slip planes. The similarity of the dislocation structure of the Fe-36% Ni alloy after ECA pressing and after cold rolling with the same degree of deformation was shown.
4. At the increasing the degree of deformation the initial grain boundaries disappeared and the ribbon structure transformed to the equiaxial fragmented structure with the fragment size about 0.2 μm and the mean value of fragment misorientation angle 10°. We can classify the final structure after 12 passes as a microcrystalline but not as a nanocrystalline one.

5. Acknowledgments

This work supported by the Ministry of Science and Technologies of Russia in frames of I.P.Bardin Russian State Science Center of Iron and Steel Industry.

6. References

1. Molotilov, B.V. (1983) *Precision Alloys* (in Russian), Metallurgia, Moscow.
2. Segal, V.M., Reznikov, V.I., Drobyshevskiy, A.E., Kopylov, V.I. (1981) Equal angular extrusion, *Russian Metally* 1, 99-105.
3. Rybin, V.V. (1986) *Severe plastic deformation and fracture of metals* (in Russian), Metallurgia, Moscow.
4. Izotov, V.I., Rusanenko, V.V., Kopylov, V.I. (1996) Structure and properties of Fe-36%Ni invar alloy after severe plastic deformation, *Fizika Metallov i Metallovedenie (in Russian)* 82, 123-135.

EVALUATION OF THE TENSILE PROPERTIES OF SEVERELY DEFORMED COMMERCIAL ALUMINIUM ALLOYS

M.V. MARKUSHEV AND M.YU. MURASHKIN
Institute for Metals Superlasticity Problems Russian Academy of Sciences, 39 Khalturin St., 450001 Ufa, Russia

Abstract

Data on ambient temperature tensile properties of commercial aluminium alloys processed from cast ingots are reviewed. The properties of alloys in the severely deformed condition and after annealing are compared with standard values for conventionally treated commercial wrought products. It is shown that with non-heat treatable alloys, processed by severe deformation, a significant increase in tensile strength is obtainable. The combination of strength and ductility achievable is comparable to, or even better than, that found in many high-strength conventionally treated precipitation-hardened alloys. For most of heat treatable alloys, the use of severe deformation for improving mechanical properties is ineffective.

1. Introduction

Understanding the mechanical behavior of submicrocrystalline (SMC) and nanostructured materials processed by severe plastic deformation (SPD) is of great importance if their potential is to be exploited in commercial applications. The issue is complex since a wide range of deformation and grain structures can be produced, depending on the alloy type, processing method and route, strain level, *etc.* In spite of the above, the majority of papers begin with one dominating statement, that severe plastic deformation results in improved and unique level of properties, especially tensile strength, to metals and alloys [1–5]. However, logical reasons make us doubt the general correctness and applicability this statement. On the one hand, it cannot be valid for all metallic materials. On the other hand, it is generally known that along with its positive effects the grain refinement also exerts a negative influence on a number of service properties. The latter is predominantly caused by less resistance to crack growth in ultra-fine grain structures [6–8].

2. Objectives

This study is aimed at checking the validity of the statement on the positive effect of SPD on mechanical properties of commercial aluminium alloys. Main non-heat treatable and heat treatable aluminium alloy compositions processed by ingot

319

T.C. Lowe and R.Z. Valiev (eds.), Investigations and Applications of Severe Plastic Deformation, 319–325.
© *2000 Kluwer Academic Publishers. Printed in the Netherlands.*

metallurgy and widely used in industry for wrought and finished productions are analysed. Ambient temperature tensile properties are compared since they are the only properties properly quoted in the literature. Such an analysis will be correct if two major conditions are followed. Firstly, the data obtained with standard measurements will be considered only[1]. Secondly, the properties of severely deformed alloy will be compared with ones for the same alloy after treatments having a similar purpose, *e.g.* after commercial methods of heat or strain hardening.

3. Tensile strength of aluminium alloys

Even a brief review of published data devoted to mechanical properties of submicrocrystalline severely deformed aluminium alloys indicates that the correct measurements of tensile strength parameters were presented in few papers only. In the majority of investigations the tensile strength has been analysed on non-standard small size specimens or obtained by recalculation of hardness. Such measurements usually result in increased values of tested characteristics, predominantly the ductility. Thus, we have also used data of non-standard tensile tests for SMC alloys that were obtained with required accuracy.

In Tables 1 and 2, one can see trustworthy data published previously and few new results on properties of severely deformed heat treatable and non-heat treatable aluminium alloys. The typical values of tensile strength characteristics for conventionally treated commercial wrought semi-products are presented for comparison.

3.1. NON-HEAT TREATABLE ALLOYS

As seen from Table 1, the as-deformed through equal channel angular extrusion or complex angular extrusion (ECAE and CAE) non-heat treatable alloys show yield and ultimate tensile strength (YS and UTS) slightly exceeding the level of conventional cold worked semi-products (treatment H1x). Besides, no sufficient reasons were found to affirm the significant improvement in alloys ductility (elongation) after treatment, involving severe plastic deformation. To increase ductility and toughness of the cold worked products a stabilizing annealing (treatment H3x) is commonly used. It is apparent that for severely deformed materials this treatment should also positively affects the plasticity. In order to serve strengthening effect on grain refinement, the post-deformation annealing of SMC billets should be conducted at temperature and duration limited by the recovery of grain and grain boundary structures [11, 14]. As a rule, such annealing leads to a slight (10–20%) decrease in tensile strength and a significant (up to two times!) improvement in alloy elongation (Table 1). Consequently, the strength of SMC severely deformed billet becomes close to that of strain hardened conventional products, but its ductility is much higher.

Thus, severe plastic deformation and post-deformation annealing of non-heat treatable aluminium alloys is quite an effective way to obtain an unusual balance of

---

[1] - at least the relation of specimens gauge height to diameter (thickness) should be not less than 5

tensile properties resulting in equivalent strength with higher ductility compared to conventional materials.

Another interesting point is whether it is possible to receive unique, otherwise unobtainable, strength in the bulk billets of non-heat treatable alloys. To answer this question the SMC billet of 1560 alloy processed by CAE have been rolled at an ambient temperature. An essential improvement in alloy strength with increasing the rolling reduction was found (Table 1). After rolling to 90% the alloy strength not only reaches, but also overcomes typical values of yield and tensile stresses of many high-strength 7xxx series alloys.

Therefore, the treatment, involving stages of submicrocrystalline structure processing and further SMC structure work hardening can provide new, extremely strengthened condition with a comparatively cheap, low alloyed material. It is possible to predict that a similar kind of processing of super-strength wrought products with SMC structure from 5xxx alloys could be quite effective for commercial application.

3.2. HEAT TREATABLE ALLOYS

The influence of SPD on properties of heat treatable alloys is more complex (Table 2). Depending on the specific alloy, it could be positive or negative.

In the first alloy group, consisting of medium-strength low-alloyed heat treatable compositions, the trends discussed above for non-heat treatable alloys, can be reproduced. For example, changes in strength and ductility of 6061 alloy after SMC structure processing (Table 2) are positive and similar to those of non-heat treatable alloys. Post-deformation annealing also exerts the same effect on improving the alloy tensile properties. According to our experience, the described influences of severe plastic deformation through ECAE and further annealing on mechanical behaviour are also observed in another 6xxx alloys. That is why, the proposal that thermomechanical treatment involving SMC structure processing, work and heat hardening has a good potential as a commercial method for enhancement properties with this class of materials, looks reasonable.

For another group of high-alloyed medium- and high-strength heat treatable alloys it can be seen that the use of severe plastic deformation for improving mechanical properties is not effective (Table 2). Moreover, for most of alloys of 2xxx, 7xxx, 8xxx compositions it leads to both less strength and ductility. The reason for this degradation of properties is that the effects of structural strengthening due to submicrocrystalline structure processing cannot surpass the effect of precipitation hardening. The peak aged heat treatment (T6) cannot be used in severely deformed products of the majority of commercial heat treatable alloy compositions. The solution treatment would lead to strong grain coarsening and lose of deformation hardening and SMC structure strengthening effects. The latter is true even for 1460 alloy with complex additions of transition metals (Zr and Sc) that forms high densities of dispersoids for preventing recovery and grain growth on heating.

TABLE 1. Ambient temperature tensile properties of non-heat treatable aluminium alloys

Alloy	Thermomechanical treatment	Processed billet, mm	Gauge of tensile specimen, mm	YS, MPa	UTS, MPa	Elong., %	Ref.
1100 (99Al)	H18	-	-	152[1]	165	5-15	[11]
	ECAE (RT, e-6)	Rod, Ø10x60	2x3x5	190	-	25	[12]
3004 (Al 1.2Mn 1.0Mg)	H38	-	-	250[1]	285	4-6	[11]
	ECAE (RT, e-8)	Rod, Ø10x60	2x3x5	370	-	15	[12]
5056 (Al 4.8Mg 0.07Mn 0.06Cr)	H18	-	-	407[1]	434	10	[11]
	H38	-	-	345[1]	414	15	[11]
	Extrusion (RT, 75%)	-	Ø3x10	385	420	10	[2]
	ECAE (RT, e-4)	-	Ø3x10	410	440	12	[2]
5083 (Al 4.4Mg 0.7Mn 0.15Cr)	H34	-	-	283/340[2]	345/405	9/6-8 (min)	[11]
	H116	Plate, 100	Ø5x25	235/240[3]	310/325	17/13	[13]
	CAE (e-14)	Plate, 25x120x120	Ø5x25	370/385[3]	420/435	11/10	[13]
	CAE+anneal. 200°C, 2hs	Plate, 25x120x120	Ø5x25	345/355[3]	370/385	20/15	[13]
	ECAE (RT, e-3)	Rod, Ø10x60	2x3x5	420	-	10	[12]
1560 (Al 6.0Mg 0.6Mn)	H14	Sheet, 2	-	320[1]	420	10	[14]
	H34 (anneal. 90°C, 10hs)	Sheet, 2	-	350	450	11	[15]
	ECAE (e-8)	Rod, Ø20x120	Ø3x15	355	435	20	[16]
	ECAE+anneal. 200°C, 8hs	Rod, Ø20x120	Ø3x15	350	415	17	[16]
	CAE (e-14)	Plate, 25x120x120	Ø5x25	375/384[3]	467/478	10.5/9.0	
	CAE+anneal. 200°C, 8hs	Plate, 25x120x120	Ø5x25	315	418	18	
	CAE+RT roll. 20%	Plate, 20	Ø5x25	432	505	6	
	CAE+RT roll. 90%	Sheet, 2	2x3x18	540	635	3	

TABLE 2. Ambient temperature tensile properties of heat treatable aluminium alloys

Alloy	Heat or thermomechanical treatment	Processed billet, mm	Gauge of tensile specimen, mm	YS, MPa	UTS, MPa	Elong., %	Ref.
6061 (Al 1.0Mg 0.6Si 0.3Cu 0.2Cr)	T6	-	-	276[1]	310	12	[11]
	ECAE (110°C, e~4.6)	Rod, 25x25x130	-	-	~400	~11	[17]
	ECAE + anneal. 210°C, 1h	Rod, 25x25x130	-	-	~330	~19	[17]
	ECAE (RT, e~6)	Rod, Ø10x60	2x3x5	280	-	17	[12]
1420 (Al 5.5Mg 2.1Li 0.1Zr)	T6	-	-	310[1]	470	10	[15]
	T6 (ageing 120°C, 12hs)	Plate, 30x100x150	Ø5x25	328	467	11	[18]
	ECAE (e~10)	Rod, Ø20x120	Ø3.5x10	350	405	9	[13]
	ECAE	-	-	265	450	13	[5]
1460 (Al 3Cu 2Li 0.1Zr 0.08Sc)	T6 (ageing 160°C, 30hs)	Sheet, 2	-	474/508[2]	543/583	7.7/7.2	[19]
	T861	Sheet, 3	-	534/527[2]	592/593	4.7/5.7	[20]
	-	Panel, ~200x470	-	470/497[2]	592/577	8.4/3.9	[20]
	ECAE (e~12)	Rod, Ø20x100	-	400	480	6	[21]
	ECAE + T6 (solution. 500°C, 0.16h; ageing 100°C, 15hs)	Rod, Ø20x100	-	470	565	5	[21]
2024 (Al 4.4Cu 1.5 Mg 0.6Mn)	T6, T651	-	-	393[1]	476	10	[11]
	T861	-	-	490[1]	517	5	[11]
	ECAE (RT, e~4)	Rod, Ø10x60	2x3x5	330	-	8	[12]
7075 (Al 5.6Zn 2.5Mg 1.6Cu 0.23Cr)	T6, T6951	-	-	503[1]	572	11	[11]
	ECAE (RT, e~6)	Rod, Ø10x60	2x3x5	470	-	~8	[12]

[1] - typical values of properties
[2] - properties in longitudinal/transverse directions

4. Conclusions

1. It has been shown that in non-heat treatable alloys, processed by severe plastic deformation through equal channel or complex angular extrusion, the unique balance of tensile strength and plasticity is obtainable.

2. For non-heat treatable alloys, processed by severe plastic deformation through angular extrusion and rolling, a significant increase in tensile strength is achievable. The combination of strength and ductility obtainable is comparable to, or even better than, that found in many high-strength conventionally processed precipitation-hardened alloys.

3. Thermomechanical treatment involving severe plastic deformation is quite effective for enhancement the tensile strength of low-alloyed heat treatable aluminium alloys.

4. For most of heat treatable alloys, especially high-strength alloys, the use of severe plastic deformation for improving ambient temperature mechanical properties is ineffective.

5. Acknowledgement

The authors are grateful to Dr. P.B.Prangnell for fruitful discussions of the data presented in the paper.

6. References

1. Valiev, R.Z., Korznikov, A.V., and Muliukov, R.R. (1993) Structure and properties of ultrafine-grained materials produced by severe plastic deformation, *Mat.Sci.Eng.* **A168**, 141–148.
2. Kawazoe, M., Shibata, T., Mukai, T., and Higashi, K. (1997) Elevated temperature mechanical properties of 5056 Al-Mg alloy processed by equal-channel-angular-extrusion, *Scr. Mat.* **36**, 699–705.
3. Gertsman, V.Y., Birringer, R., and Valiev, R.Z. (1995) Structure and strength of submicrometer-grained copper, *Phys. Stat. Sol.* **149**, 243–252.
4. Valiakhmetov, O.R., Galeev, R.M., and Salischev, G.A. (1990) Mechanical properties of titanium alloy VT8 with submicrystalline structure, *FMM* **10**, 204–206 (in Russian).
5. Horita, Z., Furukava, M., Nemoto, M., Tsenev, N.K., Valiev, R.Z., Berbon, P., and Langdon, T.G., (1996) Processing of and Al-Li-Mg alloy with ultra-fine grain size, *Mat.Sci.Forum*, **243–245**, 239–244.
6. Williams, J.C. and Starke, E.A. (1986) The role of thermomechanical processing in tailoring the properties of aluminium and titanium alloys, in Krauss, G. (ed) *Deformation, Processing and Structure*, Proc ASM Sem., 279–354.
7. Lasalmonie, A. and Strudel, J.L. (1986) Influence of grain size on the mechanical behaviour of some high strength materials, *J. Mater. Sci.* **21**, 1837–1852.
8. Rabinovich, M.Kh. and Markushev, M.V. (1994) Effect of grain size on the crack resistance of aluminium alloys, *Metals Science and Heat Treatment* **7–8**, 429–436.
9. Rabinovich, M.Kh. and Markushev, M.V. (1995) Influence of fine grained structure and superplastic deformation on the strength of aluminium alloys. Part 1. The phenomenology of the influence of fine grained structure and superplastic deformation on the strength of aluminium alloys, *J. t .Sci.* **30**, 4692–4702.
10. Rabinovich, M.Kh. and Markushev, M.V. (1996) Influence of fine grained structure and superplastic deformation on the strength of aluminium alloys. Part 2. The physical nature of the influence of fine grained structure on the strength of aluminium alloys, *J. t. Sci.* **31**, 4997–5001.

11. ASM Specialty Handbook, *Aluminium and Aluminium Alloys*, (1993) Davis, J.R. (ed).
12. Horita, Z., Fujinami, T., Nemoto, M., and Langdon, T.G. (1998) Microstructures and mechanical properties of submicrometer-grained Al alloys produced by equal-channel angular extrusion, in *Aluminium Alloys*, Proc. ICAA-6 **1**, 449–454.
13. Markushev, M.V., Bampton, C.C., Murashkin, M.Yu. and Hardwick, D.A. (1997) Structure and properties of ultra-fine grained aluminium alloys produced by severe plastic deformation, *Mat. Sci. Eng.* **A234-236**, 927–931.
14. Nikolaev, G.A., Fridlyander, I.N., and Arbyzov, I.P. (1990) Welded aluminium alloys, Metallurgia.
15. Commercial aluminium alloys (1984) Aliev, C.G., Altman M.B., Ambarcumyuan, S.M., etc. *Metallurgia* (in Russian).
16. Markushev, M.V., Murashkin, M.Yu., Prangnell, P.B., Gholinia, A., and Maiorova, O.A. (1999) Structure and mechanical behaviour of an Al-Mg alloy after equal channel angular extrusion, *Nanostructured Materials* **12**, 839–842.
17. Ferrasse, S., Segal, V.M., Hartwig, K.T., and Goforth, R.E. (1997) Development of a submicrometer-grained microstructure in aluminium 6061 using equal channel angular extrusion, *J. Mater. Res* **12**, 1253–1261.
18. Rabinovich, M.Kh., Markushev, M.V., and Murashkin, M.Yu. (1994) Influence of Superplastic Strain-Heat Treatment on the Mechanical Behaviour of Aluminium-Lithium Alloy 1420 // in T.G.Langdon (ed) *Superplasticity in Advanced Materials*, Proc.ICSAM'94, *Mat. Sci. Forum.* **170–172**, 243-248.
19. Zaharov, V.V. and Rostova, T.D. (1996) Role of shear bands in the Al-Li sheets, *Technologia legkich splavov* **5**, 35–39 (in Russian).
20. Fridlyander, J.N.and Bozich, W. (1998) The properties of semiproducts of 1460 (Al-Cu-Li-Sc) and 1421 (Al-Mg-Li-Sc) at 293K and 77K, in *Aluminum Alloys*, Proc. ICAA–6 **2**, 937–941.
21. Tsenev, N.K., Valiev, R.Z. and Kuzeev, I.R. (1997) Advanced properties of ultra fine-grained aluminium alloys, *Mat. Sci. Forum* **242**, 127–134.

SUPERPLASTICITY OF MECHANICALLY ALLOYED NANOCRYSTALLINE AND AMORPHOUS MATERIALS

O.M. SMIRNOV, JEONG SEUNG, I.V. POUSTOVALOVA
Laboratory of Superplastic Materials Deformation
Moscow State Steel and Alloys Institute
117936, Moscow, Russia

1. Introduction

A wide variety of natural and technical materials exhibit superplastic or superplastic-like flow under external or internal stress in certain temperature regimes. Phenomenological analysis of mass-transfer mechanics producing such flow gives the basis for dividing these materials into two principal groups: (1) polycrystalline superplastic materials which exhibit so called fine-structure superplasticity *i.e.* superplastic flow where mass-transfer units are crystalline grains of size in micro-, submicro- or nanometer range; and (2) materials exhibiting superplastic-like flow where the basic mass-transfer units are either single atoms (molecules) or groups of them. The latter group includes inorganic non-metallic and metallic glass-forming systems and polymers. Amorphous alloys (metallic glasses) with small volume fraction of nanocrystalline dispersed phase are known to have very high mechanical strength and low ductility at low and high temperatures compared with those for crystalline alloys. On the other hand, some recently developed bulk amorphous alloys with large glass-forming ability have shown striking superplastic-like behavior and very high formability in supercooled liquid state which seems to be promising for future development as a new type of superplastic metallic materials [1–3]. One of the most powerful tools for producing bulk amorphous alloys is severe plastic deformation (SPD) in the form of mechanical alloying. Other forms of SPD can be used to yield traditional fine-structure superplastic alloys.

2. Rheological Behavior of Superplastic and Superplastic-like Materials

Rheological behavior of superplastic and superplastic-like materials as well as advantages of industrial application of superplastic deformation (low flow stress, exceptionally high ductility and formability) can be easily explained by the ability of related materials to exhibit viscous flow. All early and recently developed physical and rheological models of superplastic deformation are based on the stress-strain-rate relationship [4–7] which is known to be the essential characteristic of viscous materials. The level of superplasticity is estimated quantitatively by strain rate-sensitivity exponent m in the equation $\sigma_e = K\xi_e^m$, where σ_e is effective flow stress, ξ_e is effective

T.C. Lowe and R.Z. Valiev (eds.), Investigations and Applications of Severe Plastic Deformation, 327–332.
© *2000 Kluwer Academic Publishers. Printed in the Netherlands.*

strain rate, and K is a coefficient. Simple analysis shows that the higher strainrate sensitivity of flow stress the more pronounced superplastic deformation advantages mentioned above. Normal metals and alloys during hot deformation show $m < 0.2$ whereas superplastic materials in optimal conditions have $m > 0.3$. Some materials behave as Newtonian-viscous liquids with m achieving a value of unity. The coefficient K is a structure-sensitive and thermally activated parameter related to the apparent shear viscosity ($\eta_a = \sigma_e/3\xi_e$) of the material as

$$\eta_a = K\xi_e^{m-1}/3. \tag{1}$$

The materials referred to the first group exhibit two resembling types of superplasticity, namely fine-structure superplasticity (FSS) and high-strain-rate superplasticity (HSRS). The nature of viscous flow of both FSS and HSRS materials as well as of other superplastic and superplastic-like materials deals generally with diffusion-based mechanisms of mass transfer at grain boundaries and in their vicinity. From the rheological point of view, viscous flow can be attributed to the liquid-like phase of the material. The behavior of this phase can be characterized by the shear viscosity $\eta_v = K_v \xi_e^{mv-1}/3$ according to Eq. 1. The calculated values of η_v for FSS alloys at optimal SPD temperatures cluster within a relatively narrow gap of two orders in magnitude. As the viscosity, a thermally activated parameter, is significantly influenced by temperature and by grain size as well. Sherby *et al.* [9] and Valiev [10] showed that submicro- or nanocrystalline alloys exhibit superplastic flow at significantly higher strain rates or lower temperatures than microcrystalline alloys of the same composition. So, for a given alloy, the smaller the grain size the lower the viscosity at the given temperature and stress.

The phenomenon of HSRS is rheologically similar to that of FSS [7]. The main difference between them is the value of strain rate. Regular FSS is known to exist at strain rates 10^{-5}–10^{-2} s^{-1}. HSRS is already obtained at strain rates up to 10^2 s^{-1} [11]. With about the same order of flow stress, the apparent viscosity of HSRS flow significantly decreases with increasing strain rate (Table 1).

TABLE 1. Values of shear viscosity η_v for some FSS and HSRS alloys

Alloy	SP type	L, μm	T_{sp}, K	T_{sp}/T_m	η_v, Pa··s	Ref.
Al-33% Cu	FSS	7.9	673	0.828	6.0×10^{10}	[13]
Al-33% Cu	FSS	7.9	723	0.889	1.9×10^{10}	[13]
IM 7475 (Al-Zn-Mg-Cu)	FSS	10.0	695	0.860	5.0×10^{10}	[14]
IM 7475 (Al-Zn-Mg-Cu)	FSS	12.3	789	0.976	5.3×10^9	[14]
IM 7475 (Al-Zn-Mg-Cu)	FSS	19.8	806	0.998	2.5×10^9	[12]
MA IN 9052 (Al-Mg-C-O)	HSRS	0.5	863	0.997	5.0×10^5	[12]
MA IN 905XL (Al-Mg-Li-C-O)	HSRS	0.4	848	0.996	2.0×10^5	[12]
MA IN 9021 (Al-Cu-Mg-C-O)	HSRS	0.5	823	0.996	1.2×10^5	[12]
MMC IN 9021- 5%SiC$_{(p)}$	HSRS	0.5	823	0.950	3.3×10^5	[12]

The realistic explanation of that phenomenon, suggested by Higashi and Mabuchi [12], assumes the presence of an isolated liquid phase at grain boundaries or liquid boundaries at high temperatures as one of the possible accommodation mechanisms for grain boundary sliding. The temperature conditions of HSRS superplasticity are very favorable for appearance of the liquid phase at grain boundaries of HSRS materials. Indeed, for aluminum based alloys of the IM, MA and MMC types, which are the most typical ones to exhibit HSRS, the optimal temperatures of SPD cluster between 0.97 T_m and T_m. Moreover, pronounced adiabatic heating as a result of high strain rates is an additional contributor to the increasing likelihood of the liquid phase at grain boundaries of HSRS materials. Very small grain sizes of HSRS materials, which are typically less than 1 µm, seem to be another contributor to likelihood of the liquid phase at grain boundaries. From the analysis of creep behavior of a two-phase material composed of rigid grains embedded in a contiguous viscous matrix made by Dryden *et al.* [15], the overall viscosity of this composite η_c can be obtained by using the following equation:

$$\eta_c = 2\, \eta_{gb}/(3\, f_l^3) \qquad (2)$$

where η_{gb} is the viscosity of the liquid phase at grain boundaries and f_l is the volume fraction of the liquid phase ($f_l \ll 1$).

Following to Eq. 2, HSRS materials exhibit quite a large effect of decreasing viscosity of two-phase system with increasing f_l. Quantitative estimation of these relationships is made by comparing the values of grain boundary material shear viscosity η_{gb} obtained from experimental data by calculation with help of Eq. 1 and Eq. 2 and of shear viscosity of the melt η_m at given temperatures [16] (Table 2).

TABLE 2. Grain boundary viscosity values of FSS and HSRS aluminum alloys

Alloy	SP type	L, µm	T_{sp}/T_m	$\dot{\epsilon}_{sp}$, s^{-1}	f_l	η_c, Pa·s	η_{gb}, Pa·s	$\eta_m(T_{sp})$,Pa·s	η_{gb}/η_m	Ref.
IM 7475	FSS	10.0	0.860	3.3 × 10^{-5}	10^{-4}	5.0 × 10^{10}	7.9 × 10^{-2}	1.7 × 10^{-3}	46.5	[14]
IM 7475	FSS	12.3	0.976	10^{-3}	8.3 × 10^{-5}	5.3 × 10^{9}	4.5 × 10^{-2}	1.3 × 10^{-3}	3.5	[14]
MAIN905XL	HSRS	0.4	0.996	20	2.5 × 10^{-3}	2.0 × 10^{5}	5.0 × 10^{-3}	1.3 × 10^{3}	3.8	[13]
MA IN9021	HSRS	0.5	0.996	50	2.0 × 10^{-3}	1.2 × 10^{5}	1.5 × 10^{-3}	1.3 × 10^{3}	1.1	[12]

The η_{gb}/η_m ratio shows that the state of grain boundaries of HSRS materials is much closer to liquid melt than that of FSS materials. This comparison gives an additional evidence of the liquid phase at grain boundaries of HSRS alloys which makes them capable of undergoing extensive tensile deformation at extremely high strain rates.

Materials that belong to the second group may be unified by similarity in the mode of mass-transfer units, which are either single atoms (molecules) or groups of them. From this point of view, inorganic glass-forming systems as well as polymers should be included into this group. Inorganic glass may be defined as a product of melt cooled down to the solid state without crystallization. There is a wide range of materials which can be easily solidified without crystallization *i.e.* pure chemical elements such as sulfur, selenium, phosphorus, as well as oxides, halogenides, chalcogenides and their composites which are used to produce regular glasses. Moreover, about 30 years ago, a new kind of glass-forming system was discovered: so called metallic glasses. These materials are known to have excellent mechanical, magnetic and anticorrosive properties. Drastic brittleness is the only negative feature of these materials. But in the supercooled liquid state they exhibit superplastic behavior. Some of these materials can be produced in the form of bulk amorphous bars. Due to that, an example of using them as preforms for producing intricate shapes has already appeared [2].

Two essential features of amorphous alloys may be used in defining potentials of using them as preforms for metal forming *i.e.* glass-forming ability and glass stability. The former is estimated by the critical cooling rate of the melt $(dT/d\tau)_c$ and the reduced glass transition temperature $T_{rg} = T_g/T_m$ where T_g is the absolute glass transition temperature and T_m is the melting point. The latter can be estimated by the complex criterion [17]: $H_r = (T_x - T_g)/(T_m - T_x)$ where T_x is the crystallization temperature. Recent developments have resulted in the creation of metallic glasses with high glass-forming ability, wide supercooled liquid range $(T_x - T_g)$ and high glass stability. These parameters approach sometimes those of regular non-metallic glasses (Table 3).

TABLE 3. Bulk amorphous alloys as compared to some non-metallic glasses

Alloy	T_m,K	T_g,K	T_x,K	T_{rg}	$(T_x - T_g)$,K	H_r	$(dT/d\tau)_c$,K/s	Ref
$Zr_{41}Ti_{14}Cu_{12.5}$-Ni_{10}-$Be_{22.5}$	937	625	770	0.67	145	0.87	<2.0	[18]
$Pd_{40}Ni_{40}P_{20}$	884	537	652	0.61	115	0.50	170.0	[19]
$Zr_{65}Cu_{15}Ni_{10}Al_{10}$	~990	652	~747	0.66	105	0.43	~1.5	[20]
$La_{55}Al_{25}Ni_{20}$	~750	465	545	0.62	80	0.39	100.0	[1]
Fluoride glasses: Ga 1	735	507	625	0.69	118	1.07	<1.0	[17]
Ga 2	729	510	631	0.70	121	1.23	<1.0	[17]
In 2	874	568	662	0.65	94	0.84	<1.0	[17]

Metallic glasses in supercooled liquid state $(T_g < T_{sls} < T_x)$ exhibit Newtonian-viscous flow with strain-rate-sensitivity coefficient $m = 1.0$ for wide range of strain rates. This should lead to extremely high formability of these materials, but up to now such information is limited to the results obtained by A. Inoue, T. Masumoto and colleagues [1, 2, 3]. Elongation up to about 20,000% was obtained by them during tension of $La_{55}Al_{25}Ni_{20}$ at 473 K with a super high strain rate of 5×10^5 s^{-1} [1]. Their results indicate a decreasing apparent viscosity of this material down to 200–10 Pa × s. Such a low viscosity, combined with obvious signs of HSRS shown by this alloy, suggest a very promising new field of metal forming technology.

3. Thermodynamical Analysis of Viscous Flow

Temperature dependence of shear viscosity of the supercooled liquid $\eta(T)$ is the essential feature that can be successfully used in analysis of mass-transfer mechanism as well as in developing technological regimes of forming processes. At temperatures below T_g the viscosity decreases slightly with increasing temperature (Figure 1) resulting in a very low apparent activation energy, which is typical for isoconfigurational viscosity. Transition from the glassy state to the supercooled liquid equilibrium state looks in the $\eta(1/T)$ diagram (see Figure 1) as narrow temperature range of sharp increase in the slope of the curve related to increased apparent activation energy. Isothermal heating of non-equilibrium amorphous material in glassy state results in its relaxation towards equilibrium state (see dotted line 3, Figure 1). Equilibrium viscosity of non-metallic glasses in the supercooled liquid region drops rapidly with temperature decrease up to the melting point. As to metallic glasses, the viscosity drops with increasing temperature up to the onset of crystallization where the viscosity starts to increase sharply and reaches the values similar to that of fine structure polycrystalline materials [7]. So, the minimum in viscosity appears very evidently at T_{vs} inside the supercooled liquid temperature range, signaling the onset of steady state Newtonian-viscous flow. The maximum in free volume fraction was found to occur at the same temperature [21].

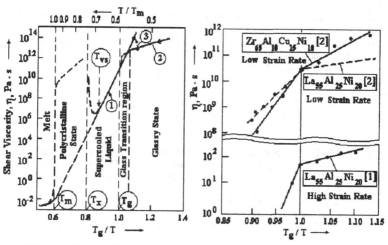

Figure 1. Schematic diagram of temperature dependence of metallic glass viscosity during heating: 1- equilibrium viscosity; 2- isoconfigurational viscosity; 3- relaxation of non-equilibrium configuration.

Figure 2. Temperature dependence of shear viscosity of metallic glasses around glass transition temperature (experimental data are taken from Ref. [1] and [2]).

Schematic representation of thermodynamic features of metallic glasses can be illustrated by a couple of examples. The experimental data of Ref. [1] and [2] on $La_{55}Al_{25}Ni_{20}$ and $Zr_{65}Cu_{15}Ni_{10}Al_{10}$ metallic glasses were recalculated and plotted onto the $\eta(T_g/T)$ diagram (Figure 2). The comparison of the curves show a difference of more than eight orders in magnitude between the values of shear viscosity of supercooled liquid at low and high strain rates. Taking into consideration that atoms move in the elementary act of mass transfer cooperatively in groups, it seems to be possible to calculate the number of atoms n_a making up the unit of mass transfer as the quotient of the value of apparent activation energy of viscous flow divided by the kinetic energy of gram-atom of the material at a given temperature. This is approximately equal to $3RT$ [22]. By estimation of n_a value from the data presented in

332

Figure 2, one can find that mass transfer units in viscous flow of supercooled liquid of $La_{55}Al_{25}Ni_{20}$ at low and high strain rates and $Zr_{65}Cu_{15}Ni_{10}Al_{10}$ at low strain rates consist of 14, 28 and 22 atoms, respectively, which relate to two to three co-ordination spheres. As to isoconfigurational viscous flow ($T<T_g$), n_a value for the same alloys at low strain rates equals to 5 and 9 atoms respectively which seems to be limited by the first co-ordination sphere.

4. References

1. Masumoto, T. (1994) Recent progress in amorphous materials in Japan, *Mater. Sci. Eng.*, A 179/ A 180, 8–16.
2. Inoue, A. and Zhang, T. (1997) Novel superplasticity of supercooled liquid for bulk amorphous alloys, *Mater. Sci. Forum* 243–245, 197–205.
3. Inoue A., Zhang, T., and Takeuchi, A. (1998) Ferrous and nonferrous bulk amorphous alloys, *Mater. Sci. Forum* 269–272, 855–864.
4. Sherby, O.D. and Wadsworth, J. (1989) Superplasticity-recent advances and future directions, *Progress in Mater. Sci.* 33 169–221.
5. Smirnov O.M. (1979) *Working of Metals under Pressure in the Superplastic State*, Moscow, (in Russian).
6. Smirnov, O.M. (1991) Superplastic material behavior affected by structure evolution, *The Japan Soc. for Res. On Superplasticity*, 813–818.
7. Smirnov, O.M. (1997) Superplasticity of metals: phenomenology based on rheo- logical properties and structural dynamics, *Mater. Sci. Forum* 245–245, 443–452
8. Reiner, M. (1958) *Rheology*, Springer Verlag., Berlin - Goettingen - Heidelberg.
9. Sherby, O.D., Nieh, T.G., and Wadsworth, J. (1997) Some thoughts on future directions and applications in superplasticity, *Mater. Sci. Forum* 243–245, 11–20.
10. Valiev, R.Z. (1997) Superplastic behavior of nanocrystalline metallic materials, *Mater. Sci. Forum* 243–245, 207–216.
11. Higashi, K. (1994) Deformation mechanisms of positive exponent superplasticity in advanced Al alloys with nano or near-nano scale grained structures, *Mater. Sci. Forum* 170–172, 131–140.
12. Higashi, K. and Mabuchi, M. (1997) Critical aspects of high strain rate superplasticity, *Mater. Sci. Forum* 243–245, 267–276.
13. Chokshi, A.H. and Langdon, T.G. (1988) The mechanical properties of the superplastic Al-33 Pct Cu eutectic alloy, *Met.Trans.* 19A 2487–2496.
14. Hamilton, C.H., Bampton, C.C., and Paton, N.E. (1982) Superplasticity of high strength aluminum alloys, *Met. Soc. of AIME*, 165–178.
15. Dryden, J.R., Kucerovsky, D., Wilkinson, D.S., and Watt, D.F. (1989) Creep deformation due to a viscous grain boundary phase, *Acta Metall.* 37, 2007–2015.
16. Battezzati, L. and Greer, A.L. (1989) The viscosity of liquid metals and alloys, *Acta metall.* 37, 1791–1802.
17. Surinach, S., Illekova, E., Zhang, G., Poulain, M., and Baro, M.D. (1997) An approach for the assessment of optical fiber drawing temperature of novel fluoride glasses, in *Rapidly Quenched Metastable Materials, Supplement*, 298–301.
18. Masuhr, A., Bush, R., and Johnson, W.L. (1998) Rheometry and crystallization of bulk metallic glass forming alloys at high temperatures, *Mater. Sci. Forum* 269–272, 779–784.
19. Wilde, G., Klose, S.G., Soelner, W., Goerler, G.P., Jeropulos, K., Willnecker, R.W., and Fecht, H.J. (1997) On the stability limits of the undercooled liquid state of Pd–Ni–P, *Mater. Sci. &Eng.* 226–228, 434–438.
20. Kato, H., Kawamura, Y., Inoue, A., and Masumoto, T. (1997) Bulk glassy Zr-based alloys prepared by consolidation of glassy alloy powders in supercooled liquid state, *Mater. Sci. & Eng.* A226–228, 458–462.
21. Myung, W.-N., Kim, H.-G., and Masumoto, T. (1994), Glass transition behavior of Zr- and Ti-based binary amorphous alloys, *Mater. Sci. & Eng.* A179–A180, 252–255.
22. Kaloshkin, S.D. and Tomilin, I.A. (1997) On evaluation of activation energy of amorphous metallic alloy crystallization while heating, *Materialovedenie* 1, 7–13, (in Russian).

STUDY OF COLLECTIVE ELECTRONIC EFFECTS CAUSED BY SEVERE PLASTIC DEFORMATION

A.N. LACHINOV, T.G. ZAGURENKO, V.M. KORNILOV, I.V. ALEXANDROV*
Institute of Physics of Molecules and Crystals, Ufa Scientific Center of Russian Academy of Sciences, K. Marks str., 6, Ufa 450025, Russia.
Ufa State Aviation Technical University, Institute of Physics of Advanced Materials, K. Marks str., 12, Ufa 450000, Russia.

1. Introduction

The severe plastic deformation (SPD) of metals is known to modify their crystal structure. Particularly, the average size of the crystalline grains considerably decreases while the extent of crystal lattice distortion and amorphous phase both increase. Such a state is characterized as non-equilibrium one. When the metal is in this state, it undergoes a polymorphous transformation at temperatures below the melting point (most intensive at $T \sim 0.4\, T_{melt}$).

Such a transformation leads to an increase in the crystalline grain size up to the equilibrium value [1]. It is obvious, that modification of the scale of the metal substructure is accompanied by a modification of its electronic properties. For example, the electrical resistance [2] and the work function increase [3, 4]. Collective electronic effects are predicted, and as a result a metal particle of nanometer size can posses a common electronic shell [5]. In all these appearances it is possible to trace correlation between the structure of a material and its electronic properties. Few experiments however are known in which such correlations are obvious.

This paper presents the results of the investigation of the transition from a non-equilibrium nanostructure in metal to the original equilibrium one by means of measuring the injection current in the metal-insulator interface.

The idea is as follows. In the metal-insulator interface a potential barrier with an area of the space charge appears. Its properties depend on the electronic properties of the contacting materials. When the electric field is applied to the contact, the injection current arises and is characterized by the energy properties of the metal-insulator interface.

Hence, changing, for example, the metal properties we thereby will affect the injection current. If this effect is rather great, it can easily be measured using of standard equipment.

T.C. Lowe and R.Z. Valiev (eds.), Investigations and Applications of Severe Plastic Deformation, 333–337.
© 2000 *Kluwer Academic Publishers. Printed in the Netherlands.*

2. Methods

The experimental cell of the metal-polymer-metal "sandwich" type was used in this work. The nanostructured (Cu and Ni) were used for one of the electrode. The nanocrystalline state in these metals is not equilibrium. Heating to $T \sim 0.4 T_{melt}$ causes structural modifications and restores the original structure. The nanostructure was identified by methods of the electronic microscopy and X-ray structure analysis. As it was shown earlier [5], an ultrafine-grained structure is formed in Cu and Ni as a result of the severe plastic deformation of torsion (SPD) with the crystallite size of about 100 nm, a high level of the crystal lattice microdistortions, non-equilibrium grain boundaries, high density of the grain-boundary dislocations, and increased static and dynamic atom displacements. Recently SPD methods has been successfully used to obtain nanostructures in various metals and alloys [5].Estimations performed according to [3, 6] show that SPD can change material work function by 0.1–0.5 eV and more. It was suggested that low-temperature annealing restores the original state, which will affect the current passing through the NM–P–M system.

A polymer in the M-P-M structure for the polyphthalidilidenbiphenylilene (PPB) film was chosen. The reasons were as follows: 1) the transition into the high conductive state (HCS) in this polymer was extensively studied earlier, 2) PPB possesses excellent film forming properties, which allows us to obtain homogeneous high quality films down to thickness of 100 Å [7], and 3) PPB reveals no temperatures singularities up to the softening temperature (360°C in air), which allows us to use it in the temperature range of most intensive structural modifications in nanostructural copper and nickel. The polymer film was obtained by the centrifugation method from a solution in cyclohexanone.

A 0.5 μm thick film was produced on a flat polished metal electrode of vanadium or copper. The polymer film quality and homogeneity were inspected by TEM and SPM methods.

The measurements were carried out in an electrical circuit containing a large ballast resistance. The experimental cell was placed in a heater allowing heating at a constant rate up to a temperature of 350°C.

3. Experimental results

Figure 1 shows the temperature dependence of the current flowing through the NM–P–M system. When the copper is used as NM, the increase in the temperature does not change the current up to $T_{min} = 146°C$. In the temperature range of 146 °C –173 °C I(T) dependence has a trapezoid shape. When the nickel is used as NM, I(T) dependence is constant up to $T_{min} = 206°C$ and has the trapezoid shape in the range of 206°C –227°C. In the control experiments the samples with nanocrystal structure were substituted with the samples of Cu and Ni with the equilibrium grain structure. The following samples were used in these experiments: 1) The initial large-grained material which was used as raw material for nanocrystal samples obtaining. 2) The nanocrystal samples heat-treated at 500°C during 30 minutes. Such a treatment is known to completely destroy nanocrystal structure [1]. The experiments reveal no current singularities in the investigated temperature range. The evolution of a microstructure of nanostructural Cu and Ni during heating was explicitly investigated in works [8] using mentioned above methods, and also the method of differential scanning calorimetry. The results obtained indicate intensive relaxation processes associated with a reorganization of the defect structure in a course of low-temperature heat-treatment. These processes are observed in temperature intervals of 150–200°C for copper and 180–220°C for nickel. It is important to note that these temperature intervals are close to those in which the above described changes in the conductivity were detected. It allows us to conclude that the mechanisms initiating these transformations are the same and are caused by the temperature transformation of the metal sample microstructure.

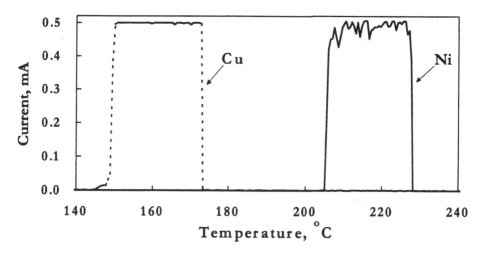

Figure 1. Temperature dependence of current flowing through the polymer film into system nanocrystal metal-polymer film - metal. Solid line – upper electrode is nanocrystal Cu, dashed line – nanocrystal Ni. Polymer film thickness – 2 µm; lower electrode –V; applied voltage – 5 V; heating rate – 8° C/min.

As compared to the results of work [9] there is one significant distinction. When the conductive state appears as a result of the electrode melting, it persists in the whole investigated temperature range. When the nanocrystal material is used as the electrode, HCS exists only in a narrow temperature range. The explanation may be that the contact improvement resulting from the electrode melting promotes HCS stabilization, whereas the electrode structural modifications do not lead to the contact improvement.

The measurements of the current flowing through the NM–P–M system during sequential "heating-cooling" cycles show that the increase in the cycle number leads to the decrease in the electrical charge passed through the sample down to a minimum value that corresponds to the current flowing through the sample at room temperature (Figure 2). It, apparently, reflects that fact, that at identical measurement conditions the relative variation of the work function decreases as the sample approaches the equilibrium state.

It is necessary to pay attention to some peculiarities of the dependence represented in Figure 2. First, during one cycle of heating the sample does not pass into the equilibrium state. The complete transition occurs during several cycles. Secondly, as a

336

rule, the maximum temperature at which current singularities are registered is observed during the first cycles of measurements. Thirdly, as the cycle number increases there is a tendency of decreasing of temperature at which current singularities are measured. Simultaneously, broadening of the temperature range of the effect occurs.

The electrical field must influence injection current. This influence was estimated using an integral I_U (T)dT over (T$_1$, T$_2$) as a function of the voltage applied. This integral is proportional to the net charge passed through the sandwich in the given temperature range. T$_1$ and T$_2$ are the boundary temperatures of the measuring interval, I_U (T) is the temperature dependence of the current when the voltage U is applied to the electrodes.

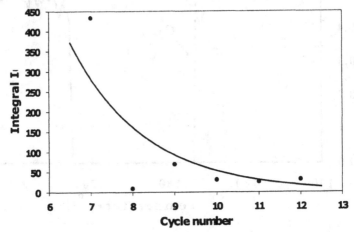

Figure 2. Integral I(T) – cycle number dependence. Film thickness – 2 μm; upper electrode – nanostructural Cu, lower electrode – V, applied voltage – 5V, heating rate – 8° C/min.

4. Conclusion

The analysis of the experimental results showed that for U < 0.1 V the magnitude of the charge flowing through the sample is constant. In the range of voltages of 0.1 V < U < 5 V superlinear increase in the net charge passing through the polymer is observed. Above 5 V the dependence approaches linear. A similar dependence is observed practically in all samples.

The sharp growth of conductivity, when a polymer film of a submicron thickness is used in M–P–M structure, may have trivial explanations such as film inhomogenities, microbreakdowns, piercing of the film resulting from the nanostructural electrode grain size growth, *etc.* Such explanations can be excluded bearing in mind the following facts (besides direct control by TEM and SPM methods): 1) trapezoid shape of the temperature dependence of current; 2) dependence of the switching effect on the voltage applied; 3) PPB modulus of elasticity is equal to 3 GPa whereas the electrode is pressed to the polymer film with the pressure of no more than 100 g/cm^2; 4) the net charge carried through the film decreases with. The dependence of the effect on the voltage applied and the decrease in the net charge carried through the M–P–M system with each consequent heating cycle are in accordance with the familiar model of the injection charge transport. The existence of the threshold voltage, apparently, indicates an essential role of double injection. As a rule the main charge carriers in polymers are holes, therefore the effects observed are caused by the electron injection. This conclusion is in agreement with the results of the experiments on direct injection of electrons into the polymer film [10].

Thus, it was established that structural transformations in nanocrystal metal samples cause collective electronic excitations of large intensity, which can be measured by the modification of injection parameters of metal - polymer contact. Commonly, the form of current - temperature dependence reflects the state of a metal crystal structure. So, the technique of metal state identification can be developed.

5. Acknowledgment

Work was particularly supported by Russian Foundation of Basic Research grant # 98–03–33322

6. References

1. Valiev, R.Z., Alexandrov, I.V., Islamgaliev, R.K. (1998) Processing and properties of nanostructured materials prepared by severe plastic deformation, in G.M. Chow and N.I. Noskova (eds.), *Nanostructured Materials*, Kluwer Academic Publishers, Dordrecht, pp. 121–142.
2. Islamgaliev, R.K., Chmelik, F., Kuzel, R., (1997) Thermal structure changes in copper and nickel processed by sever plastic deformation, Mater. Sci. Eng. A **234–236**, 335 .
3. Craig P.P. (1969) Direct observation of stress-induced shifts in contact potentials, Phys.Rev.Letters, **22**, 700,
4. Gertsman, V.Yu., Birringer, R., Valiev, R.Z., Gleiter, H. (1994) On the structure and strength of ultrafine-grained copper produced by severe plastic deformation, *Ser. Met. Mater.* **30**, 229–234 .
5. Kresin, V.V., Knight, W.D., (1988) Quantized electronic states in metal microclusters: electronic shells, structural effects, and correlations, in V.Z.Kresin (eds), *Pair Correlation in Many-Fermion Systems*, Plenum Press, New York, pp. 245–261.
6. Beams, I.W. (1968) Potentials on rotor surfaces, *Phys. Rev. Letters,* **21**, 1093–1096.
7. Rasmusson, J.R., Kugler, Th., Erlandsson, R., Lachinov, A., Salaneck, W.R. (1996) Thin poly(3,3'-phthalidylidene-4,4'-biphenylene) films studied by scanning force microscopy, *Synthetic Metals*, **76**, 195–201.
8. Krasil'nikov, I.A., Raab, G.I., Kil'mamatov, A.R., Alexandrov, I.V., Valiev, R.Z. (1998) Production and investigation of nanostructured copper, *Physics of Metals and Metallography*, **86**, 491–496.
9. Kornilov, V., Lachinov, A. (1997) Metal Phase in electroactive polymer indused by change in boundary conditions, *Synth. Metals*, **84**, 893–894.
10. Kornilov, V.M., Lachinov, A.N. (1995) Electron beam stimulated phenomena in poly(phthalidylidenarylene)s, *Synth. Metals*, **69**, 589–590.

INFLUENCE OF GRAIN BOUNDARY DIFFUSION FLUXES OF ALUMINUM ON STRENGTH PROPERTIES AND CREEP OF COPPER AND CU- 0.9%VOL AL₂O₃ NANOCOMPOSITE

YU.R. KOLOBOV, K.V. IVANOV, G.P. GRABOVETSKAYA
Institute of Strength Physics and Materials Science, Russian Academy of Science Pr,. Academichesky 2/1, 634021, Tomsk, Russia

1. Introduction

The practical application of nanostructured materials obtained by severe plastic deformation involves two main problems at least. Firstly, there is the instability of non-equilibrium structure connected with high energy accumulated by severe plastic deformation [1]. The instability is the reason for intense softening at relatively low temperature. This temperature is close to room temperature for several nanostructured materials, for example, copper [2]. Secondly, there is the increased diffusivity of nanostructured material state. This leads to high sensitivity to the influence of surrounding. One of the most important type of this action is grain boundary fluxes from an external source, for example, coating [3, 4].

Investigation on possibility to create nanostructured composites in which dispersed particles of oxides, carbides or nitrides are distributed in a nanostructured matrix (having grain size of 0.1 – 0.3 5 μm) are intensively developed at present. It is supposed that nanocomposites may have not only increased mechanical properties relative to nanostructured materials without particles but essentially higher resistance to temperature and surrounding action [5, 6].

Taking into account the above, it is interesting to investigate the influence of temperature and grain boundary diffusion fluxes from a surrounding (coating, for example) on the strength properties of nanocomposite under tension and creep and compare them with respective properties of nanostructured and coarse grained materials

2. Experimental

The specimens of coarse grained (grain size of 10 μm) and nanostructured (grain size of about 0.3 μm) copper as well as Cu-0.9 vol.% Al₂O₃ nanocomposite (matrix grain size of about 0.6 μm, particles size of about 20 nm) were investigated. The coarse grained specimens were obtained by rolling reduction up to 90% at room temperature followed by recrystallization during annealing at 773 K, 30 min. The nanostructure in copper was obtained by equal-channel angular pressing method [7]. The nanocomposite was produced by the internal oxidation method (Glide Cop® material) followed by extrusion and rolling up to 70%. Specimens having dimension of 10·2.5·0.3 mm³ were cut out

T.C. Lowe and R.Z. Valiev (eds.), Investigations and Applications of Severe Plastic Deformation, 339–344.
© *2000 Kluwer Academic Publishers. Printed in the Netherlands.*

340

by electro-spark method. The specimens were coated with aluminum layer of 10 μm in thickness by ion-plasma deposition method.

Transmission electron microscope studies of the Cu-0.9%Al$_2$O$_3$ nanocomposite were carried out by carbon extraction replicas method using EM-125 electron microscope under 100 kV. Tensile tests for the nanostructured copper and nanocomposite were carried out in temperature interval of 293-673 K under strain rate of 3,3·10^{-3} s^{-1}. Creep tests in vacuum and under the influence of grain boundary diffusion fluxes of aluminum from the surface were carried out at 373-473K for the nanostructured copper, at 423 K for the nanocomposite and at 573-673 K for the coarse grained copper. All mechanical tests were performed in vacuum 10^{-2} Pa using PV-3012M machine. Temperature was controlled using a single Pt-Pt(Rh) thermocouple. The temperature drift during testing was less then 1 degree per an hour.

3. Results and Discussion

Structure studies of the Cu-0.9vol%Al$_2$O$_3$ nanocomposite by the replicas method have shown that it the has fine-grained structure (average grain size of 0.6 μm). Annealing of the Cu-0.9vol%Al$_2$O$_3$ nanocomposite at temperature higher than 773 K leads to recrystallization and grain growth of copper matrix. Aluminum oxide particles having size of about 20 nm are uniformly distributed in the nanocomposite volume. Annealing at temperatures less than 1073 K does not change particles size and theirs distribution in the matrix. It should be noted that there are recrystallization and grain growth in the nanostructured copper even after annealing upper 423 K [2].

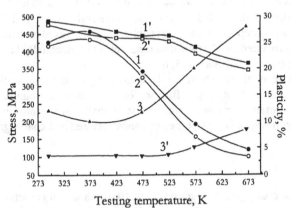

Figure 1. Dependence of strength (1, 1'), yield strength (2, 2') and plasticity (3,3') on tensile temperature for the nanostructured copper (1,2,3 curves) and Cu-0.9 vol.% Al$_2$O$_3$ nanocomposite (1',2',3' curves)

The tensile testing for the nanostructured copper and for the Cu-0.9 vol.% Al$_2$O$_3$ nanocomposite have shown that the flow curves in temperature interval of 293-673 K have ordinary form which are similar to flow curves for coarse grained copper. Figure 1 represents a temperature dependence of strength σ$_B$, yield strength σ$_{0.2}$ and deformation up to failure δ for the nanostructured copper (1, 2, 3 curves, respectively) and for the Cu-0.9 vol.% Al_2O_3 nanocomposite (1', 2', 3' curves, respectively). From figure 1 one can see that at ambient temperature σ_B and $\sigma_{0.2}$ values for the nanostructured copper are close to those for the Cu-0.9 vol.% Al_2O_3 nanocomposite and come to σ_B – 425 and 490 MPa, $\sigma_{0.2}$ – 415 and 476 MPa, respectively. These values are 3 – 4 times more than those for coarse grained copper. The indicated values σ_B and $\sigma_{0.2}$ monotonously

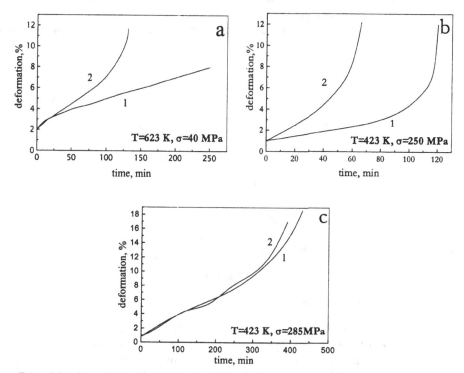

Figure 2. Specimens deformation as a function of time for the coarse grained (a), nanostructured (b) copper and for the Cu-0,9%vol.Al_2O_3 nanocomposite (c) in vacuum (curves 1) and under the grain boundary diffusion fluxes of aluminum atoms (curves 2).

decrease with temperature increase while plasticity δ enhances. It should be noted that σ_B and $\sigma_{0.2}$ for the Cu-0.9 vol.% Al_2O_3 nanocomposite decrease lower than for the nanostructured copper with temperature increase. High thermal stability of mechanical properties of the Cu-0.9 vol.% Al_2O_3 nanocomposite is caused, probably, by Al_2O_3 dispersed particles existence in volume of the nanocomposite. Al_2O_3 particles may block dislocations and grain boundary migration and so resist to materials recrystallization which leads to softening.

Figure 2a shows the creep curves of the coarse grained Cu both in vacuum and under diffusion of aluminum atoms along the grain boundaries at 623 K. It is easy to see that the deformation rate during creep is significantly increased by the grain boundary diffusion fluxes of aluminum (by approximately two times). Creep rate increase of the coarse grained copper under grain boundary diffusion of aluminum is observed in temperature interval of 573-673 K. As it was established in [8,9] this effect is caused by grain boundary sliding activation and increase of its contribution to total deformation

The creep curves of the nanostructured copper and the Cu-0.9 vol.% Al_2O_3 nanocomposite at 373-473 K have usual three-stage shape characteristic for plastic deformation of polycrystalline materials (figure 2b, 2c, curves 1) except the curves at 473K and load less than 140 MPa which have five stages (figure 3). Herewith plasticity of the copper is risen strongly (from 10 up to 50%). One may assume that such creep character is determined by kinetic singularity of process like dynamic recrystallization development.

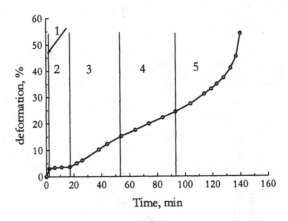

Figure 3. Specimen deformation as a function of time. Creep under σ=127 MPa, T=473 K.

Let us discuss the influence of grain boundary diffusion fluxes of aluminum on the creep of the nanostructured copper and Cu-0.9 vol.% Al_2O_3 nanocomposite. The penetration of aluminum into copper at 373-473 K may be by grain boundaries only because of volume diffusion is "frozen" (extrapolation the aluminum volume diffusion coefficient in copper using Ahrrenius equation gives the D_V=1,8·10^{-25} m^2/s at 473 K. The penetration depth of aluminum into the volume of copper grains does not exceed 1,8·10^{-11} m). The studies have shown that the creep curves of the nanostructured copper under the coating (aluminum layer) and in vacuum are the same. The creep acceleration effect determined as the ratio $\dot\varepsilon_2/\dot\varepsilon_1$ ($\dot\varepsilon_1$ - the creep rate of Cu in vacuum; $\dot\varepsilon_2$ - the creep rate of Cu under conditions of the diffusive contact with Al) for the nanostructured copper under the influence of aluminum grain boundary diffusion fluxes takes place at 423-473K. For coarse grained copper the present effect is observed at 573-673K. The decrease of the effect manifestation temperature is supposed to be caused by the significant increase of diffusion coefficients of aluminum in nanostructured copper beside coarse grained one. It should be noted that the analogous shift of the creep acceleration effect manifestation temperature for nanostructured nickel as compared to coarse grained one is found recently in [3]. Maximum increase of the nanostructured copper creep rate under the action of grain boundary diffusion fluxes (by five times) takes place at 423 K (figure 2b). Plasticity of the nanostructured copper under existence of the diffusion fluxes of aluminum does not change and comes to about 10%. At 473 K plasticity of the nanostructured copper under existence of the diffusion fluxes of aluminum does not increase in comparison with the pure copper tested under the same conditions and comes to 40-50%. However the creep curves of the nanostructured copper under aluminum coating is three-stage in contradiction to the creep curves of pure copper which as mentioned above have five stage shape. Hence the same value of the copper plasticity may be caused by different reasons. For example plasticity of the pure copper is determined by dynamic recrystallization but plasticity of the copper under the influence of grain boundary diffusion aluminum fluxes may increase because of increase of grain boundary sliding contribution to total deformation.

Figure 2c represents creep curves for the Cu-0.9vol%Al$_2$O$_3$ nanocomposite in vacuum (curve 1) and under aluminum grain boundary diffusion (curve 2). Comparing the curves 1 and 2 one can see that grain boundary diffusion fluxes of aluminum practically not influence on the curve shape and creep rate of the Cu-0.9vol%Al$_2$O$_3$ nanocomposite which points to the depressing of creep activation effect by the action indicated.

As mentioned above the creep activation effect of polycrystals by grain boundary diffusion fluxes of impurity from an external source is connected with grain boundary sliding activation and increase of its contribution to total deformation during creep. One may assume that depressing of the effect indicated during the Cu-0.9vol%Al$_2$O$_3$ nanocomposite creep is caused by well known braking action of dispersed particles on grain boundary sliding.

4. Conclusions

1. The creep acceleration effect for the nanostructured copper under the action by grain boundary diffusion fluxes of aluminum atoms from the surface takes place in the interval of lower temperatures as compared to respective interval for the creep of coarse grained copper. Apparently it is caused by the increase of aluminum grain boundary diffusion coefficient in copper being in nanostructured state relative to coarse grained state.

2. Using the Cu-0.9vol%Al$_2$O$_3$ nanocomposite by way of example (Al$_2$O$_3$ particles size ~ 10-20 nm) it is established that the dispersed particles presence leads to the depressing of the effect connected with the action by grain boundary diffusion fluxes.

5. Acknowledgments

The authors would like to express their gratitude to Prof. R.Z. Valiev (UGATU, Ufa, Russia) for supplying materials for the experiments and fruitful discussion. The work is financially supported by Russian Foundation of Fundamental Research (grant N98-02-16517) and Europe commission (grant INTAS N97-1243).

344

6. References

1. Valiev R.Z. (1997) Structure and mechanical properties of ultrafine-grained metals, *Materials Science and Engineering A* **234-236**, 59-66.
2. Korznikov A., Dimitrov O., Korznikova G. (1996) Thrmal evolution of the structure of ultrafine-grained materials produced by severe plastic deformation, *Ann. Chim. Fr.* **21**, 483- 491.
3. Kolobov Yu.R., Grabovetskaya G.P., Ratochka I.V. *et al.* (1996) Effect of grain boundary diffusion fluxes of copper on the acceleration of creep in submicrocrystalline nickel, *Ann. Chim. Fr.* **21**, 483-491.
4. Kolobov Yu.R., Grabovetskaya G.P., Ratochka I.V., Ivanov K.V. (1998) Creep features and diffusion parameters of submicrocrystalline metals, *Russian Physics Journal* **3**, 77-82.
5. Sauer C., Weißgärber T., Dehm G. *et al.* (1998) Dispersion strengthening of copper alloy, *Z. Metallkd.* **2**, 119 –125.
6. Sekino T., Niihara Q. (1995) Nanostructural characteristics and mechanical properties for Al2O3/metal nanocomposites, *Nanostructured Materials* **5-8**, 663-666.
7. Utyashev F.Z., Enikeev F. U., Latysh V.V. (1996) Comparison of deformation methods for ultrafine-grained structure formation, *Ann. Chim. Fr.* **21**, 379-389.
8. Kolobov Yu.R. (1998) Diffusion controlled processes on grain boundaries and plasticity of metal polycrystals, Nauka, Novosibirsk.
9. Kolobov Yu.R., Ratochka I.V., Marvin V.B. (1998) Diffusion controlled processes on grain boundaries and plasticity of metal polycrystals, in V.E. Panin (ed.), *Physical mesomechanics of Heterogeneous Media and Computer*, pp.249-264.

V. FUTURE HORIZONS FOR SEVERE PLASTIC DEFORMATION MATERIALS: APPLICATIONS AND COMMERCIALIZATION

OVERVIEW AND OUTLOOK FOR MATERIALS PROCESSED BY SEVERE PLASTIC DEFORMATION

T.C. LOWE[1], Y.T. ZHU[1], S.L. SEMIATIN[2], D.R. BERG[1]
[1]Los Alamos National Laboratory, Los Alamos, New Mexico, USA
[2]Air Force Research Laboratory, Wright-Patterson AFB, Ohio USA

Abstract

The status and prospectus for severely deformed materials are overviewed from the perspective of interrelated trends in materials characterization, modeling, and processing. Technological drivers for accelerating the development of these materials are summarized. Technological barriers to understanding and applying severely deformed materials are highlighted. Recommendations are made to direct future work, including facilitation of more rapid commercialization.

1. Introduction

The science and technology of materials subject to severe plastic deformation (SPD) have evolved largely during the past 30 years. During this period substantial advances have been made, yet they do not appear to have been systematically overviewed, evaluated, and reported in the open literature. The purpose of this overview is to establish a clear understanding of the status of the severe plastic deformation field, analyze recent trends, identify barriers to further advancements, and propose key areas of opportunity for additional research and technology commercialization. We expect this review to better enable researchers and their sponsors to set priorities for future work. We also will elucidate the maturity of severe plastic deformation technologies so that those interested in applications or commercialization will better understand near term and long term potentials.

The recent emergence of research on nanocrystalline materials and their remarkable properties has accentuated the interest in severe plastic deformation as a process by which the grain size of conventional metals and alloys can be made submicrocrystalline (grain diameter of 100 nm – 1000 nm) or nanocrystalline (grain diameter of 1-100 nm). This overview is limited in scope to the science and technology of severe plastic deformation to produce these classes of materials. Because of the brevity of this particular work, only key highlights and selected references will be discussed.

Large strain deformation has been studied in a variety of contexts including: metal forming, machining, ball milling of powders, wear of high friction surfaces, localized phenomena near the tips of propagating cracks, fatigue, and superplasticity. Research in any of these areas has contributed to the overall knowledge of how severe plastic deformation evolves the microstructures of metals and alloys, leading to the novel

T.C. Lowe and R.Z. Valiev (eds.), Investigations and Applications of Severe Plastic Deformation, 347–356.

properties that we are beginning to understand and exploit. Though large deformations have been employed to shape and form metals for centuries, only recently have various facets of deformation processing, materials characterization, and computational modeling advanced sufficiently together to enable systematic development of these materials.

The unfolding of scientific research typically follows certain patterns and occurs in identifiable stages. At any given time, scientists and technologists work simultaneously over a range of these stages, from conception or discovery, to initial probing and inquiry, to expanding investigation and analysis, all the way through to applied research, application, technology development, and in some cases, to commercialization. The distribution of effort and relative levels of activity within each phase is indicative of the maturity of a field of research and suggestive of what kinds of work are most critical to the ongoing emergence of knowledge and technology development. Thus, this overview will summarize historical contributions in the archival literature, patents, and the most recent results that are being published today.

2. Procedure for literature analysis

The principal method of investigation in this work is literature acquisition and analysis using computer-based knowledge discovery tools. These tools have been developed at Los Alamos National Laboratory to explore, analyze, and utilize published scientific information [1]. Bodies of knowledge published within the past 25 years in the international patent literature and in over 200 international technical journals were examined. Additional insight, plus synthesis and conclusions were guided by the collective experience of the authors' work in the field.

Our results are presented in five parts: summary of overall body of knowledge, SPD processing methods, characterization of microstructures and properties of SPD materials, modeling of SPD processes and microstructures, and applications and commercialization of SPD.

3. Results

3.1. SUMMARY OF OVERALL BODY OF KNOWLEDGE

The historical origins of many technologies can usually be traced back to a series of incubating discoveries or developments, rather than a single event. For example, aspects of SPD materials processing appear in the work of Bridgman in the 1940s and 1950s [2]. This work helped establish the basic knowledge of material response to combinations of high pressure and severe deformation. Modified Bridgman dies were used for the first time to fabricate ultrafine-grained microstructures in the late 1980s [3,4]. Another catalyzing contribution that gave birth to severe plastic deformation processing of materials was the work of Segal et al.[5], inventor of equal channel angular extrusion (ECAE). Segal proposed and demonstrated the concept of subjecting large volumes of material to simple shear in order to modify their microstructure and enhance properties. The most highly cited work on severe plastic deformation that has spawned the development of ECAE and related techniques for producing ultrafine grain materials is that of Valiev, *et al.*[6]. Thus this review focuses on research since Segal published his concept in the international archival literature in 1981, with special attention to the majority of the research published since the contribution of Valiev, *et al.* [6] in 1993.

Research on topics directly or indirectly related to producing ultrafine grain materials by severe deformation was published in this same time period. This worked helped further establish the framework for advancing development the field. However, we will not address this body of knowledge in this review. For instance, research by Cohen and coworkers on strain hardening and microstructure evolution during wire drawing [7,8] typified efforts within the past 30 years to examine the behavior of materials subject to increasingly larger uniform strains. Efforts to model severe plastic deformation also emerged in the early 1980s, for example research to develop finite element methods suited for simulating metal forming processes such as backwards extrusion [9].

3.1.1. Analysis of the literature

At least 197 articles on severe plastic deformation have been published in archival journals since 1981. The annual rate of publication has grown steadily in recent years, as shown in Figure 1. This accelerating growth indicates the increasing significance and success of severe plastic deformation technology.

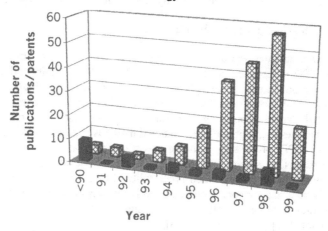

Figure 1. Growth in publications (cross-hatch) and patents (solid black) on severe plastic deformation

The greatest portion of SPD research has been conducted during the past 10 years. The most comprehensive and widely cited work on the structure and properties of ultrafine grain materials produced by SPD is that of Valiev, Korznikov, and Mulyukov [9] in 1993. In this research, a compendium of experimental techniques were applied to determine the important role of grain boundaries on elastic, strength, superplastic, damping, and magnetic properties of ultrafine grain materials. As we will discuss later, progress in this field critically depends on the ability to create samples under carefully controlled conditions and meaningfully characterize their properties.

The distribution of research effort between the following subfields illustrates the current level of development of severe plastic deformation technology. These subfields are: 1) synthesis and processing, 2) characterization of microstructures, 3) characterization and measurement of properties, 4) modeling, and 5) application of SPD materials. At the time of writing of this review, the number of publications addressing five sub-topical areas pertaining to severe plastic deformation materials was distributed as shown in Table 1.

TABLE 1. Distribution of topical emphasis in publications on severe plastic deformation

Topic	Number of publications
Microstructures	144
Properties	78
Synthesis and processing methods	44
Modeling	18
Applications and productization	18

The largest amount of published research has addressed observation and analysis of microstructure. These observations are usually connected with measurement of associated properties. Mechanical properties have been the most widely studied, followed by physical properties such as density, thermal conductivity, thermal stability, corrosion resistance, electrical conductivity, and magnetic properties.

3.2. SPD PROCESSING METHODS

Equal channel angular extrusion or pressing is currently the single most widely researched SPD process, being the focus of 31% of all SPD publications. Alternative processing methods using severe plastic deformation are increasingly being explored. Variants of Segal's equal channel angular extrusion method are most common, for example [10]. Novel processing techniques are emerging, for example hybrids of powder techniques [9], high pressure torsional methods [12], roll bonding [13], and ball milling [14]. SPD processing is being investigated in an ever increasing range of materials systems, as shown in Table 2. Aluminum and aluminum alloys have been most widely studied because of their great utility of fine grain materials for superplastic forming. Though the concepts underlying SPD processing were developed for metals and alloys, they have been applied for processing of intermetallics [15], polymers [16], semiconductors [17] and composites [18].

3.3. CHARACTERIZATION OF MICROSTRUCTURES AND PROPERTIES OF SPD MATERIALS

Characterization of microstructures and properties of SPD materials has been the most pervasive focus of SPD research. The microstructural characteristics most studied include grain size, grain boundary structure, dislocation substructure, texture, and local atomic structure. X-ray diffraction and optical and electron microscopy are the most commonly used tools for microstructural analysis of SPD materials. However, neutron scattering [19], Raman [20], Mössbauer [21], and XAFS [22] techniques have also been employed.

TABLE 2. Distribution of publications on severe plastic deformation by material studied

Material	Number of publications
Aluminum or alumium alloys	58
Copper or copper alloys	48
Nickel or nickel alloys	28
Iron or iron alloys	24
Titanium or titanium alloys	10
Intermetallics	11
Semiconductors	7
Composites	7
Palladium or palladium alloys	7
Magnetic materials & rare earths	5
Polymers	3
Other metals: Ag, Nb, Co, Zn, Mg, Li,	11

3.3.1. Microstructure

The microstructures of SPD-processed materials depend on the SPD processing method. Two most common variants of the SPD technique are ECAE and High Pressure Torsion (HPT) straining [10, 16]. For ECAE, multiple passes are often used to produce the ultrafine-grained microstructures. The rotation of material billets around its longitudinal axis between sequential ECAE passes affects the development of microstructures. Four rotation routes have been used, for example as demonstrated in [21]. These routes are: route A: the billet is not rotated; route B_A: the billet is rotated 90° clockwise and counterclockwise alternatively; route B_C: the billet is rotated 90° clockwise; and route C: the billet is rotated 180°. For all routes except route C, shear bands are observed in optical micrographs. The shear bands become denser with increasing number of passes. For route C, the shear bands disappear at each even numbered pass.

Submicron-sized grains are observed under TEM within the shear bands [23]. After the first ECAE pass, subgrains with low-angle grain boundaries form inside the original coarse grains. These subgrains are refined and their misorientations increase with increasing number of passes, forming high-angle grain boundaries. The fraction of high angle grain boundaries increases with increasing number of passes. The grain boundaries are usually not clearly defined, indicating that they are in a non-equilibrium state [23-25]. High internal stress exists inside grains. Dislocation arrangement and density inside grains depends on the grain size [23, 25]. Smaller grains have fewer dislocations in them while large grains may contain subgrains with low angle misorientations.

The effectiveness of grain refinement increases from routes A to B_A to C to B_C. In all cases, grains become more equiaxed with increasing passes because of dynamic recrystallization. For the same number of passes, varying ECAE route from A to B_A to C to B_C generates grains that become increasingly equiaxed [23]. The grain size distributions generally follow log-normal statistics, with the average grain size determined by the dynamic balance of grain refinement and recrystallization. Overall, the microstructures are very uniform through the cross-section of an ECAE-processed billet. However, local microstructural variation may exist [24].

Similar to ECAE [26], the HPT process initially produces cellular microstructures with low-angle misorientations inside initial large grains. The misorientation across some low-angle grain boundaries then increases with increasing strain, forming high-angle grain boundaries, whose quantity increases with increasing strain. However, the HPT process is more effective in reducing the grain size and can result in higher defect density than ECAE. At large plastic strains, HPT processing led to non-distinct microstructures [26] with a large quantity of dislocation debris inside grains. In addition, HPT produces strong textures and microstructures that may not be uniform along the radius of sample disc [12, 18].

3.3.2. Mechanical Properties

The microhardness and strength of SPD-processed materials increase with increasing number of ECAE passes or HPT turns. However, the ductility usually decreases with increasing strength. For example, ultrafine grained Ti processed by ECAE has a tensile strength of 790 MPa, which is 72% higher than coarse-grained Ti, while its elongation to failure is reduced from 27% to 15% [18]. In compression the SPD-processed materials behave much like cold-worked materials under a wide range of strain rate [27], exhibiting high strain-rate sensitivity and little work-hardening.

3.4.   MODELING OF SPD PROCESSES AND MICROSTRUCTURES

Modeling SPD materials has focused largely on the mechanics of processing methods and to a much lesser extent on describing the resulting microstructures.

3.4.1.   Process modeling

A number of approaches have been taken to model the deformation that occurs during severe plastic deformation processes such as equal channel angular extrusion (ECAE) and HPT. These include descriptions based on slab, slip-line field, finite-element, and physical models, among others. Predictions of deformation patterns, loads, and the effect of tooling design on metal flow have been developed from such analyses.

Segal, the inventor of ECAE, performed the first analysis of the kinematics and loads involved in the process using the slip line field method [5]. Iwahashi, Furukawa, and their coworkers [28,29] extended the work of Segal, et al. [5] for the frictionless case in which the channels intersect gradually such that deformation occurs as material elements pass through a centered-fan of specified angle. They also explored the formation of texture and other effects of multi-pass deformation. The finite-element-modeling (FEM) work of Prangnell, et al. [30] and DeLo and Semiatin [31] provided further insight into the effect of die-geometry and friction on metal flow. DeLo and Semiatin also modeled ECAE under nonisothermal conditions.

We have yet to see sufficiently quantitative analysis of severe torsional deformation under high pressure. The approximations used in the few existing models for this SPD process (discussed in [30]) have limited ability to address the complexities of variation in pressure distribution and radial material flow between the confining anvils.

3.4.2.   Microstructure modeling

Though microstructures produced by large deformations have been studied extensively, efforts to model SPD-induced microstructures have been fairly limited, for example [33]. This work by Valiev, et al. addresses defect structures and long range stress fields near grain boundaries in ultrafine grain materials. There have been no quantitative models of how microstructures evolve during SPD. Similarly, though texture evolution has been studied experimentally, there are no published simulations of texture evolution.

Deformation-induced microstructures resulting from very small strains have been modeled [34], but the capability to model the large deformations and complex deformation histories typical of SPD has yet to be developed.

3.5. APPLICATIONS AND COMMERCIALIZATION OF SPD

Review of United States, European, and Japanese patent activity on fine grain materials illustrates the commercial prospects for SPD materials. The earliest relevant patent activity began in Russia in the mid-1970's with the work of Segal and coworkers. However, the largest body of relevant patents lies in the domain of superplastic forming. The commercial value of superplasticity has long been established for alloys with micron-sized grains. Technological developments in superplasticity were extensively patented during the 1980's and these patents continue to grow in number. Outside of Russia, substantial growth in patent activity specific to SPD materials and processing didn't occur until the early-1990's (see Figure 1), again beginning with the U.S. patents of Segal and coworkers in the United States. Though nine (non-Russian) patents relevant to SPD were issued prior to 1990, none of these claim grain refinements below 1000 nm. The key areas that have been covered include the basic concepts of equal channel angular forming, design and construction of apparatus for ECA processing, extensions or variants of ECA processing, material-specific developments of ECA processing, application-specific adaptations of ECA processing, and superplasticity of ECA materials. In total there are fewer than 40 patents relevant to SPD materials.

Overall, the growth in SPD-specific patent activity parallels the growth in published SPD-specific research. This fact suggests that the value of the intellectual property in the field is well recognized by contributing individuals and institutions. Since commercialization of patented technologies typically lags patents by several years, it is reasonable to expect that the earliest significant commercial products may begin to appear soon. However, as material-specific and application-specific patents are emerging, we can optimistically anticipate reduction to commercial practice. Initial applications will exploit the exceptional strength of SPD materials or their superplastic formability. From a pessimistic viewpoint one could speculate that it could be yet another 10 years before we see many uses of SPD materials. This may be attributable to the observation that new materials often require 10-20 years of development before they appear in widespread commercial use. Though the pace of technology development has been accelerated by the information age, the cycle time for evolving new materials has not yet been reduced substantially.

One indicator of technology maturity is the extent to which technology marketing companies have begun promoting an emerging technology area in which they see notable potential. This phenomenon has not yet begun for SPD materials. However, venture capitalists and a few large corporations are currently investing in SPD technology.

4. Discussion

The field of SPD materials has several defining characteristics that provide a partial basis for directing future research. First, it is notable that the technology appears to have been dormant for the first ten years since its introduction in the archival literature in 1981. It appears that the emergence of Russian scientists following the collapse of the Soviet Union has been a significant factor in the expansion of SPD technology. Also, there is a substantial portion of SPD research that appears only in conference proceedings and formats other than archival journals. Though these formats are useful for communication to restricted audiences, they do not address the broader research communities necessary to develop cross-institutional coordination of research or alliances with industrial interests.

The distribution of emphasis of SPD reported in Table 1 suggests that development of SPD technology is still in an early stage. The greatest amount of work is still on characterizing SPD material microstructures and properties. Results of early characterization efforts have been limited by the lack of samples of sufficient size or homogeneity to conduct systematic studies of SPD process variables and their effects on microstructure and properties. Improved process controls and research on the heterogeneity in SPD-processed materials has resolved these early issues.

Though the next greatest research effort has been on developing processing methods, much of this work still focuses on the original ECAE and severe plastic torsional straining techniques. Fortunately, process modeling efforts have been particularly helpful in increasing the understanding of alternative ECAE processing methods. The further development of finite element process modeling capabilities to better quantify the effects of die temperature, die friction, temperature-dependent flow, deformation rate effects, microstructure evolution, and fracture will greatly accelerate the development of SPD materials and applications.

Table 2 illustrates the very limited effort that has been directed towards composites. Further development of this class of materials may require advancement of hybrid processing methods, for example using combinations of powder compaction and ECAE. It is not yet clear which combinations of deformable constituent phases may be effectively processed by SPD to obtain properties that are superior to monolithic materials.

Models to interpret and predict SPD material microstructures are critical to further understand the dynamics of microstructure refinement. In most cases, the grain structure of SPD materials is ultrafine, but rarely truly nanocrystalline. Models of microstructure and microstructure evolution may guide the optimization of grain size and other microstructural features.

Commercial applications demand more research into manufacturing engineering issues pertinent to scale-up and evaluation of economic feasibility. For instance, because of the nature of the less highly deformed ends in ECAE-processed billets, product yield and process economics are an important challenge. Similarly, tooling design must be developed for ECAE to accommodate the effects of large hydrostatic stresses. Efforts directed toward die life extension may become critical to ensuring the economic viability of SPD material processing. Severe torsional deformation offers certain advantages for obtaining greater refinement of grain size, but methods for scaling this technique to production scale require additional development. The earliest applications of SPD materials will necessarily be in areas which benefit significantly from enhanced performance. Niche markets which can sustain high initial material and manufacturing costs will likely develop first. Current examples include ultrafine grain metals for sputtering targets, nearly-nanocrystalline titanium for medical prosthetics, and production of flat ultrafine grain billets appropriate for high rate superplastic forming.

5. Conclusion

SPD materials and processing methods have evolved substantially during the past thirty years. However, the field is still young and offers significant challenges. The most critical research contributions have been published within the past ten years. For the field to advance substantially, a greater array of research tools and methods will need to be applied. Prominent among these is computational modeling of SPD processing and of SPD material microstructure evolution. Because of the complexity of SPD processing, greater coordination of research capabilities across research institutions is needed to conduct more comprehensive analyses of SPD materials. The preponderance of research that has been directed towards aluminum and its alloys make this material system a more likely candidate for cross-institutional research.

The commercial viability of SPD materials is supported by a substantial patent portfolio, but has not been substantially realized in commercial products. The economics of SPD processing needs to be addressed. This is particularly true in view of the potential for competing processes for making nanocrystalline metals to emerge more quickly than the present rate at which SPD manufactuing processing is evolving.

6. References

1. Luce, R.E. (1994) Shaping the library of the future: digital library developments at Los Alamos National Laboratory's research library, *Los Alamos National Laboratory Technical Report*, LA-UR-94-3307.
2. Bridgman, P.W. (1952) Studies in Large Plastic Flow and Fracture, McGraw-Hill, New York.
3. Smirnova, N.A., Levit, V.I., Pilyugin, V.P., Kuznetsov, R.I., and Degtyarev, M.V. (1986) Low temperature recrystallization pecularities of nickel and copper, Metal Physics and Metallography, 65, 3, 566-570 (in Russian)
4. Abdulov, R.Z., Valiev, R.Z., Krasilnikov, N.A. (1990) Formation of submicrometre-grained structure in magnesium alloy due to high plastic strains. Mater. Science Letters,9,1445-1447.
5. Segal, V.M., Reznikov, V.I., Drobyshevskiy, A.E., and Kopylov, V.I. (1981) Plastic working of metals by simple shear, *Russ. Metall* 1, 99-105.
6. Valiev, R.Z., Korznikov, A.V., and Mulyukov, R.R. (1993) Structure and properties of ultrafine-grained materials produced by severe plasti c deformation, *Matls Sci Engg A*,168, 141-148.
7. Langford, G. and Cohen, M. (Sept 1969) Strain hardening of iron by severe plastic deformation, *ASM Transactions* 62, 623-638.
8. Biswas, C., Cohen, M., and Breedis, J.F. (1973) Strain hardening of titanium by severe plastic deformation, *Institute of Metals Monogr Rep Ser*, 1, 16-20.
9. Mori, K. , Osakada, K., and Fukuda, M. (1983) Simulation of severe plastic deformation by finite element method with spatially fixed elements, *Intl. J. of Mechanical Sciences*, 25, 775-783.
10. Segal, V.M., Hartwig, K., and Goforth, R.E. (1997) In situ composites processed by simple shear, *Matls Sci Engg A* 224, 107-115.

356

11. Xiang, S. Matsuki, K., Takatsuju, N., Tokizawa, M, Yokote, T., Kusui, J., and Yoko, K. (1997) Microstructure and mechanical properties of PM 2024Al-3Fe-5Ni alloy consolidated by a new process, equal channel angular pressing, *J Matls Sci Letters* **21**, 1725-1727.

12. Alexandrov, I.V., Zhu, Y.T., Lowe, T.C., Islamgaliev, R.K., and Valiev, R.Z. (1998) Consolidation of nanometer-sized powders using severe plastic torsional straining, *Nanostructured Matls* **10**, 45-54.

13. Saito, Y, Tsuji, N., Utsunomiya, H., Sakai, T., Hong, R.G. (1998) Ultra-fine grained bulk aluminum produced by accumulate roll-bonding (ARB) process, *Scripta Materialia* **39**, 1221-1227.

14. Murty, B.S. and Ranganathan, S. (1998) Novel materials synthesis by mechanical alloying/milling, *International Materials Review* **43**, 101-141.

15. Semiatin, S. L., Segal, V.M., Goetz, R. L., Goforth, R.E., Hartwig, T. (1995) Workability of a gamma titanium aluminide alloy during equal channel angular extrusion, *Scripta Met et Mater* **33**, 535-540.

16. Sue, H.J., Li, C.K.Y. (1998) Control of orientation of lamellar structure in linear low density polyethylene via a novel equal channel angular extrusion process, *J Matls Sci Letters* **17**, 853-856.

17. Islamgaliev, R.K., Kuzel, R., Mikov, S.N., Igo, A.V., Burianek, J., Chmelik, F., and Valiev, R.Z. (1999) Structure of silicon processed by severe plastic deformation, *Matls Sci Engg A* **266**, 205-210.

18. Alexandrov, I.V., Zhu, Y.T., Lowe, T.C., Islamgaliev, R.K., and Valiev, R.Z. (1998) Microstructure and properties of nanocomposites obtained through SPTS consolidation of powders, *Metallurgical and Materials Transactions A* **29**, 2253-2260.

19. Maier, Ch., Blaschko, O., and Pichl, W. (1997) Influence of uniaxial deformation on the martensitic phase transformation in Li and LiMg, *Physica B* **234-236**, 126-128.

20. Islamgaliev R.K, Kuzel R., Obraztsova E.D., Burianek J., Chmelik F., and Valiev R.Z. (1998) TEM, XRD and Raman-scattering of germanium processed by severe deformation, *Matls Sci Engg A* **249**, 152-157.

21. Teplov, V.A., Pilyugin, V.P., Chernyshev, E.G., Gaviko, V.S., Klejnerman, N.M., and Serikov, V.V. (1997) Formation of Fe-Cu and Fe-Bi nonequilibrium solid solutions under large plastic deformation and subsequent heating, *Fizika Metallov I Metallovedenie* **84**, 82-94.

22. Babanov, Y.A., Blaginina, L.A., Mulyukov, R.R., Musalimov, R.S., Shvetsov, V.R., Ryazhkin, A.V., Sidorenko, AF, and Fadyushina N.V (1998) EXAFS study of short-range order in submicron-grainedcopper produced by severe plastic deformation, *Fizika Metallov I Metallovedenie* **86**, 47-52.

23. Ferrasse, S., Segal, V.M., Hartwig, K.T., and Goforth, R.E. (1997) Microstructure and Properties of Copper and Aluminum Alloy 3003 Heavily Worked by Equal Channel Angular Extrusion, *Metall. Mater. Trans.* **28A**, 1047-57.

24. Furukawa M., Iwahashi Y., Horita,Z., Nemoto, M., Tsenev, N.K., Valiev, R.Z. and Langdon, T.G. (1997) Structural Evolution and the Hall-Patch Relationship in an Al-Mg-Li-Zr Allor with Ultra-fine Grain Size, *Acta Mater.*, **45**, 4751-57.

25. Valiev, R.Z., Kozlov, E.V., Ivanov, Yu.F., Lian, J., Nazarov, A.A., and Baudelet, B. (1994) Deformation Behavior of Ultra-fine-grained Copper, *Acta Metall. Mater.* **42**, 2467-75.

26. Valiev, R.Z., Ivanisenko, Yu.V., Rauch, E.F., and Baudelet B. (1996) Structures and Deformation Behavior of Armco Iron Subjected to Severe Plastic Deformation, *Acta Mater.* **44**, 4705-12.

27. Gray III, G.T., Lowe, T.C., Cady, C.M., Valiev, R.Z., and Alexandrov, I.V. (1997) Influence of Strain Rate & Temperature on the Mechancial Response of Ultrafine-Grained Cu, Ni, and Al-4Cu-0.5Zr, *NanoStructured Mater.* **9**, 477-80.

28. Iwahashi, Y., Wang, J., Horita, Z., Nemoto, M., and Langdon, T.G. (1996) Principle of equal channel angular pressing of ultra-fine grained materials, *Scripta Mater* **35**, 143-146.

29. Furukawa, M., Iwahashi, Y., Horita, Z., Nemoto, M., and Langdon, T.G. (1998) The shearing characteristics associated with equal-channel angular pressing, *Mater Sci Eng A* **A257**, 328-332.

30. Prangnell, P.B., Harris, C., and Roberts, S.M. (1997) Finite element modeling of equal channel angular extrusion, *Scripta Mater* **37**, 983-989.

31. DeLo, D.P. and Semiatin, S.L. (1999) Finite-element modeling of nonisothermal equal-channel angular extrusion, *Metall Mater Trans A* **30A**, 1391-1402.

32. Kuhlmann-Wilsdorf, D., Cai, B.C., and Nelson, R.B. (1991) Plastic flow between Bridgman anvils under high pressures, *J Mater Res* **6**, 2547-2564.

33. Valiev, R.Z., Pshenichnyuk, A.I., and Nazarov, A.A. (1994) Structural model of ultrafine grained materials produced by severe plastic deformation, *Key Engineering Materials* **97-99**, 59-64.

34. Devincre, B. and Kubin L.P. (1997) Mesoscopic simulations of dislocations and plasticity, *Matls Sci Engg A* **234**, 8-14.

INFLUENCE OF SEVERE PLASTIC DEFORMATION ON THE STRUCTURE AND PROPERTIES OF ULTRAHIGH-CARBON STEEL WIRE

DONALD R. LESUER[1], CHOL K. SYN[1], OLEG D. SHERBY[2]
[1] Lawrence Livermore National Laboratory, Livermore, CA 94551
[2] Stanford University, Stanford, CA 94305

Abstract

Ultrahigh-carbon steel wire can achieve very high strength after severe plastic deformation, because of the fine, stable substructures produced. Tensile strengths approaching 6000 MPa are predicted for UHCS containing 1.8%C. This paper discusses the microstructural evolution during drawing of UHCS wire, the resulting strength produced and the strengthening mechanisms active in these materials. Drawing produces considerable alignment of the pearlite plates. Dislocation cells develop within the ferrite plates and, with increasing strain, the size normal to the axis (d) decreases. These dislocation cells resist dynamic recovery during wire drawing and thus extremely fine substructures can be developed (d < 10 nm). Increasing the carbon content reduces the mean free ferrite path in the as-patented wire and the cell size developed during drawing. For UHCS, the strength varies as d^{-n}, where n is 0.5 to 1. The influence of processing and composition on achieving high strength in these wires during severe plastic deformation is discussed.

1. Introduction

Extensive plastic deformation resulting from wire drawing is commonly used to produce steel wires with high strengths. Typically these steels are eutectoid and hypereutectoid steels and drawing strains up to 4 are used during processing. At Lawrence Livermore National Laboratory, we have been studying the processing, structure and properties of steels with very high carbon content. These steels, which are called ultrahigh-carbon steels (UHCSs), have a hypereutectoid composition up to the limit of carbon solubility in austenite (approximately 2.1%C) [1]. The potential for increasing the strength of wires is illustrated in Fig. 1, which shows wire strength as a function of carbon content. The wire diameter (0.28 mm) and the amount of cold work is constant for all data points given as solid circles. The UHCSs are projected to have wire strengths in excess of 4000 MPa. These ultrahigh strength UHCS wires are expected to be significantly stronger than the 3400 MPa wire currently used in premium tires. At 1.8% C, the UHCS wires are projected to have a strength approaching 6000 MPa. This paper discusses the microstructural evolution during drawing of UHCS wire, the resulting strength produced and the strengthening mechanisms active in these materials.

T.C. Lowe and R.Z. Valiev (eds.), Investigations and Applications of Severe Plastic Deformation, 357–366.
© 2000 Kluwer Academic Publishers. Printed in the Netherlands.

358

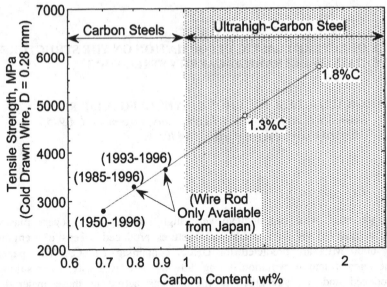

Figure 1. Influence of carbon content on the strength of cold drawn steel wire. Current premium radial tires use high carbon steel wire with a strength of 3400 MPa. The projected strengths for ultrahigh carbon steels containing 1.3% and 1.8%C are shown in the figure by the open circle symbols

2. Processing and microstructural evolution

2.1. PROCESSING BEFORE WIRE DRAWING

Achieving high strength and good ductility in UHCS requires breaking up the deleterious pro-eutectoid carbide network and developing ultra-fine carbides in pearlitic form. The formation of this pro-eutectoid carbide network can be influenced by composition, control over phase transformations and thermo-mechanical processing. Ochiai and his co-workers [2-5] have investigated the influence of composition and cooling rate during patenting on the conditions which lead to a proeutectoid carbide network. Patenting involves rapid heating to austenite followed by rapid transformation to a fully pearlitic structure. Alloying additions that were found to inhibit the formation of a network include Si and Co. In the cooling rate studies, the tendency to form a proeutectoid carbide network was studied in a 0.2Si-0.5Mn steel as a function of carbon content and cooling rate. High cooling rates tend to suppress the formation of a network. In addition, as the carbon concentration in a hypereutectoid steel is increased, faster cooling rates are required to obtain pearlitic structures without a network. In this study, most of the cooling rates were achieved by lead patenting. The authors concluded that hypereutectoid steels with carbon content up to 1.10% could be patented to obtain a network-free microstructure using commercially-available cooling equipment at cooling rates of 10-15°C/s.

2.2. MICROSTRUCTURAL EVOLUTION DURING WIRE DRAWING

Several investigators have studied the evolution of microstructure during drawing of mildly hypereutectoid steels [6-8] and iron [7,9]. The starting microstructure for wire drawing of eutectoid and hypereutectoid steels is fine pearlite with randomly oriented lamellae. The fine pearlite results from a patenting heat treatment. The mean free ferrite path in a eutectoid steel after patenting is typically 70 nm. Increasing the carbon content will decrease the mean free ferrite path.

The evolution of this microstructure during wire drawing is shown schematically in Fig. 2. Drawing produces considerable alignment of the pearlite plates parallel to the drawing direction and reduction in the mean free ferrite path. The ferrite also develops a <110> wire texture typical for BCC metals. Throughout the wire drawing process, the carbide plates deform to strains comparable to the ferrite plates and the carbide plates are also observed to fracture but not as much as might be expected. Recent work has also shown that partial dissolution of the cementite phase can occur during severe plastic deformation of pearlitic steels [10]. Dense dislocation tangles form in the ferrite and, at small drawing strains (approximately .25), dislocation cell walls form. These cell walls contain fragmented carbide particles. Transmission electron microscopy studies reveal that very few dislocations are found within the cells and a high dislocation density is found in the cell walls. With continuing deformation, these cells become thinner and resist extensive dynamic recovery. Embury et al. [7] have shown that the width of the cells scales as the diameter of the wire. The result is the development of a fine, stable substructure with very fine cell dimensions normal to the wire axis (e.g. < 10 nm).

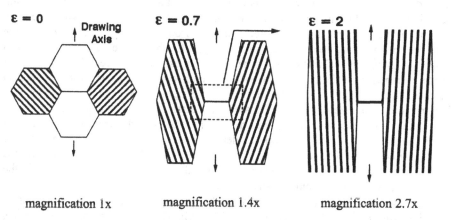

magnification 1x   magnification 1.4x   magnification 2.7x

Figure 2. Schematic drawing showing the development of microstructure during drawing of eutectoid composition steel. Typical drawing strains are indicated in the figure.

3. Properties and Strengthening Mechanisms

The dominant deformation resistance in UHCS wire depends on the structure and substructure developed during processing. Previous studies of eutectoid and hypereutectoid steels [11, 12] with pearlitic microstructures have shown that, in the absence of severe plastic deformation, the yield strength is derived from hardening contributions associated with pearlite colony size, interlamellar carbide spacing and solute additions. The pearlite colony size and the interlamellar carbide spacing represent the dimensions of microstructural features that impose barriers to dislocation motion. These strengthening mechanisms contribute to the yield strength in an additive manner and, for eutectoid and hypereutectoid steel with pearlitic microstructure, the following equation has been derived.

$$\sigma_y = (\sigma_o)_{ss} + 145(D_s^*)^{-1/2} + 460L^{-1/2} \tag{1}$$

where σ_y is the yield strength, $(\sigma_o)_{ss}$ is the resistance to dislocation motion resulting from solid solution atoms, $(D_s^*)^{-1/2}$ is the carbide spacing (interlamellar spacing) and L is the pearlite colony size.

After severe plastic deformation, additional strengthening mechanism(s) are introduced. A number of investigators have studied the increase in strength that results from cold drawing of eutectoid and hypereutectoid steel wires [3, 6, 13-17]. Most of these studies have reported strength as a function of wire diameter. Data from seven such materials are shown in Fig. 3. Five of these materials are hypereutectoid composition wires and two of the materials are eutectoid composition wire (piano wire and the data of Kim). The starting strength in all these studies is derived from a fully pearlitic microstructure produced as a result of a patenting treatment. The strengthening produced by patenting to obtain these microstructures represents an important starting point for the very high strengths typically observed in severely drawn wire. The influence of cold wire drawing on the strength of hypereutectoid steels can be understood through further analysis of the data in Fig. 3. The increase in strength resulting from cold wire drawing was calculated as a function of drawing strain by subtracting the strength in the patented condition from the strength of the wire in the cold-drawn condition. Results are shown in Fig. 4 for three hypereutectoid steels with similar compositions. These materials had different as-patented strengths and different starting diameters. The results fall on a common curve suggesting that a common mechanism is responsible for the increase in strength by wire drawing. It is also important to recognize that, even after a drawing strain of 4, the strength of the wire is continuing to increase with increasing strain. These high work hardening rates at high strains are in contrast to the work hardening rates in FCC metals [7], which are substantially lower. The reduced work hardening rates in FCC metals at large strains result from more extensive dynamic recovery and thus a slower rate of reduction in the size of dislocation substructures relative to iron.

Embury and Fisher [6] have studied the principal strengthening mechanisms resulting from cold wire drawing in a Fe-.93C-.2Si-.37Mn wire. As discussed above, wire drawing develops a stable, cellular substructure consisting of narrow, oriented dislocation cells which are the dominant source of strengthening in these materials. In Fig. 5, the variation of stress with cell size (measured normal to the drawing axis) is shown. The wire drawing produced very small cell sizes (10 nm and less). The yield strength of the wire was found to vary as the inverse square root of the width (d) of the cells. Thus, in general terms,

$$\sigma_y = \sigma_o + k_y d^{-n} \tag{2}$$

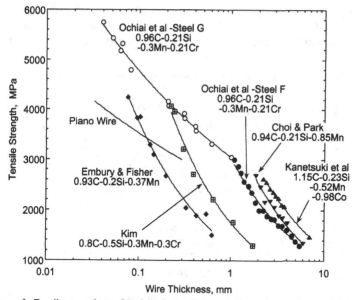

Figure 3. Tensile strength as a function of wire diameter during wire drawing for eutectoid and hypereutectoid steels

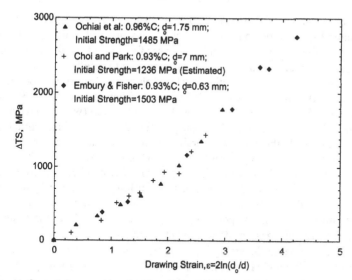

Figure 4. Strength increment as a function of wire drawing strain in hypereutectoid steels with similar composition

362

where σ_y is the yield strength, σ_o is the strength from all sources other than dislocation cells, and k_y and n are constants. The strengthening in these severely drawn wires with stable substructures resulted from the cell walls acting as barriers to dislocation motion.

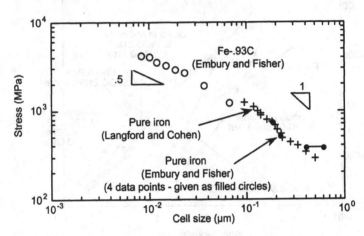

Figure 5. Stress versus cell size for drawn Fe-.93C (data of Embury and Fisher[6]) and pure iron. The pure iron data is from the two investigations (Embury and Fisher[6] and Langford and Cohen[9]).

The contribution of a stable cellular structure to strengthening has also been studied by Langford and Cohen [9] for pure iron. In contrast to the work of Embury and Fisher, the yield strength was found to vary as an inverse linear function of the dislocation cell size (d^{-1}). The dependence of flow stress on d^{-1} was theorized to result from the work of deformation required to generate the length of dislocation line necessary to produce the imposed deformation. A comparison of the Embury and Fisher data on a Fe-.93C steel (which showed $d^{-1/2}$) and the Langford and Cohen data on pure iron (which showed that strength varied as d^{-1}) is shown in Fig. 5. Clearly different slopes are appropriate for the two data sets, although there is an overall continuity in the data for the two investigations. It is important to note, however, that, in addition to the Fe-.93C alloy, Embury and Fisher also studied a commercially pure iron (Ferrovac E) and concluded that the yield strength varied also as $d^{-1/2}$. Despite these differences, the results in Fig. 5 suggest a change in the dominant deformation resistance or significant contributions from other deformation mechanisms upon decreasing the cell size. Over the range of cell sizes studied, the strength varies as d^{-n} with n between 0.5 and 1.

Two observations relative to the strength levels shown in Fig. 5 are relevant to the various mechanisms of strengthening. First, the initial (as patented) strength before wire drawing is higher in the hypereutectoid steel than in pure iron. Clearly this difference arises from the strengthening effects of the carbide plates in the pearlitic steel. By analogy with the experimental work of Taleff *et al.* [11, 12], one might expect that the strength of a pearlitic steel results from the sum of strengthening contributions from different barriers to dislocation motion. Thus for a severely worked pearlitic steel,

$$\sigma_y = (\sigma_o)_{ss} + \sigma_{pearlite} + \sigma_{colony} + \sigma_{cell} \qquad (3)$$

where $(\sigma_o)_{ss}$, $\sigma_{pearlite}$, σ_{colony} and σ_{cell} represent the resistance to dislocation motion resulting from solid solution additions $((\sigma_o)_{ss})$, pearlitic plate spacing $(\sigma_{pearlite})$, pearlite colony size (σ_{colony}) and dislocation cell size (σ_{cell}). The $(\sigma_o)_{ss}$, $\sigma_{pearlite}$, σ_{colony} terms for hypereutectoid steels are given in equation (1). The σ_{cell} term is the dominant source of strengthening in Fig. 4 (assuming in a severely drawn wire that the yield strength equals the tensile strength). The data in Fig. 4 suggests that for severely drawn wire the cell size dominates the strength of the wire, and the pearlite spacing and pearlite colony size are secondary contributors. Furthermore, this view agrees with the continuous nature of the data shown in Fig. 5, which compares a pearlitic structure wire with a totally ferritic structure wire. The σ_{cell} data in Fig. 4 is for a single composition hypereutectoid steel. It is informative to examine the strength increment due to drawing for all the steels shown in Fig. 3. The results are shown in Fig. 6, which shows a good correlation between the incremental increase in strength and the drawing strain. There is some scatter in the data; the general trend, however, is that steels with higher carbon content have higher strength increments for a given amount of drawing strain.

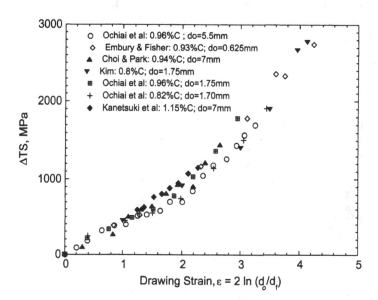

Figure 6. Strength increment as a function of wire drawing strain for the eutectoid and hypereutectoid steels shown in Fig. 3.

The second observation that can influence the strength levels in Figs. 3, 5 and 6 is the size of the initial cells that form. As noted by Embury and Fisher, who compared dislocation substructures produced in commercially pure iron and Fe-.93C (with both coarse and fine pearlite), the presence of carbide plates produced a smaller initial cell size in the ferrite. In addition, reducing the interlamellar spacing through suitable heat treatment, reduced the initial cell size. Thus, by analogy, increasing the carbon content for a given carbide-to-carbide distance in pearlite will reduce the mean free ferrite path and thus the initial cell size. Since the cell thickness scales with wire diameter during drawing [6], smaller cell sizes (and higher strengths) are possible for a given drawing strain. These conclusions are consistent with experimental observations by Ochiai [3, 5] on eutectoid and hypereutectoid steel wires in that both the as-patented strength and the work hardening rate increased with increasing carbon content.

3.1. COMPARISON WITH OTHER HIGH STRENGTH FIBERS

The wire strengths predicted here (*e.g.* the 5000 MPa wire strength shown in Fig. 1) represents a very high value - exceeding 20% of the theoretical cohesive strength of steel. Figure 7 compares the strength of these highly drawn UHCS wires with the strength of bulk materials as well as other fibers that are used as reinforcement in composite materials. The highly drawn UHCS wires are predicted to be significantly stronger than bulk metals used in structural applications but more importantly the strength of these wires compares favorably with those of other reinforcing fibers such as S-glass, Kevlar-49 and carbon fiber. The figure also shows the cost of the reinforcing fibers; UHCS wire (with an estimated as-processed cost of $.60/pound) is significantly less expensive than other reinforcing fibers. Thus one can expect significant commercial applications with the ultrahigh strength hypereutectoid steel wires described here.

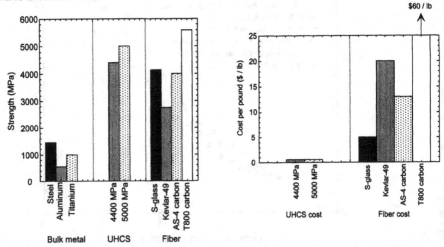

Figure 7. Comparisons of strength and cost of UHCS wire with other bulk structural metals and reinforcing fibers used in composite materials.

4. Summary and Concluding Remarks

This paper has reviewed the processing, microstructure evolution and the predicted resulting strength of UHCS subjected to severe plastic deformation during wire drawing. Mechanisms of strengthening have been discussed. Important conclusions are as follows.

• Very high strengths are possible in hypereutectoid steel wire. At 1.8% C, hypereutectoid steel wires are projected to have a strength approaching 6000 MPa.

• Severe plastic deformation during wire drawing of hypereutectoid steels results in considerable alignment of the pearlite plates and the development of a substructure that resists dynamic recovery.

• The yield strength in hypereutectoid steels has been shown to result from additive strengthening contributions from solid solution additions, pearlite spacing, pearlite colony size and dislocation cell size. For severely drawn wire, the strengthening due to dislocation cells can dominate the strength of the wire. The strength varies as the inverse square root of the cell size.

• Increasing the carbon content reduces the mean free ferrite path in the as-patented wire and the initial cell size developed during drawing. This results in a higher flow stress in the as-patented wire and a higher work hardening rate.

• Achieving high strength requires a) eliminating the continuous carbide network that can form during cooling from temperatures in the austenite phase field, b) a starting microstructure of fine pearlite, c) development of a stable dislocation substructure within the ferrite and d) avoiding fracture.

5. Acknowledgments

This work was performed under the auspices of the U. S. Department of Energy by Lawrence Livermore National Laboratory under contract No. W-7405-ENG-48. Support was provided in part by the Laboratory Directed Research and Development Program.

6. References

1. Lesuer, D.R., Syn, C.K., Goldberg, A., J. Wadsworth, and Sherby, O.D., (1993) The case for ultrahigh carbon steels as structural materials, *JOM* **45** (8), 40-45.
2. Ochiai, I., Ohba, H., Yohji, Y. and Nagumo, M. (1988) Effect of central segregation on drawability of high carbon steel wire rod manufactured from continuously cast blooms, *Tetsu-to-Hagane* **74**, 1625-1632.
3. Ochiai, I., Nishida, S., Ohba, H. and Kawana, A. (1993) Application of hypereutectoid steel for development of high strength steel wire, *Tetsu-to-Hagane (J. of Iron and Steel Inst. of Japan)* **79**, 89-95.
4. Ochiai, I., Nishida, S. and Tashiro, H. (1993) Effects of metallurgical factors on strengthening of steel tire cord, *Wire Journal International* **26**, 50-61.
5. Ochiai, I., Nishida, S., Ohba, H., Serikawa, O. and Takahashi, H. (1994) Development of ultra-high strength hypereutectoid steel wires, *Materia Japan (Bulletin of Japan Inst. of Metals)* **33**, 444-446.
6. Embury, J.D. and Fisher, R.M. (1966) The structure and properties of drawn pearlite, *Acta Metallurgica* **14**, 147-159.
7. Embury, J.D., Keh, A.S. and Fisher, R.M. (1966) Substructural strengthening in materials subject to large plastic strains, *Transactions of the Metallurgical Society of AIME* **236**, 1252 - 1260.
8. Langford, G. (1970) A study of the deformation of patented steel wire, *Metallurgical Transactions* **1**, 465 - 477.
9. Langford, G. and Cohen, M. (1969) Strain hardening of iron by severe plastic deformation, *Transactions ASM* **62**, 623-638.
10. Languillaume, J., Kapelski, G. and Baudelet, B. (1997) Cementite dissolutiobn in heavily cold drawn pearlitic steel wire, *Acta Materiala*, **45** (3), 1201 - 1212.
11. Taleff, E.M., Syn, C.K., Lesuer, D.R. and O.D. Sherby, (1997) A comparison of mechanical behavior in pearlitic and spheroidized hypereutectoid steels, in *Thermomechanical Processing and Mechanical Properties of Hypereutectoid Steels and Cast Irons*, D.R. Lesuer, C.K. Syn, and O.D. Sherby, Eds. Warrendale, PA: TMS,.

366

12. Taleff, E.M., Syn, C.K., Lesuer, D.R. and Sherby, O.D. (1996) Pearlite in ultrahigh carbon steels: heat treatments and mechanical properties, *Metallurgical Transactions* **27A**, 111-120.
13. Lesuer, D.R., Syn, C.K., Sherby, O.D. and Kim, D.K. (1996) Processing and mechanical behavior of hypereutectoid steel wire, in *Metallurgy, Processing and Applications of Metal Wires*, H. G. Paris and D. K. Kim, Eds. Warrendale, PA: TMS, 109-121.
14. Lesuer, D.R., Syn, C.K., Sherby, O.D., Kim, D.K. and Whittenberger, W.D. (1997) Mechanical behavior of ultrahigh strength, ultrahigh carbon steel wire and rod, in *Thermomechanical Processing and Mechanical Properties of Hypereutectoid Steels and Cast Irons*, D. R. Lesuer, C. K. Syn, and O. D. Sherby, Eds. Warrendale, PA: TMS, 175-188.
15. Kanetsuki, Y., Ibaraki, N. and Ashida, S. (1991) Effect of cobalt addition on transformation behavior and drawability of hypereutectoid steel wire, *Iron and Steel Inst. of Japan International* **31**, 304-311.
16. Kim, D.K. and Shemenski, R.M., US Patent 5,167,727, 1992.
17. Choi H.C. and Park, K.T. (1996) The effect of carbon content on the hall-petch parameters in the cold-drawn hypereutectoid steels, *Scripta Materialia* **34**, 857-862.

THE DEVELOPMENT OF ULTRAFINE-GRAINED TI FOR MEDICAL APPLICATIONS

V.V. STOLYAROV[1], V.V. LATYSH[1], R.Z. VALIEV[1], Y.T. ZHU[2], T.C. LOWE[2]
[1]Ufa State Aviation Technical University, Institute of Physics of Advanced Materials, St. K. Marks, 12, Ufa, 450000, Russia
[2]Materials Science and Technology Division, Los Alamos National Laboratory, Los Alamos, NM 87545, USA

Abstract

Bulk samples of commercially pure (CP) titanium were processed by severe plastic deformation (SPD), namely, equal channel angular (ECA) pressing in combination with thermal mechanical treatment, to produce ultrafine-grained (UFG) microstructures. It is shown that by altering SPD processing parameters one can form distinct types of UFG microstructures, which differ in grain shape, size and intragrain defect density and defect distribution. These UFG microstructures in CP titanium result in strength and fatigue properties similar to those of Ti-alloys. We are thus able produce CP titanium suitable for replacing Ti-6Al-4V in biomedical and orthopaedic applications such as dental implants, medical instruments, and trauma fixation devices (nails, plates, screws, and wires).

1. Introduction

Ti and Ti alloys are widely used in medical applications, particularly in traumatology [1,2,3]. The medical community throughout the world has great interest in these materials because of their superior physical, chemical and mechanical properties. First, Ti and Ti alloys have excellent corrosion resistance in many corrosive mediums, including human body fluids. The excellent corrosion resistance results from the dense oxide film that forms on the surface of Ti and Ti alloys. The oxide film resists both uniform and localized corrosion, providing corrosion resistance superior to many other materials, including stainless steels. Second, Ti and Ti alloys have an outstanding biocompatibility with human tissues. They do not cause allergic reactions as do stainless steels and Co-Cr alloys. Ti dissolves very little in bio-fluids. More importantly, dissolved Ti ions are not toxic and remain localized in the tissue near the Ti implants. The international medical community generally considers Ti as a non-toxic metal [1,4]. Third, Ti and Ti alloys have high specific strength and low elastic modulus, which are highly favorable for orthopaedic and traumatology applications. The elastic modulus and specific density of Ti and Ti alloys are about half of that of steels and Co alloys.

T.C. Lowe and R.Z. Valiev (eds.), Investigations and Applications of Severe Plastic Deformation, 367–372.

Currently, Ti-6Al-4V is the most common alloy used for medical applications. It was first developed for the aerospace industry. Although it is very biocompatible as compared with other alloys, it is not without shortcomings. Its alloy elements, Al and V, are known to be toxic [5], and may potentially cause series of ailments including cancer, causing great concerns among medical experts. It is also found experimentally that Ti alloy implants do not fix to tissues as well as pure Ti. Though pure titanium is chemically inert, coarse-grained pure titanium lacks the strength needed for implants. Therefore, the development of new technologies to enhance mechanical properties of CP Ti is very important for medical applications.

One way to improve the mechanical properties of metallic materials such as CP-Ti is to refine the grain size. Our recent work has demonstrated the great potential of SPD processing for refining the grain-size and improving the mechanical properties of metallic materials [6, 7, 8]. The objective of this paper is to report on the results of our investigation using SPD techniques to process CP-Ti and to discuss possible medical applications of ultrafine grained CP-Ti.

2. Material and experimental procedure

Hot-rolled CP Ti rods, with a weight percentage of impurities as 0.12% O, 0.18%Fe, 0.07%C, 0.04%N, 0.01%H, were used as the starting material. The initial CP Ti rods had a diameter of 40 mm and mean grain size of 15 μm. UFG nanostructured states in CP Ti were obtained by a combination of ECA pressing [6-8] and other thermal mechanical treatments [9]. Samples obtained following these process steps had dimensions up to 32 mm in diameter and more than 100 mm in length. Longitudinal and transverse sections of these samples were characterized by TEM and X-ray analysis. Tensile tests were conducted on samples with a gage length of 25 mm and a gage section diameter of 5 mm at strain rate $4 \times 10^{-2} s^{-1}$. Microhardness was measured using a PMT-3 indentor under a load of 100 g and a duration of 10 sec. Fatigue properties were determined by an accelerated method using step loading [10] with fully reversed rotating bending (R = -1) at a frequency of 100Hz for 2×10^7 cycles in air. The fatigue samples had a gage dimension of $\varnothing 7.5 \times 25$ mm

3. Experimental results

3.1. MICROSTRUCTURE

Three UFG nanostructured states (1, 2, 3), significantly different in their microstructures and mechanical properties, were obtained using specific processing parameters (temperature, strain rate, strain degree, number of ECAP passes and routes) and subsequent thermomechanical treatments. They had different grain size and shape, dislocation density, volume fraction of high angle grain boundaries and mechanical properties.

In UFG nanostructured state 1 the CP Ti has an equiaxed grain structure with a mean grain size 0.25 μm and dislocation density up to 10^{12}-$10^{13} m^{-2}$. In a typical selected area diffraction (SAD) pattern taken from a $2 \mu m^2$ area one can see a large number of diffraction spots uniformly arranged in circles, which is typical for UFG materials processed by SPD. This indicates the existence of high fraction of high-angle grain boundaries. The azimuth spreading of spots indicates high internal stresses.

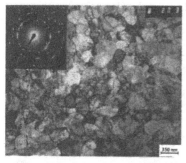

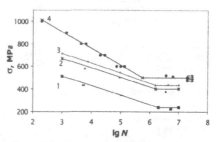

Figure 1. TEM microphotograph of UFG microstructure in CP Ti (state 3)

Figure 2. S - N curves for CP Ti with (1) coarse grained, (2) UFG equiaxed, (3) fibrous and (4) fragmentary microstructures.

The UFG nanostructured state 2 has a fibrous grain structure in the longitudinal direction of the billet, a rather small mean grain size of 0.2 μm, and a higher dislocation density 10^{14}-10^{15}m^{-2}. SAD patterns show more diffraction spots than the UFG state 1 and indicate the existence of both high and low angle boundaries. The UFG nanostructured state 3 (Fig. 1) has the smallest grain size of 0.15 μm and the highest strength (table 1). It has a dislocation density of 10^{13}-10^{14} m^{-2}. Some small grains are free of dislocations. Both high-angle and low-angle grain boundaries exist. Fringes are shown near some grain boundaries (Fig. 1), which indicates the relief of internal stresses and the transition of non-equilibrium grain boundaries to a partly equilibrated state. The grains in UFG state 3 are more equiaxed than UFG states 1 and 2.

3.2. MECHANICAL PROPERTIES

Mechanical properties including microhardness, tensile properties and fatigue limit are presented in Table 1. Microhardness in the initial coarse-grained CP Ti is 1800 MPa. The microhardness is much higher in all UFG states than in the initial coarse-grained state, especially in the UFG state 3. The UFG state 3 also has a very high ultimate strength of 1100 MPa, which is 139% higher than in the initial coarse-grained state, while the maintaining decent elongation to failure of 9%. This ductility is sufficient for most medical applications of CP-Ti. Note that CP Ti in the UFG nanostructured state 3 has higher strength and ductility, comparable with the widely used Ti-6Al-4V ELI alloy [3, 11].

3.3. FATIGUE PROPERTIES

As seen from Table 1, the fatigue limit of the UFG CP-titanium strongly depends on its microstructural state. The UFG nanostructured state 3 has a fatigue limit of 500 MPa, which is 100% higher than coarse-grained Ti. This fatigue limit is comparable to that of high strength titanium alloy Ti-6Al-4V ELI (515 MPa) [3, 11]. The increase of the fatigue limit correlates with the increase of microhardness and ultimate and yield strength. One can see from the S-N curves (Fig. 2) that UFG nanostructured titanium, has higher strength than the coarse-grained Ti in both multiple cyclic and low cyclic fatigue tests.

TABLE 1. Microhardness, tensile mechanical properties and fatigue limit of titanium in different states.

State (structure type)	Hv, MPa	US, MPa	YS, MPa	δ, %	RA, %	Fatigue limit, MPa
Coarse-grained (15μm)	1800	460	380	26	60	238±10
UFG nanostructured state 1 (equiaxed)	2700	710	625	14,0	60	403±8
UFG nanostructured state 2 (fibrous)	2821	960	725	10	45	434±5
UFG nanostructured state 3 (fragmentary)	2850	1100	915	9	40	482±8 500*
Ti-6Al-4V ELI (annealed)**		965	875	10-15	25-47	515

NOTE: * - tested under cyclic conditions at constant stress amplitude.
**- upper limit of mechanical properties [3, 11]

3.4. APPLICATIONS IN TRAUMATOLOGY AND ORTHOPAEDICS

The excellent mechanical properties of UFG Ti processed by SPD make it suitable for fabricating biocompatible medical implants such as hip and knee prostheses; trauma fixation devices such as nails, plates, screws and wires; orthopaedic and dental implants; and medical instruments. For example, in trauma operations, metal plates and screws are widely used to fix broken bones. Such metal implant plates must have high compressive and bending strength and sufficient plasticity. Fatigue strength does not play a significant role because of the short duration (1-3 months) of these plates in human body. The analyses of different implant-plate constructions for boned osteosynthes has resulted in the design and processing of series of implant plates from UFG CP Ti (Fig.3).

Another application of UFG CP Ti is a special conic screw for spine fixation (Fig.4). Due to the great cyclic stresses during its service, the screw must have high fatigue strength. In this case, all advantages of UFG CP Ti, namely, the high static and fatigue strength and excellent biological compatibility, are fully utilized. Note also that the fatigue properties of the screw can be further improved by redesigning the screw to better match the unique properties of UFG CP Ti. The same sort of redesign is appropriate for manufacturing UFG-Ti dental implants.

With the advance of the SPD techniques to process larger Ti rods, larger medical implants may be made of the UFG Ti. Shown in Fig. 5 are some examples.

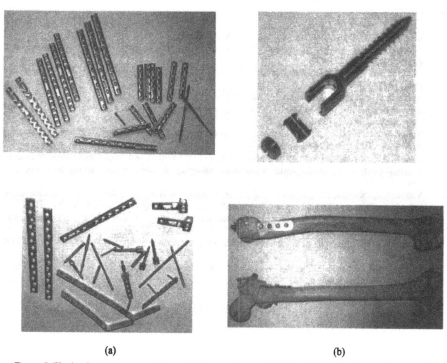

(a)   (b)

Figure 5. The implants (a) processed from UFG nanostructured CP Ti and its application at hip break (b).

4. Summary

UFG microstructures were formed in CP Ti processed by SPD. These microstructures significantly improved the static strength and fatigue strength of commercial purity Ti. The enhancement in properties was sufficient to makes SPD-processed CP Ti a viable alternative to Ti-6Al-4V for medical implants applications. The chief advantage of this new material is the absence of toxic Al and V alloying elements that are currently present in Ti alloy prosthetics. Future efforts will focus on further enhancing properties, optimizing processing parameters, and minimizing processing cost.

5. Acknowledgement

The authors are grateful for funding for this work provided by the U.S. Department of Energy NIS-IPP program, and executed through a joint collaboration between the Institute for Physics of Advanced Materials at Ufa State Aviation Technical University and Los Alamos National Laboratory (Contract #87880017-35).

372

6. References

1. Igolkin, A.I. (1993) Titanium in medicine, *Titanium* 1, 86-90.
2. Wang K. (1996) The use of titanium for medical applications in the USA, *Materials Science and Engineering* A213, 134-137.
3. Niinomi, M., (1998) Mechanical properties of biomedical titanium alloys, *Materials Science and Engineering* A243, 231-236.
4. Agins, H.J., Alcock, N..W., Bansal, M., Salvati, E.A. (1988) *J.Bone Joint Surg.* 70 (3), 347-356.
5. Steinemann S.G., Perren S.M., (1984) Titanium alloys as metallic biomaterials - *Proc. of the fifth world conf. on titanium*, v.2, 1327-1334.
6. Valiev, R.Z., (1996) ed., Ultrafine-grained materials prepared by severe plastic deformation, *Special Issue, Ann. Chim. Fr.* 21 (6-7), 369-480
7. Valiev, R.Z., Korznikov, A.V., and Mulukov, R.R. (1993) Structure and properties of ultrafine-grained materials produced by severe plastic deformation, *Materials Science and Engineering* A168 (2), 141-148.
8. Valiev, R.Z., Kozlov, E.V., Ivanov, Yu.F., Lian, J., Nazarov, A.A., Baudelet, B., (1994) Deformation behaviour of ultrafine-grained copper, *Acta Metallurgica and Materialia*, 42 (7), 2467-2475.
9. Valiev, R.Z. *et al.* (1998)Nanostructured Materials for Applications in Medicine, Reports submitted to Los Alamos National Laboratory.
10. Lokati, L. (1995) Le prove di fatica come ausilio alla prodettazione ed alla predusone, *Met. Ital.*, 47 (9) 832-837.
11. Brown, S.A., Lemons, J.E., (eds) (1996) *Medical applications of titanium and its alloys*, ASTM STP 1272, ASTM, Coshohochen, PA.

COMPRESSIVE BEHAVIOR OF SEVERELY PREDRAWN STEEL WIRE

ETIENNE AERNOUDT
Department of Metallurgy and Materials Engineering
Katholieke Universiteit Leuven – Belgium

Abstract

Severely drawn steel wire is nanostructured. This leads to very high strengths thus opening a number of interesting engineering applications. Several of these applications require a change in strain path subsequent to the drawing strain. This has important influences on material flow, on the occurrence of shear bands and on fatigue behavior, hence on residual ductility during subsequent forming and on product performance. The underlying reasons are the very strong crystallographic and structural anisotropy as well as the polarity of the dislocation arrangements introduced by the drawing process. The present contribution focuses on the first mentioned aspect, namely the influence of structure induced by a previous strain mode on material flow in a subsequent strain mode.

1. Introduction: ductility

Ductility in general is a function of both the metal being deformed and the process being used to deform it. Of all the ductility parameters determined in a tensile test – uniform elongation, fracture strain, reduction of area - only the last one appears to give a reliable prediction of formability in processes such as rolling, cold forging and drawing [1]. But even that is not valid for all metals and alloys and certainly not valid at all if one is interested in knowing the ductility in a second -subsequent- process with different strain mode.

In what follows, poor ductility resulting from material defects *–e.g.* leading to central bursting- will not be treated. Attention will only be paid to the residual ductility and flow of prestrained material with *recognized surface and interior soundness.*

The paper treats the particular case of *carbon steel wire*, doubtless the material family with the largest variation in room temperature ductility behavior of all metallic materials.

2. The structural state of drawn carbon steel wire

Severely drawn steel wire is a "commodity" *nanostructured* material. Low carbon steel on the one hand is known to have its dislocation substructure gradually transformed by drawing into subgrain fibers, continuously decreasing in size until the largest drawing

T.C. Lowe and R.Z. Valiev (eds.), Investigations and Applications of Severe Plastic Deformation, 373–381.
© *2000 Kluwer Academic Publishers. Printed in the Netherlands.*

strains hitherto obtained and which are of the order of $\varepsilon \approx 10$ [2]. At strains of $\varepsilon \approx 4$ the mean transverse subgrain size d has already been reduced to about 0.1 µm and its steady further refinement is reflected by the experimentally measured strength which continues to increase linearly with strain thereby following the structure-based relationship [3].

$$\sigma = \sigma_o + k\left(M\tau_{cf}\right)d^{-m} \quad \text{with } m = 0.5....1 \tag{1}$$

with M the Taylor factor in tension of the <110> textured material and τ_{cf} the critical shear stress of the ferrite.

This strongly contrasts with face-centered cubic metals, which reach a subgrain size of the order of 0.2 µm at strains of $\varepsilon \approx 4$, but which remains practically unchanged on further deformation, the corresponding flow stress also having reached a stationary value [3]

The particular behavior of body-centered materials during axisymmetric drawing is linked to the so-called "curling" effect, which is the consequence of a plane strain "relaxed-constraints" deformation of the <110> oriented grains [4, 5], inducing a continuous generation of redundant dislocations, which are only partially annihilated [6].

High carbon pearlitic steel on the other hand is characterized by an aligned structure of pearlite colonies, equally curled and with the mean interlamellar distance λ decreasing down to 20...30 nm already after drawing strains of $\varepsilon \approx 3$, as was recently confirmed by AFM and high resolution SEM [7]. Strain hardening during wire drawing has been described by an exponential Hall-Petch relationship of the form:

$$\sigma = \sigma_o + k_p\left(M^*\tau_{cp}\right)\lambda_r^{-m} \tag{2}$$

with $\lambda_p = \lambda_o e^{-\varepsilon/2}$ the strain induced change of interlamellar distance as soon as the pearlite has been oriented –which is the case after $\varepsilon \approx 1.5$ -, M^* the Taylor factor of the ferrite in its pearlite environment and τ_{cp} its critical resolved shear stress [8][9], indicating that the interlamellar distance is directly determining the tensile flow stress of the wire.

Both low and high carbon steel can be drawn to strengths of more than 3000 MPa, the first one after having been deformed into hair thin fibers, the second one reaching a tensile strength of nearly 4000 MPa already after a drawing strain of $\varepsilon \approx 3...4$, e.g. a wire having been drawn from 1.75 mm down to 0.2 mm diameter.

The latter material is characterised by a pronounced yield stress anisotropy: the $\sigma_{0.2}/\tau_{0.35}$ ratio (equivalent stresses in case of Von Mises behavior) amounts to 2.4 [10], compared to 1.73 for an isotropic material, to about 2 for a normal <110> fibre texture, 2 for the cyclic {110}<110>texture and 1.4 for the also observed {110}<001> cyclic texture components ({hkl}<uvw> representing shear plane normal and shear direction respectively).

Although a detailed study of the dislocation patterning in severely deformed material is still lacking, it can be assumed that their configuration will be characterised by a pronounced polarity, which will influence flow stress and slip behavior wherever the shear stress on the slip systems changes sign.

3. Ductility and material flow in subsequent forming

A particular but most important aspect of steel wire ductility is the behavior of the drawn wire material when changing strain path. Indeed, along the processing route of steel wire products, often a change in strain path takes place. Known examples are cold heading and roll flattening which are done on predrawn wire, and spring and cable manufacturing which are processes of combined plastic torsion and bending executed on -often severely- predrawn material.

One of the first consequences of the presence of a stretched dislocation subgrain structure in the case of low carbon steel and of an aligned pearlite structure in the case of medium and high carbon steel is a marked effect on the *formability* in subsequent forming processes.

As an example: industrial practice shows that the cold heading formability of steel wire is improved when the material has been cold drawn after annealing/patenting [11].

Fig. 1 shows the effect of wire drawing strain on the compressibility of a low-alloyed 0.07%C steel, compressive ductility being defined as the strain at which the first crack can be seen with a stereoscope at a magnification of 20. Experimental conditions can be found in [12] a.o. making clear that care was taken to work with rod material with a good internal soundness and surface quality. Similar improvements of headability (compressibility) were found for medium carbon steel.

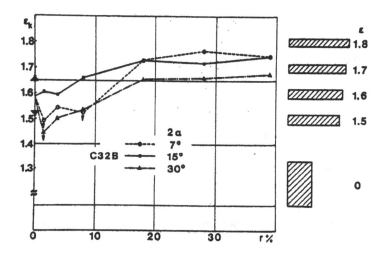

Figure 1. Wire drawing prestrain favourably affects cold headability ε_k of a 0.07%C steel [11]

It has been suggested that this improvement is due to a combined effect of a further improvement of the surface quality by drawing and to a strain softening effect occurring during cold-heading, an effect which will be discussed in what follows.

The compressive flow curves of predrawn wires after increasing amounts of wire drawing strain are shown in Fig. 2 respectively for a low, a medium and a high carbon steel [13].

376

Reversing in all cases induces a Bauschinger yield stress drop $\Delta\sigma_b$, which increases with the amount of prestrain. The reverse yield stress is then followed by a steep strain hardening over a strain range of only a few percent, after which the flow curve levels off. This levelling off leads to strain hardening values much smaller than those of the forward curves at the same equivalent strain. After moderate prestrains, the transition region of very low strain hardening is followed by a gradual restoration of the inclination of the forward curve. After large prestrains, flow curves of the three materials show a maximum after a reverse strain of about $\varepsilon = 0.05$, followed by a strain softening towards a plateau region. Reverse strains of between 20 and 50% are needed before a new strain hardening stage starts, which finally leads to a flow curve characterized by a permanent softening $\Delta\sigma_p$. It is seen in Fig. 2 that the softening effect increases with drawing prestrain and is larger when the carbon (pearlite) content of the material is higher.

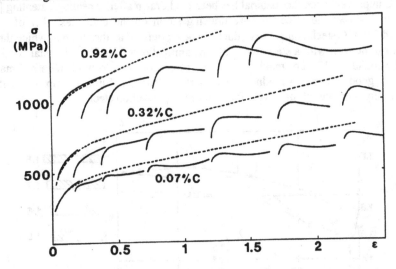

Figure 2: Compressive flow curves of predrawn wire of different carbon content, after various amounts of drawing strain [15]

The same effects as described above have also been reported by other authors, in tension-compression sequences as well as in alternating torsion, in ferrous as well as in non-ferrous metals [14, 15, 16].

It was shown that the softening *in ferritic steel* when reversing the strain mode is not the result of the residual macro-stress state in the drawn wire, nor of a negative slope of the Taylor factor (dM/dε) but is linked to the collapse of the dislocation substructure formed during drawing, which obviously is unstable when the shear stresses on the slip systems are changing sign [17].

The gradual replacement of a substructure aligned with the drawing axis by an equiaxed structure with larger subgrain size and followed by a new substructure oriented perpendicular to the compression axis is clearly illustrated in Figure 3a, b and c. The dislocation annihilation and repatterning mechanisms, active during that process, have still to be elucidated. Also the structural effects when compressing predrawn *pearlitic steel* wire are not yet known and they present without doubt an interesting topic for further investigation.

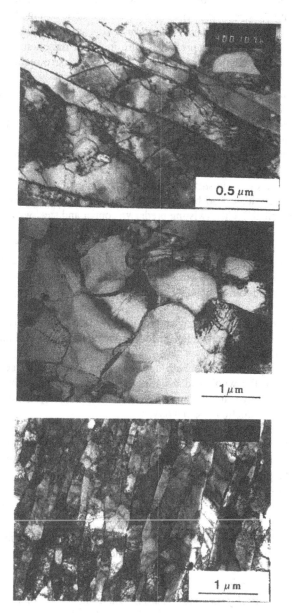

Figure 3. Dislocation subgrain structure: a. after ε=2.5 drawing strain, b. after ε=0.4 and c. after ε = 0.8 subsequent compressive strain. Wire axis is shown for a. and c.

When the change in strain mode is not as dramatic as to change the sign of all the components of the strain tensor, still a strain softening effect is visible possibly preceded by a cross effect (a transient yield stress increase) at small offset strains, which is not of relevance at larger subsequent strains. A practical example is encountered in the roll flattening of steel wire, where the strain tensor is changed from $[\varepsilon_{11}(+),\varepsilon_{22}(-),\varepsilon_{33}(-)]$ to $[\varepsilon_{11}(+), \varepsilon_{22}(-), \varepsilon_{33}(+)]$, provided shear strains are neglected.

Indeed, when compressing predrawn wire in radial direction, the compressional strain results in a markedly higher elongation in the transverse direction of the wire than in the former drawing direction [18].

Similarly, a continuous roll flattening operation leads to a broadening of the wire product which is itself not only controlled by roll gap geometry, friction and the longitudinal back stress that is sometimes applied, but equally well by the structural and hence plastic anisotropy of the material, as illustrated in Fig. 4 [19].

Because of the lower flow stress, material flow obviously is easier in directions in which the strain tensor changes sign. This behavior can be visualized in terms of the evolution of anisotropic yield surfaces when changing strain path.

Softening effects have also been reported when compressing tubes after moderate prestrains in torsion and when twisting predrawn copper and steel wires [20, 21].

Effects on material flow were also measured when bending specimens machined from a steel wire pre*compressed* from a diameter of 9 mm and a height of 18 mm down to a thickness of 3.5 mm and a diameter of 20 mm[12]. Specimens had a square cross section of 3.5x3.5 mm, a length –in radial direction d of the compressed specimen- of 20 mm and an effective free bending length in the three point bending test of 10 mm (Fig. 5)

Precision bend tests, with small and large deflection respectively, were done first with the radial direction b, perpendicular to r, as the bending axis and then with the wire axis z as the bending axis.

At the extension side of the bent specimen, the extension is compensated by a shortening strain in the width and thickness directions; on the compression side, the longitudinal compressive strain equals the sum of the two compensating elongations in width and thickness direction respectively.

Because of these strain conditions, the neutral line is shifted towards the extension side, which causes an overall thickness reduction during bending of narrow specimens.

In the four tests, the width changes Δb were measured on the extension side and on the compression side respectively, as well as the overall thickness change Δd of the specimen. The results are shown in Table 1. The ration $\Delta d/\Delta b$- illustrates the important difference in flow behavior induced by the structural anisotropy of the predrawn material.

Completely in accordance with the previous result, material flow is facilitated in the specimen directions for which the sign of the strain tensor component has been reversed.

All the examples treated teach us that in developing finite element codes for predicting the material flow and resulting dimensions in precision forming wire products from predrawn wire, structural anisotropy has to be taken into account.

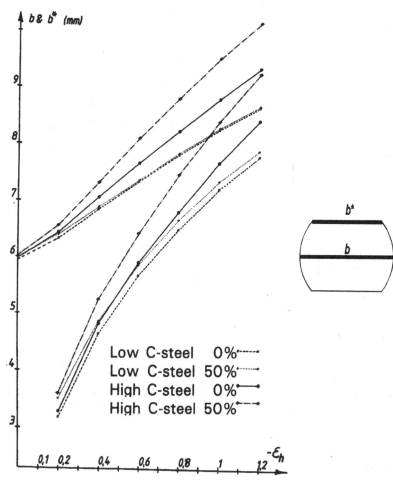

Figure 4. Broadening during the flat rolling of annealed viz. patented steel wire on the one hand and of the same materials after predrawing on the other hand [19]. The meaning of b (upper curves) and b* (lower curves) is represented in the accompanying drawing.

Table 1: Dimensional changes in precision bending tests on specimens machined from precompressed steel wire. Compressive strain ε = 1.6. (F):=forward straining, (R) = reverse straining

Direction of bending	Dimensional changes (in mm)			
	$\Delta b+$	$\Delta b-$	Δd	$\Delta d/\Delta b-$
b small deflection	0.26 (F)	0.29 (R)	0.10 (F)	0.34
b large deflection	0.41 (F)	0.45 (R)	0.16 (F)	0.36
z Small deflection	0.40 (R)	0.30 (F)	0.30 (R)	1.00
z Large deflection	0.64 (R)	0.51 (F)	0.45 (R)	0.88

380

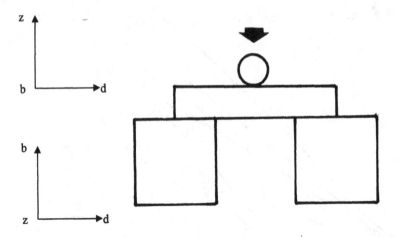

Figure 5. Experimental set up for measuring dimensional changes during the free bending of small specimens machined from a pre*compressed* steel wire ($\varepsilon = 1.6$) ; z = wire axis direction, d and b : two mutually perpendicular radial directions. [12].

4. Acknowledgements

The author likes to thank N.V. Bekaert R&D for continuing support of wire drawing research at MTM. Part of the results presented was obtained in the framework of the IUAP initiative of the Federal Services for Scientific, Technical and Cultural affairs, financed by the Belgian Ministry of Science Policy. We gratefully acknowledge that support.

5. Reference

1. Dieter, G., (1967) Ductility, ASM, Chapman and Hall Ltd, London
2. Leslie, W.C. (1972) Iron and its dilute substitutional solid solutions *Metallurgical Trans.* **3**, 5-26
3. Gil-Sevillano, J., Van Houtte, P., Aernoudt, E. (1981) Large strain work hardening and textures, *Progr. In Materials Science Series* **25, 2**
4. Hosford, W.F. (1964) Microstructural changes during deformation of <110> fiber-textured metals, *Trans. of the metallurgical society of AIME* **230**, 12-15
5. Van Houtte, P. (1988), A comprehensive mathematical formulation of an extended Taylor-Bishop-Hill Model featuring relaxed Constraints, the Renouard-Winterberger Theory and a strain Rate sensitivity Model., *Textures and Microstructures* **8-9**, 313-350
6. Gil-Sevillano, J., Garcia-Rosales, C., Flaquer-Fuster, J. (1998) Texture and large-strain deformation microstructure - The Royal Society - Proc. Discussion Meeting on "Deformation Processing of Metals", London
7. Delrue, H. (1999) "Ageing and ductility of steel wire" *PhD thesis in progress, K.U.Leuven*
8. Gil-Sevillano, J. (1991) Substructure and strengthening of heavily deformed single and two-phase materials, *Journal of Physics* **III** **1**, pp. 967-998
9. Watté, P., Van Houtte, P., Aernoudt, E., Gil-Sevillano, J., Van Raemdonck, W., Lefever, I. (1994). The work hardening of pearlite during wire drawing, *Material Science Forum* **157-162**, pp. 1689-1694

10. Delrue, H., Van Humbeeck, J., Aernoudt, E., Timmerman, H., Lefever, I., (1998) IWT project 960120, Anisotropy and Inhomogeneity in drawn steel wire – *Progress Report Autumn 98*
11. Strauven, Y, Bouwman, J., Aernoudt, E., (1981) Über das Stauchen vorgezogener Drähte, *Draht* **32**, pp. 224-295
12. Strauven Y., (1982), Cold heading of steel wire. Ductility and Bauschinger effects; (in Dutch) *PhD thesis K.U.Leuven*
13. De Bleser W. (1979) Bauschinger-effect in steel wire after large drawing strains (in Dutch) *Final Thesis, K.U.Leuven*
14. Polakowski, N.H., (1951) Softening of metals during cold-working, *Journal of the Iron and Steel Institute*, pp. 337-342
15. Hecker, F.W. (1971) Die Wirkung des Bauschinger-Effektes nach grossen plastischen Verformungen im Verdrehversuch, *Archiv für das Eisenhüttenwesen* **42**, pp. 819-824
16. Langford, G., Cohen, M., (1969) Strain hardening of Iron by severe plastic deformation, *Trans. ASM* **62**, pp. 623-627.
17. Strauven, Y., And Aernoudt, E., (1987) Directional strain softening in ferritic steel, *Acta metall.* **25-5**, pp. 1029-1036.
18. Aernoudt, E., Gil-Sevillano, J., Van Houtte, P., (1987) Structural background of yield and flow, *Constitutive Relations and their Physical Basis* (Proc. 8th Risø Intern. Symp. on Metallurgy and Materials Science – ed. S.I. Anderson *et al.* – Risø National lab. Roskilde (Denmark), 1-38.
19. Tytgat, J., (1984)) Broadening of flat-rolled steel wire (in Dutch), *Final thesis – K.U.Leuven.*
20. Lee, P.W., Kuhn, H.A., (1973) Fracture in cold upset forging –a criterion and model, *Metallurgical Transactions* **4**, pp. 969-974.
21. Polakowsky, N.H., Mostovoy, S., "Transient and destructive instability in torsion" (1961) *Trans. ASM* **54**, pp. 567-570.

WORKSHOP PARTICIPANTS

Etiene Aernoudt
Department of Metal and Material
Engineering
Katholieke University Leuven
K.U. Leuven de Croylaan 2 B-3001
Heverlee-Leuven
Belgium

Igor Alexandrov
Ufa State Aviation Technical University
K. Marks str., 12
Ufa 450000
Russia

Mikhail Alymov
Baikov Institute of Metallurgy
Leninsky Prospect 49
Moscow 117334
Russia

N. Amirhanova
Ufa State Aviation Technical University
K. Marks str., 12
Ufa 450000
Russia

Rostislav Andrievski
Institute for New Chemical Problems
Russian Academy of Sciences
Institute Prospect 16
Chernogolovka 142432
Russia

Vladimir Bengus
National Academy of Sciences of Ukraine
B.Verkin Institute for Low
Temperature Physics & Engineering
Lenin's Avenue 47
Kharkov-164, 310164
Ukraine

Irina Brodova
Institute of Metal Physics, Ural Division of
Russian Academy of Sciences
GSP-170, S. Kovalevskaya Str.18
620219 Ekaterinburg
Russia

Walter Buchgraber
Osterreichisches Forschungszentrum
Seibersdorf
2444 Seibersdorf
Austria

Igor Budilov
Ufa State Aviation Technical University
K. Marks str., 12
Ufa 450000
Russia

Hugh Casey
ONRIFO
223 Old Marylebone Road
London NW1 5TH
United Kingdom

Sergey Dobatkin
Moscow State Steel and Alloys Institute
Leninsky prospekt,4
Moscow 117936
Russia

Frank De Bruyne
N.V. Bekaert S.A.
Bekaertstraat 2
Zwevegem B-8550
Belgium

Nariman Enikeev
Ufa State Aviation Technical University
K. Marks str., 12
Ufa 450000
Russia

Vladimir Farber
Urals State Technical University
Mira, 19
Ekaterinburg 620002
Russia

Sergiy Firstov
Francevich Institute For Problems of
Materials Science
3, krzhizhanovskoho str.
Kyiv 252142
Ukraine

Vasiliy Gaviko
Institute of Metal Physics
18, S.Kovalevskaya
Ekaterinburg 620219
Russia

Amit Ghosh
The University of Michigan
2300 Hayward St
Ann Arbor, MI 48109-2136
USA

Herbert Gleiter
Forschngsrentrum
Karlsruhe-INT, Postfach 3640
D76021 Karlsruhe
Germany

A. M. Glezer
Institute of Metal Physics
2nd Baymonstaya St.
9123; 107005 Moscow
Russia

Bella Greenberg
Institute of Metal Physics, Ural Branch c
Russian Academy of Sciences
18 S.Kovalevskaya str.
Ekaterinburg GSP-170
Sverdlovsk Region 620219
Russia

K. (Ted) Hartwig
Texas A&M University
Department of Mechanical Engineering
College Station, TX 77843-3123
USA

Thomas Hebesberger
Erich Schmid Institut F. Materialwissens
Oesterr.Akademie D. Wissensch
Jahnstrasse 12
Leoben, Steiermark A-8700
Austria

Zenji Horita
Kyushu University
Dept. of Materials Science & Engineerir
Faculty of Engineering 36
Fukuoka 812-81
Japan

Rinat Islamgaliev
Ufa State Aviation Technical University
K.Marks str., 12
Ufa 450000,
Russia

Eugene Ivanov
Tosoh SMD, Inc.
3600 Gantz Rd
Grove City, OH 43123
USA

Ascar Kilmametov
Ufa State Aviation Technical
University
K. Marks str., 12
Ufa 450000
Russia

Yuri Kolobov
Institute of Strength Physics and
Material Science
pr. Academichesky 2/1
634021 Tomsk
Russia

Lembit Kommel
Tallinn Technical University
5 Ehilajate St.
Tallinn EE-0026
Estonia

Vladimir Kondratyev
Institute of Metal Physics, Ural Branch
Of the Russian Academy of Sciences
18, S. Kovalevsaya st.
Ekaterinburg 620219
Russia

Nina Koneva
Tomsk State University of Architecture
and Building
2 Solyanaya Sq.
Tomsk 634003
Russia

Vera Konenkova
Moscow State Steel and Alloys
Institute
Leninsky prospekt,4
Moscow 117936
Russia
.

Vladimir I. Kopylov
Phusical-technical Institute of National
Academy of Science of Belarus
Akademika Kuprevicha str., 10
Minsk, 220141
Belarus

Alexander Korshunov
Russian Federal Nuclear Center
All-Russian Research Institute of
Experimental Physics
Technology Department
607190, Sarov (Arzamas-16)
Nizhny Novgorod Region
Russia

Boris Koutcheryaev
Moscow State Steel and Alloys
Institute
Leninsky prospekt,4
Moscow 117936
Russia

Eduard V. Kozlov
Tomsk State university of Architecture
and Building
2 Solyanaya sq.
Tomsk 634003
Russia

Nikolay Krasilnikov
Ulyanovsk State University
L. Tolstoy Str., 42
Ulyanovsk 432700
Russia

Alexei Lachinov
Institute of Physics of Molecules and
Crystals, Ufa Scientific Centre of
Russian Academy of Sciences
Pr. Oktyabrya, 151
Ufa 450075
Russia

Terence Langdon
University of Southern California
Dept. of Materials Science
Los Angeles, CA 90089-1453
USA

Rimma Lapovok
CSIRO, Manufacturing Science and
Technology
Locked Bag N9
Preston, Vic 3072
Australia

Vladimir Latysh
Ufa State Aviation Technical
University
K.Marks str., 12,
Ufa 450000
Russia

Don Lesuer
Lawrence Livermore National
Laboratory
L-342, P.O. Box 808
Livermore, CA 94550
USA

Terry Lowe
Materials Science & Technology
Division
MS G754
Los Alamos National Laboratory
Los Alamos, NM 87545
USA

Michael Markushev
Institute for Metals Superplasticity
Problems
Russian Academy of Sciences
39 Khalturin St.
Ufa 450001
Russia

Hael Mughrabi
University of Erlangen-Nuernberg
Martensstr.5 D-91058
Erlangen
Germany

Amiya Mukherjee
University of California
Dept. of Chemical Engg. & Materials
Science
One Shields Ave.
Davis, CA 95616
USA

Oleg Naimark
Institute of Continuous Media
Mechanics, Russian
Academy Of Sciences, Head of the
Division
1 Acad.Korolev str., 614013
Perm
Russia

Nina Noskova
Institute of Physics Metals UD RAS
18, S.Kovalevskay Str., GSP-170
Ekaterinburg, 620219
Russia

Victor Panin
Institute of Strength Physics and
Material Science
8, Academicheskii Pr., 634055, Tomsk
Russia

Philip Prangnell
UMIST / Manchester Materials
Science Centre
Grosvenor St
Manchester M1 7HS
United Kingdom

Virgil Provenzano
Naval Research Laboratory
Physical Metallurgy Branch, Code
6320
Washington, DC 20375-5343
USA

Igor Yu. Pyshmintsev
Urals State Technical University
Ekaterinburg, 620002
Russia

Georgy Raab
Ufa State Aviation Technical
University
K.Marks str., 12
Ufa 450000
Russia

Edgar Rauch
GPM2 - INPG
ENSPG BP46
38402 Saint Martin d'Hères
France

Gennady Salishchev
Institute for Metals Superplasticity
Problems
Russian Academy of Sciences
Khalturina 39
Ufa 450001
Bashkortostan
Russia

Shankar Sastry
Washington University
One Brookings Drive
St. Louis, MO 63130
USA

Erhard Schafler
Inst. F.Materialphysik, Universitaet
Wien
Strudlhofgasse 4
Wien A-1090
Austria

Marc Seefeldt
Department MTM, K.U. Leuven
De Croylaan 2
B-3001 Heverlee
Belgium

Oleg Senkov
University of Idaho
321 Mines Bldg.
Moscow, ID 83844-4284
USA

Oleg Smirnov
Moscow State Steel and Alloys
Institute
Leninsky prospekt,4
Moscow 117936
Russia

Vladimir Stolyarov
Ufa State Aviation Technical
University
K. Marks str., 12
Ufa 450000
Russia

Tatyana Sviridova
Moscow State Steel and Alloys
Institute
Leninsky pr., 4
Moscow 117936
Russia

Elena Tabachnikova
B. Verkin Institute for Low
Temperature
Physics and Engineering
Ukraine Academy of Sciences
47, Lemin's Avenue
Kharkov
Ukraine

Anya Tarakanova
Ufa State Aviation Technical
University
K. Marks str., 12
Ufa 45000
Russia

Laurent Taravella
DGA/DCE/CTA
16 bis, avenue Prieur de la Cote d'Or
Arcueil 94114 cedex
France

Victor Tcherdyntsev
Moscow State Steel and Alloys
Institute
Leninsky prospect,4
Moscow 117936
Russia

Ellen Thiele
Institute of Physical Metallurgy
Technical University Dresden
Dresden D-01062
Saxony
Germany

Igor Tomilin
Moscow State Steel and Alloys
Institute
Leninsky prospect,4
Moscow 117936
Russia

Alexander Tyumentsev
Institute of Strength Physics and Materia
Technology, RAS
2/1 Akademichesky Ave.
Tomsk 634055
Russia

Tamas Ungar
Eotvos University Budapest, Hungary
Pazmany Peter setany 1/A
Budapest H-1518, P.O.B. 32
Hungary

F.Z. Utyashev
Institute for Metals Superplasticity Prob
Khalturina 39, Ufa 450001
Russia

Ruslan Valiev
Ufa State Aviation Technical University
K. Marks str., 12
Ufa 450000
Russia

Vener Valitov
Institute of Metals Superplasticity Prob
Khalturina 39
Ufa 450001
Russia

Alexei Vinogradov
Osaka City University
Sugimoto 3-3-138, Sumiyoshi-ku
Osaka 558-8585
Japan

Takashi Yamasaki
Department of Mechanical Engineering
Faculty of Engineering
Doshisha University
Kyotanabe 610-0321
Kyoto
Japan

Reza Alain Yavari
Institut National Polytechnique, Grenoble
France
LTPCM CNRS, BP 75, Domaine Universitaire
38402 St-Martin-d'Heres
France

Michael Josef Zehetbauer
Institut fuer Materialphysik
Universitaet Wien
Strudlhofgasse 4
Wien A-1090
Austria

Sergey Zherebtsov
Institute for Metals Superplasticity Problems
Russian Academy of Sciences
Khalturina str. 39
Ufa 450001
Bashkortostan
Russia

SUBJECT INDEX

A

activation energy 10, 27, 47, 228, 236, 262, 331, 301
all-round multiple forging 21
aluminum 32, 84, 110, 151, 155, 163, 242
alternating bending 21
aluminum alloys 9, 29, 65, 152, 156, 185, 221, 232, 242, 319, 328, 350
Al 6061 303
Al-Mg alloys 32, 156, 189
aluminum powder 3, 7, 109
amorphous alloy 115, 330,
annealing 10, 27, 43, 46, 115, 124, 152, 153, 155, 174, 205-207, 222, 251, 255, 262, 268, 273, 283, 298, 300, 319, 334, 340,
analytical methods 4, 5, 6
angular domains 37
application and commercialization 353
atomic displacement 94, 106

B

ball milling 139
bauschinger effect 222, 228, 247, 290, 292, 375
bend strength 285
biocompatible material 361
biomedical and orthopaedic applications 364

C

cells 165, 197, 247, 350
Cr 16, 281
clear ECAP 23
cold rolling 164
collective electronic effects 333
combined deformation 74, 87
conformal mapping 38
corrosion resistance 16, 350, 361
creep 10, 115, 261, 271, 294, 327, 340
crystallographic and structural anisotropy 373

crystallographic texture 58, 62, 94, 106, 221
crystal lattice distortion 128, 223
Cu 103, 127, 133, 156, 202, 205, 221, 241, 255 261, 273, 297, 333, 339
Cu-Nb metal matrix composite 9
Cu-Al$_2$O$_3$ 273
Cu + 0.5%Al$_2$O$_3$ nancomposite 267
cyclic extrusion compression method 87
cyclic plastic deformation 174, 248, 289
cyclic stress-strain response 174, 248, 289

D

damage 61, 241, 304
Debye-Waller parameter 106
deformation induced vacancies 86, 204
deformation mechanism 231
die-wall sticking 29
coherent scattering domains 101
diffusion 9, 26, 43, 52, 86, 118, 131, 149, 158, 215, 228, 237, 262, 289, 328, 342
disclinations 26, 127-130, 197, 204, 315
dislocation cell wall 124, 165
dislocation contrast factor 95
dislocation density 30, 88, 93, 103, 123, 133, 137, 156, 164, 176, 180, 199, 247, 294, 362
dislocation structure 82, 100, 121, 179, 235, 255, 313
disordering 49, 216, 223
distorted crystal lattice 221
ductile-to-brittle transition temperature 281
ductility 54, 149, 303, 375
dynamic recovery 13, 30, 360
dynamic recrystallization 13, 51, 58, 173, 208, 340, 352